Lecture Notes in Biomathematics

Lecture Notes in Biomathematics

Managing Editor: S. Levin

23

Geometrical Probability
and Biological Structures:
Buffon's 200th Anniversary

Proceedings of the Buffon Bicentenary Symposium
on Geometrical Probability, Image Analysis,
Mathematical Stereology, and Their Relevance
to the Determination of Biological Structures,
Held in Paris, June 1977

Edited by R. E. Miles and J. Serra

Springer-Verlag
Berlin Heidelberg New York 1978

Editors

Roger E. Miles
Department of Statistics
Institute of Advanced Studies
Australian National University
P.O. Box 4
Canberra, A.C.T. 2600/Australia

Jean Serra
Centre de Morphologie Mathématique
Ecole des Mines de Paris
35 rue St. Honoré
77305 Fontainebleau/France

Library of Congress Cataloging in Publication Data

Buffon Bicentenary Symposium on Geometrical Probability,
    Image Analysis, Mathematical Stereology, and Their
    Relevance to the Determination of Biological Struc-
    tures, Paris, 1977.
    Geometrical probabilities and biological structures,
Buffon's 200th anniversary.

    (Lecture notes in biomathematics ; 23)
    Bibliography: p.
    Includes index.
    1. Morphology--Mathematics--Congresses. 2. Geo-
metric probabilities--Congresses. 3. Stereology--
Mathematics--Congresses. I. Buffon, Georges Louis
Leclerc, comte de, 1707-1788. II. Miles, Roger
Edmund, 1935-   III. Serra, Jean, 1940-   IV. Title.
V. Series.
QH351.B83 1977   519.2   78-18234

AMS Subject Classifications (1970): 60 D 05, 01 A 50, 53 C 65, 92-00

ISBN-13: 978-3-540-08856-1   e-ISBN-13: 978-3-642-93089-8
DOI:10.1007/978-3-642-93089-8

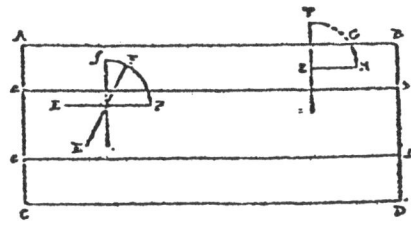

cm 7

cm 5

cm 9

erson

1977

Adam 21.5.77

P R E F A C E

The idea for this Buffon bicentenary Symposium was due to E. Weibel, and realized by an organizing committee, made up of H. Haug, G. Matheron, J.L. Binet, R. Miles (programme chairman) and J. Serra (organization chairman). In the year 1777, Buffon published the famous study of a needle dropped at random on a grid of lines, which we could consider as the starting point of geometric probability. This publication is all the more remarkable because it comes from one of the founders of the natural sciences. We wanted to celebrate this, two hundred years later, by gathering together in the same Symposium, about thirty biologists and the same number of mathematicians, in particular specialists in integral geometry and geometric probability.

The Symposium was held in June 1977, in the buildings of the Jardin des Plantes de Paris, the very gardens that Buffon created, and where he spent much of the last years of his life. This memorial aspect of the Symposium appears in the introduction which follows, where we survey the germination of the basic idea of "Buffon's needle", from Buffon to the beginning of the twentieth century. During the Symposium itself, the homage to Buffon was manifested also by an exhibition of paintings in the Salpêtrière organized by J.L. Binet, and by the publication of a book where J. Roger and J.L. Binet collected together the major works of Buffon.

In these proceedings, we have not changed the chronological order of the papers and we have respected the alternation between mathematical and biological presentations. Both are written in a deliberate everyday manner, avoiding difficult technicalities.

We would like to acknowledge the Museum d'Histoire Naturelle, the Centre of Mathematical Morphology of the Paris School of Mines, The Anatomy Institute of Berne, the Hospital of La Salpêtrière, the International Association for Stereology, and the University of Paris VI, for their moral and financial help.

R.E. MILES, Canberra
J. SERRA, Fontainebleau

LIST OF THE PARTICIPANTS
-=-=-=-=-=-=-=-=-=-=-=-=-

*	R. AMBARTZUMIAN	Armenian Academy of Science EREVAN (S.U.)
	A. BADDELEY	18 Price Avenue MONTMORENCY Victoria 3094 (Australia)
*	G. BERNROIDER	UNIVERSITAT SALZBURG ZOOLOGISCHES INSTITUT AKADEMIESTRASSE 26 A-5020 SALZBURG
*	S. BEUCHER	CENTRE DE MORPHOLOGIE MATHEMATIQUE E.M.P. 35 Rue St-Honoré 77305 FONTAINEBLEAU
	S. BIDIKHOF	Bolchoï Vouzovski per. 3/12 M/I/E/M/ Chaire de Cybernétique MOSCOU
* +	J.L. BINET	PITIE SALPETRIERE 91 Bld de l'Hôpital 75013 PARIS
*	J.C. BISCONTE	C.H.U. DE BOBIGNY Dept de Biologie Av. Marcel Cachin 93000 BOBIGNY
	H. BOLEY	Department of Histochemistry and Cytochemistry University of LEIDEN Wassenaarserveg 72 LEIDEN
	H. BRAUN	STATISTICS Princeton University PRINCETON NJ 08540
	G. BRUGAL	Laboratoire de Zoologie et Biologie Animale U.S.M. B.P. 33 CENTRE DE TRI 38041 GRENOBLE CEDEX
	J.M. CHASSERY	Laboratoire de Zoologie et Biologie Animale U.S.M. B.P. 33 CENTRE DE TRI 38041 GRENOBLE CEDEX
*	R. COLEMAN	Department of Mathematics Imperial College Queen's Gate LONDON SW7 2BZ
	F. CORELLA	CENTRE DE MORPHOLOGIE MATHEMATIQUE E.M.P. 35 Rue St-Honoré 77305 FONTAINEBLEAU
	M. COSTER	UNIVERSITE DE CAEN Labo. Chimie Minérale Industrielle 14032 CAEN CEDEX
	M. COURTOIS	Laboratoire de Microscopie Quantitative U.E.R. de Médecine 93000 BOBIGNY
	L.M. CRUZ ORIVE	UNIVERSITAT BERN ANATOMISCHES INSTITUT Postfach 139, Bühlstrasse 26 3000 BERN 9
*	P. D'ATHIS	PITIE SALPETRIERE 47 Bld de l'Hôpital 75634 PARIS CEDEX 13
	P. DEBRE	PITIE SALPETRIERE 47 Bld de l'Hôpital 75634 PARIS CEDEX 13
*	R. DEHOFF	College of Engineering Department of Matérials Science and Engineering, University of Florida GAINESVILLE Florida 32611

P. DELFINER   CENTRE DE MORPHOLOGIE MATHEMATIQUE E.M.P.
35 Rue St-Honoré 77305 FONTAINEBLEAU

* H. DIGABEL   CENTRE DE MORPHOLOGIE MATHEMATIQUE E.M.P.
35 Rue St-Honoré 77305 FONTAINEBLEAU

G. DIGIERO   PITIE SALPETRIERE 47 Bld de l'Hôpital
75634 PARIS CEDEX 13

* A. FELLOUS   39 Rue Lamarck 75018 PARIS

W.M. FISCHER   Laboratorium für Elektronen mikroskopie der Theoretisch-
Medirinischen Institute der Universität Innsbrück
Fritz Pregl Strasse 3 A-6020 INNSBRUCK

M. GAUTHIER   CENTRE DE MORPHOLOGIE MATHEMATIQUE E.M.P.
35 Rue St-Honoré 77305 FONTAINEBLEAU

G. GIRAUT   PITIE SALPETRIERE 47 Bld de l'Hôpital
75634 PARIS CEDEX 13

* J. GRANARA   153 Fbg St-Denis 75010 PARIS

* H.J.GUNDERSEN   Arhus Kommunehospital 2nd University Clinic of Internal Me.
DK-8000 ARHUS C

E. HARDING   UNIVERSITE DE CAMBRIDGE STATISTICAL LABORATORY
16 Mill Lane   CAMBRIDGE CB2 1SB

* + H. HAUG   Abteilung für Anatomie der Medizinischen Hochschule Lübeck
Ratzeburger Allee 160 D-2400 LUEBECK

D. JEULIN   I.R.S.I.D. 57210 MAIZIERES LES METZ

H. KELLER   Anatomisches Institut Bühlstrasse 26 CH-3012 BERN

J.C. KLEIN   CENTRE DE MORPHOLOGIE MATHEMATIQUE E.M.P.
35 Rue Saint-Honoré 77305 FONTAINEBLEAU

* K. KRICKEBERG   U.E.R. de Mathematique Logique Formelle et Informatique
Sorbonne 12 Rue Cujas 75005 PARIS

* Ch. LANTUEJOUL   CENTRE DE MORPHOLOGIE MATHEMATIQUE E.M.P.
35 Rue St-Honoré 77305 FONTAINEBLEAU

H. MAITRE   ECOLE DES TELECOMMUNICATIONS 45 Rue Barrault 75013 PARIS

* B. MANDELBROT   P.O. Box 218 Yorktown Heights NEW YORK 10598

* J.C. MARIANNA   I.N.R.A. NOUZILLY 37380 MONNAIE

B. MATERN   SKOGSCHOGSCKOLAN Royal College of Forestry S-104 05 STOCKHOLM

* + G. MATHERON   CENTRE DE MORPHOLOGIE MATHEMATIQUE E.M.P.
35 Rue St-Honoré 77305 FONTAINEBLEAU

*   F. MEYER   CENTRE DE MORPHOLOGIE MATHEMATIQUE E.M.P.
35 Rue St-Honoré 77305 FONTAINEBLEAU

*   J. MIGNOT   HOPITAL AMBROISE 9 Av. du Gal de Gaulle 92100 BOULOGNE

* + R. MILES   THE AUSTRALIAN NATIONAL UNIVERSITY THE RESEARCH SCHOOL OF
SOCIAL SCIENCES   Box 4 P.O. CANBERRA ACT 2600

     F. De MONTAUX   PITIE SALPETRIERE 47 Bld de l'Hôpital   75634 PARIS CEDEX 13

*   R. OSTERBY   Arhus Kommunehospital 2nd University Clinic of Internal Med.
DK-8000 ARHUS

     W. PFALLER   Institut für Physiologie der Universität Innsbruck
Fritz Pregl Strasse 3   A-6020 INNSBRUCK

     F. PIEFKE   Institut B für Mathematik Technische Universität
Pokelstrasse 14 - Forum D-330 BRAUNSCHEWEIG

     J. PIRET   CENTRE DE MORPHOLOGIE MATHEMATIQUE E.M.P.
35 Rue St-Honoré   77305 FONTAINEBLEAU

     Y. POULIGUEN   PITIE SALPETRIERE 47 Bld de l'Hôpital   75634 PARIS CEDEX 13

     J.P. RASSON   Facultés Universitaires Département de Mathématiques
8 Rempart de la Vierge   5000 NAMUR   Belgique

*   A. REITH   Norsk Hydro's Institute for Cancer Research
The Norwegian Radium Hospital Montebello   OSLO 3

*   J. ROGER   13 Rue de Mézières   75006 PARIS

*   H. RUBEN   Department of Mathematics Mc Gill University
P.O. Box 6070, Station A MONTREAL Québec H3C 3GI

*   L.A. SANTALÓ   Cochabamba 780 1150 BUENOS AIRES

     H.E. SCHROEDER   Zahnärztliches Institut der Universität
Plattenstrasse 11   8000 ZURICH

* + J. SERRA   CENTRE DE MORPHOLOGIE MATHEMATIQUE E.M.P.
35 Rue St-Honoré 77305 FONTAINEBLEAU

     B.W. SILVERMAN   UNIVERSITY OF CAMBRIDGE STATISTICAL LABORATORY
16 Mill Lane   CAMBRIDGE CB2 1SB

     A. SOLARI   I.N.R.A.   78350 JOUY EN JOSAS

*   F. STREIT   Section de Mathématique Université de Genève
2-4 Rue du Lièvre Case Postale 124 1211 GENEVE 24

*   T. TAKAHASKI   Department of Pathology Tohoku University School of Medecine
Seiryomachi, SENDAI

     J.F. THERY   ECOLE DES TELECOMMUNICATIONS 45 Rue Barrault 75013 PARIS

     H. VROLYK   Department of Histochemistry and Cytochemistry
University of Leiden Wassenaarserveg 72 LEIDEN

*   G.S. WATSON   Department of Statistics Fine Hall Princeton University
PRINCETON, New Jersey

* + E.R. WEIBEL   Universität Bern Anatomisches Institut
Bühlstrasse 26   3000 BERN 9

*   Speakers

+   Scientific Committee

CONTENTS
-=-=-=-=-=-=-=-

OPENING SPEECH
-=-=-=-=-=-=-=-=

Jean DRY

President of the P.M. Curie University

-:-

Several reasons lead the President of the Pierre et Marie Curie University to agree to sponsor this Symposium, which I had the privilege to open on 21 June 1977 :

- the fact that Professor J.L. BINET, who asked me to open the Symposium, belongs to one of the three Medicine Faculties of this University.

- the bonds that link together the Natural History Museum (where the Symposium was held) and the Pierre et Marie Curie University.

- the variety of scientific fields represented in the Symposium is comparable to that observed in a University.

- the very topic of the proposed double feature - geometrical probability for the Symposium, and its relation to graphical arts (through Buffon) for the painting exposition. This suggests binding thought and expression, its formal translation, and once more raises the problem of sign interpretation, well known by medical scientists as the problem of semeiology.

I am especially keen on that idea of linking a scientific manifestation - this Symposium - to an art show - BUFFON 1977.

The participant is in fact introduced to two different worlds, or better yet, is placed in a different situation with regard to two closely linked realms.

By his work of research scientist, and by his participation in the Symposium, he takes part in the analysis - and building - of an intellectual system, starting with signs that are the keys which allow him to decipher the system.

As a spectator, he is confronted with another realm, which can be very similar. Painting is, in fact, a reorganization of space, at least of the space of the canvas, whether we consider lines, colors, or even a figurative scene. This organization is once again part of a coherent system, through which the spectator looks for a meaning, from the signs at his disposal.

The confrontation of these experiences can only be fascinating, especially if the orginizer has been able to find systems that are both close enough and sufficiently different to provide new keys, and that the two states of actor and spectator shed enough light on each other.

This could actually be a justification of the plurality of university curriculae (purely intellectual).

It is by all means an experience worth renewing.

Jean DRY

EN MATIERE D'INTRODUCTION...

R.E. MILES

Dept of Statistics (I.A.S.), Australian National University,
P.O. Box 4, Canberra, A.C.T. 2600, Australia

J. SERRA

Centre de Morphologie Mathématique
Ecole Nationale Supérieure des Mines de Paris
35 Rue St. Honoré, 77305 - Fontainebleau - France

Nous avons cherché à savoir comment BUFFON aurait réagi devant notre symposium, comment il voyait lui, les relations entre mathématiques et sciences naturelles. Et voici son point de vue, critique, et empreint d'un positivisme aux résonances extraordinairement contemporaines. Il l'énonce dès le début de son oeuvre, dont il consacre la fin du discours d'introduction à s'interroger sur le bon usage des mathématiques, et sur leurs abus.

Mais cet abus n'eft rien en comparaifon des inconvéniens où l'on tombe lorfqu'on veut appliquer la Géométrie & le calcul à des fujets de Phyfique trop compliquez, à des objets dont nous ne connoiffons pas affez les propriétés pour pouvoir les mefurer; on eft obligé dans tous ces cas de faire des fuppofitions toujours contraires à la Nature, de dépouiller le fujet de la plûpart de fes qualités, d'en faire un être abftrait qui ne reffemble plus à l'être réel, & lorfqu'on a beaucoup raifonné & calculé fur les rapports & les propriétés de cet être abftrait, & qu'on eft arrivé à une conclufion toute auffi abftraite, on croit avoir trouvé quelque chofe de réel, & on tranfporte ce réfultat idéal dans le fujet réel, ce qui produit une infinité de fauffes conféquences & d'erreurs.

C'eft ici le point le plus délicat & le plus important de l'étude des fciences : fçavoir bien diftinguer ce qu'il y a de réel dans un fujet, de ce que nous y mettons d'arbitraire en le confidérant, reconnoître clairement les propriétés qui lui appartiennent & celles que nous lui

prêtons, me paroît être le fondement de la vraie méthode
de conduire fon efprit dans les fciences ; & fi on ne
perdoit jamais de vûe ce principe, on ne feroit pas une
fauffe démarche, on éviteroit de tomber dans ces erreurs
fçavantes qu'on reçoit fouvent comme des vérités, on
verroit difparoître les paradoxes, les quèftions infolubles
des fciences abftraites, on reconnoîtroit les préjugés &
les incèrtitudes que nous portons nous - mêmes dans les
fciences réelles, on viendroit alors à s'entendre fur la Mé-
taphyfique des fciences, on cefferoit de difputer, & on
fe réuniroit pour marcher dans la même route à la fuite
de l'expérience , & arriver enfin à la connoiffance de
toutes les vérités qui font du reffort de l'efprit humain.

Deux siècles ont passé.

Mais l'ambiguité de l'interférence des mathématiques avec les autres
sciences, si clairement perçue par BUFFON demeure, comme si la progression
des mathématiques et celle des sciences biologiques, chacune pour soi, l'ai-
guisait indéfiniment sans la résorber. Et la voici aujourd'hui, cette question, au centre des débats dans notre symposium.

C'est que communiquer avec autrui n'est simple ni pour les individus,
ni encore moins pour les corps de doctrine organisés. Au point qu'il y a
quelque chose d'antinomique à vouloir développer une science, et à lui deman-
der, en même temps, d'en reconnaître une autre. A les voir de l'extérieur, on
pourrait penser que les mathématiques organisent la vision logique de l'esprit
par un discours purement déductif, et tautologique. En réalité, ce point de
vue est faux, et chacun le sent bien. Ou plutôt, il est exact, mais ne marque
qu'un moment dans la pensée mathématique, celui où un problème est re-formulé
dans les catégories mentales du mathématicien, à une époque donnée.

Le moteur premier, et plus profondément actif, se trouve au contraire
dans les limites mêmes de l'imaginaire des mathématiciens (ou des biologistes)
pris à un moment de leur histoire. A l'opposé du discours tautologique, ce
moteur autorise la communication entre disciplines, ou dans une science donnée,

entre générations. On peut même dire qu'il en vit. Ce sont les limites du monde
imaginaire de CAUCHY, où toutes les surfaces sont dérivables, qui, par contre
coup ont engendré celui de MINKOWSKI puis de l'anneau convexe. C'est parce
qu'il a omis d'introduire des hypothèses sur ses polyèdres, qu'EULER a provoqué
suffisamment de remous pour permettre à RIEMANN de découvrir le rôle fondamental
de l'analysis situs en mathématiques. Relisez dans cette optique le dialogue de
sourds entre SERRET et CROFTON que nous présentons un peu plus loin. C'est en
s'appuyant sur les limites de l'autre que ce dernier a affiné sa propre pensée.

La communication entre mathématicien et biologistes procède de la même
manière. Certes il y a les modèles tout prêts, la formule de CAUCHY par exemple,
immédiatement utilisables en histologie quantitative. Mais ce sont là mathématiques mortes, dont la diffusion, bien que souhaitable, ne justifierait pas à
elle seule un symposium tel que le nôtre. La vraie communication vient de ce que
nos deux mondes imaginaires ont besoin de se heurter l'un à l'autre pour connaître leurs propres limites. Car les biologistes aussi, se construisent un monde
d'images. Entre un foelgen et un golgi, entre une micrographie optique et une
autre, électronique, lesquelles des histologies sont plus proches du tissu
"vrai" ? Imaginez simplement qu'on veuille décrire les anisotropies d'un coeur
en train de battre, ou les techniques d'un globule blanc façocytant une amibe,
ou encore le réseau des connexions nerveuses d'un plexus ; sommes nous vrai-
ment sûrs que nos méthodes mathématiques, nées de la physique du XIX° siècle,
seront adaptées à ces objets biologiques ? Et sont-ils toujours certains, les
biologistes, que les caractéristiques majeures des objets qu'ils présentent
sont effectivement celles qu'ils en donnent ?

C'est l'avenir qui progressivement apportera des réponses à ces questions,
et les méthodes d'analyse d'image en biologie semblent encore trop jeunes pour
qu'on puisse en dresser un historique. Ce qui est dommage. La façon dont chacun
des deux mondes imaginaires dérange l'autre nous aurait sûrement beaucoup
instruits. Mais il est possible par contre - et nous allons le faire - de voir
comment par le même mécanisme, ont pris tournure, petit à petit, les théories
mathématiques mises en oeuvre dans les méthodes actuelles d'analyse d'image.
Faute de place, nous achèverons à l'aube du XX° siècle ce survol un peu rapide,
mais déjà révélateur, avec ses tête-à-queue, et ses erreurs nourricières.

Oui, depuis deux siècles le vaisseau de la géométrie intégrale et des probabilités géométriques se meut à coup de gaffes, mais il avance quand même...

———————

Avant de laisser parler les textes, quelques mots de commentaires. Dans la mesure du possible, nous avons surtout retenu les premiers textes de chaque auteur, même lorsqu'ils ne sont pas les plus complets, car ils sont de loin les plus vivants (chez CROFTON, c'est typique). De même, quand c'était possible nous avons photocopié les éditions originales de leurs écrits, aux typographies si suavement vieillottes...

Si nous avons indiqué les nationalités des auteurs cités, c'est parce qu'elles révèlent des tendances profondes. Il est extraordinaire de découvrir que, dès leurs débuts, les probabilités géométriques sont d'inspiration franco-anglaise, alors que la géométrie intégrale s'annonce comme beaucoup plus germanique (les Suisses EULER et STEINER furent d'expression allemande). Et comme l'esprit d'une culture ne se dément pas, le XX° siècle n'a fait qu'affirmer ces tendances. La chose méritait d'être soulignée.

Le rideau se lève sur le XVIII° siècle. Entrent en scène le Français BUFFON, pour les probabilités géométriques, et le Suisse EULER pour la géométrie intégrale. Et pendant plus d'un siècle, chacun va se développer à peu près indépendamment de l'autre. Explorons d'abord le phylum des probabilités géométriques.

Dès 1733, BUFFON résoud le problème de l'aiguille, et le signale oralement à l'académie, dans une communication dont on peut trouver aujourd'hui un bref commentaire dans l'Histoire de l'Acad... Paris, année 1733. C'est en 1777 seulement qu'il publie son étude, dans le supplément à son histoire naturelle intitulé "Essai d'arithmétique morale", signant l'acte de naissance de la géométrie aléatoire (réf. a). Il y calcule la probabilité qu'un écu circulaire, lancé au hasard sur des pavés carrés, hexagonaux, ou losanges, tombe à l'intérieur d'un carreau. Puis il remplace l'écu par une aiguille et les carreaux par des lates de plancher infinies, et enfin celles-ci par un pavage rectangulaire, calculant à chaque fois la probabilité d'éviter un joint. Non sans erreur du reste, car la formule qu'il donne pour son dernier cas de figure est fausse, et il faudra attendre le texte de LAPLACE, 34 ans plus tard, pour en avoir un calcul exact.

Ce dernier (réf. b) reprend, mais sans le citer, l'analyse de BUFFON dans sa théorie des probabilités de 1812. En fait le changement est notable. D'abord il ne se trompe pas (son dernier résultat $\frac{4(a+b)r-4r^2}{ab\ \pi}$, valait chez BUFFON, avec les mêmes notations pour a = b, $\frac{2(a-r)r}{\pi a^2}$). D'autre part, par son maniement du calcul intégral, par ses notations (la constante s'appelle c chez BUFFON) LAPLACE est moderne. Mais surtout, il apparaît dès le premier paragraphe comme l'initiateur de la méthode de Monte-Carlo. L'optique très pascalienne de jeux de hasard a disparu et l'on commence à s'interroger sur le calcul de l'intégrale d'une courbe ou d'une surface.

Il faut attendre ensuite un peu plus d'un demi-siècle pour voir se construire, avec l'anglais M.W. CROFTON, la première théorie des probabilités géométriques digne de ce nom. Les préoccupations de l'époque ont changé. On se demande maintenant (réf. c) quelles sont les caractéristiques probables d'un triangle dont les trois sommets sont pris au hasard dans un cercle. C'est l'époque om SYLVESTER pose son célèbre problème, de la probabilité pour quatre points au hasard de donner un quadrilatère convexe. Nous ne résistons pas au plaisir de citer en détail la première communication de CROFTON, elle est pleine de trouvailles, et de réflexions profondes. Il tient du prodige que cet homme ait abouti à tant de résultats avec des outils aussi simples sans même utiliser la notion de probabilité conditionnelle (qui pourtant apparaît en filigrane dans sa densité $\theta - \sin \theta$!) (réf. d).

La première communication de CROFTON donna lieu à un intéressant échange de note. J.A. SERRET redémontra l'un de ses théorèmes, par des

moyens géométriques élémentaires, pour les polygones (réf. e). D'une
certaine manière c'est une génération, qui va bien dans le sens de la
géométrie intégrale, mais c'est aussi une régression, car SERRET ne reprend pas les modes de raisonnement, si probabilistes, de CROFTON. D'où
la réplique de ce dernier (réf. f) qui lui fait parvenir quelques mois
plus tard, un nouveau théorème, indémontrable (au moins pour lui, CROFTON)
autrement que par ses probabilités géométriques !

Il y aurait encore bien des textes à citer. Nous ne retiendrons que
le plus poétique, où l'on voit Captain O.C. FOX lancer son bourre-pipe
par terre cinq cent quatre vingt dix fois... (réf. g).

Passons à la géométrie intégrale.

C'est en latin que le Suisse EULER publie en Russie, à l'époque de
BUFFON, son célèbre théorème sur les polyèdres (1753), que l'on s'accorde
à considérer à la fois comme un point de départ de la géométrie intégrale,
et celui de la topologie algébrique.

Nous avons reproduit (réf. h) les deux pages essentielles de son
texte. Il est à noter que DESCARTES avait démontré un siècle plus tôt le
théorème selon lequel "Dans tout polyèdre la somme des angles solides est
égale à huit angles solides droits" qui équivaut au premier théorème
d'EULER. Mais il n'a pas vu le second, qui est de loin le plus important,
et qui a fait couler beaucoup d'encre. En effet, la démonstration d'EULER
est erronnée. Son principe de troncature n'aboutit pas nécessairement à un
seul polyèdre, et de plus il a omis de considérer la présence de trous dans
ses solides. Félix culpa ! Comme aurait dit EULER ; car de 1809, où CAUCHY
a donné une autre démonstration du même théorème, fausse elle aussi, mais
débouchant sur les graphes planaires, jusqu'à la synthèse finale de cette
question par LEBESGUE au XX° siècle, LUILLIER (1813) a pu montrer les erreurs
de ses prédécesseurs, VON STAUDT en fonder la démonstration sur des hypothèses
convenables (1847) et SCHLÄFLI (1850) la généraliser à l'espace à n dimensions !
Et nous ne citons que les principaux auteurs ; les avatars du théorème d'EULER
pourraient a eux seuls remplir un livre entier (*)...

* On peut lire à ce sujet les deux intéressant chapitres que leur consacre
  J.C. PONT dans sa "Topologie algébrique des origines à POINCARE", PUF 1975.

Près d'un siècle après EULER et indépendamment de son travail apparaissent à quelques mois d'écart deux études brèves, mais que le recul du temps fait apparaître comme les deux piliers sur lesquels s'est édifiée toute la géométrie intégrale. Ce sont le théorème des projections du français CAUCHY (réf. i) et la formule des dilatations du suisse STEINER (réf. j). Beaucoup plus tard, au XX° siècle, lorsque grâce à MINKOWSKI et ses successeurs l'étude des convexes aura acquis ses lettres de noblesse, on ne pourra l'étendre à des ensembles plus généraux, comme l'anneau convexe, qu'en choisissant chaque fois entre l'un ou l'autre des deux théorèmes, renonçant à s'appuyer sur l'autre. Ou bien ce sera la formule de CAUCHY qu'on gardera, avec toute sa puissance stéréologique (voir BLASCHKE, SANTALO), ou bien celle de STEINER, avec ses possibilités d'interprétation locale, en termes de mesures (MATHERON). Les deux points de vue, révélés vers 1840, décelaient en fait une fissure profonde.

Dans la lignée de CAUCHY se situe le texte du jeune Français BARBIER (réf. k) qui étend le travail du maître aux cas stéréologiques courants du plan qui coupe une courbe dans l'espace, ou de la droite qui transperce une surface.

Ceci dit, BARBIER, comme CAUCHY, ne s'inquiétaient guère de savoir si leurs surfaces étaient mesurables. Il était sous-entendu que toutes les fonctions mises en jeu admettaient toujours et partout des dérivées. Pour les mathématiciens du XX° siècle, ce petit côté archaïque irrite un peu, même s'il n'est pas très gênant en pratique. Les modèles sont maintenant plus fins, et pensés différemment. Ce changement de style, général aux mathématiques, fit son apparition vers la fin du siècle, et pour la géométrie intégrale, sous l'impulsion de l'allemand MINKOWSKI. Dans notre parallèle ce dernier joue un rôle équivalent à celui dévolu à CROFTON dans les probabilités géométriques. Il fait basculer la méthode, l'investit de préoccupations modernes. Pour la première fois, il utilise systématiquement les fonctions d'appui, établit les inégalités isopérimétriques, et découvre, pratiquement, le théorème sur les mesures de surface. Il y en aurait long à dire sur son oeuvre qui se classe plutôt comme un début qu'une fin. Mais cela nous aurait trop entrainé vers le XX° siècle.Nous nous contenterons ici (réf. l) de ne retenir que les quatre premières pages de son texte "Volumen und Oberfläch" (1903) qui résument ses principaux résultats.

Pour la fin, nous avons gardé le captain FOX de la géométrie intégrale, à savoir le français DELESSE (1847) (réf. m) cher aux stéréologistes, qui, s'il ne jetait pas son bourre-pipe, envoya bien des gouttes d'huile dans bien des cailloux.....

ESSAI

Je suppose que dans une chambre, dont le parquet est simplement divisé par des joints parallèles, on jette en l'air une baguette, & que l'un des joueurs parie que la baguette ne croisera aucune des parallèles du parquet, & que l'autre au contraire parie que la baguette croisera quelques-unes de ces parallèles; on demande le sort de ces deux joueurs. *On peut jouer ce jeu sur un damier avec une aiguille à coudre ou une épingle sans tête.*

Pour le trouver, je tire d'abord entre les deux joints parallèles A B & C D du parquet, deux autres lignes

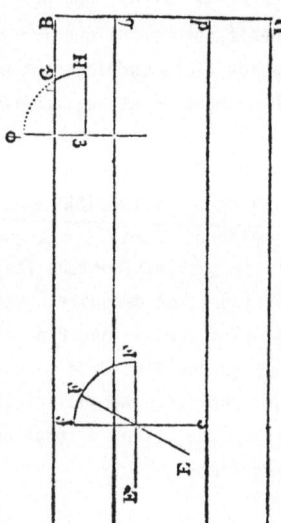

parallèles a b & c d, éloignées des premières de la moitié de la longueur de la baguette E F, & je vois évidemment que tant que le milieu de la baguette sera entre ces deux secondes parallèles, jamais elle ne pourra croiser les premières dans quelque situation E F, e f, qu'elle puisse se trouver; & comme tout ce qui peut arriver au-dessus de a b arrive de même au-dessous de c d, il ne s'agit que de déterminer l'un ou l'autre; pour cela je remarque

que toutes les situations de la baguette peuvent être représentées par le quart de la circonférence du cercle dont la longueur de la baguette est le diamètre: appelant donc 2 a la distance C A des joints du parquet, C le quart de la circonférence du cercle dont la longueur de la baguette est le diamètre, appelant 2 b la longueur de la baguette, & f la longueur A B des joints, j'aurai $f\left(\overline{a-b}\right)$ c pour l'expression qui représente la probabilité de ne pas croiser le joint du parquet, ou ce qui est la même chose, pour l'expression de tous les cas où le milieu de la baguette tombe au-dessous de la ligne a b & au-dessus de la ligne c d.

Mais lorsque le milieu de la baguette tombe hors de l'espace a b d c, compris entre les secondes parallèles,

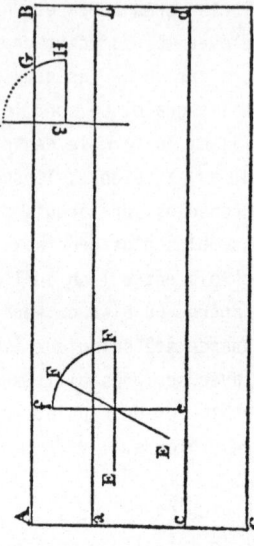

elle peut, suivant sa situation, croiser ou ne pas croiser le joint; de sorte que le milieu de la baguette étant, par exemple, en s, l'arc φ G représentera toutes les situations où elle croisera le joint, & l'arc G H toutes

BUFFON

(réf. a)

103   E S S A I

celles où elle ne le croisera pas, & comme il en fera de même de tous les points de la ligne ιφ, j'appelle dx les petites parties de cette ligne, & y les arcs de cercle φ G, & j'ai $f(\int y\,dx)$ pour l'expression de tous les cas où la baguette croisera, & $f(bc-\int y\,dx)$ pour celle des cas où elle ne croisera pas; j'ajoute cette dernière expression à celle trouvée ci-dessus $f(a-b)c$, afin d'avoir la totalité des cas où la baguette ne croisera pas, & dès-lors je vois que le fort du premier joueur est à celui du second, comme $ac-\int y\,dx : \int y\,dx$.

Si l'on veut donc que le jeu soit égal, l'on aura $ac=2\int y\,dx$ ou $a=\dfrac{\int y\,dx}{\frac{1}{2}c}$, c'est-à-dire, à l'aire d'une partie de cycloïde, dont le cercle générateur a pour diamètre $2b$ longueur de la baguette; or, on fait que cette aire de cycloïde est égale au carré du rayon, donc $a=\dfrac{bb}{\frac{1}{2}c}$, c'est-à-dire, que la longueur de la baguette doit faire à peu-près les trois quarts de la distance des joints du parquet.

La solution de ce premier cas nous conduit aisément à celle d'un autre qui d'abord auroit paru plus difficile, qui est de déterminer le fort de ces deux joueurs dans une chambre pavée de carreaux carrés, car en inscrivant dans l'un des carreaux carrés, un carré éloigné par-tout des côtés du carreau de la longueur b, l'on aura d'abord $c\,\overline{(a-b)}^2$ pour l'expression d'une partie des cas où

D'ARITHMÉTIQUE MORALE.   104

la baguette ne croisera pas le joint; ensuite on trouvera $\overline{(2a-b)}\int y\,dx$ pour celle de tous les cas où elle croisera, & enfin $cb\overline{(2a-b)}-\overline{(2a-b)}\int y\,dx$ pour le reste des cas où elle ne croisera pas; ainsi le fort du premier joueur est à celui du second, comme $c\,\overline{(a-b)}^2+cb\overline{(2a-b)}-\overline{(ca-b)}\int y\,dx : \overline{(2a-b)}\int y\,dx$.

Si l'on veut donc que le jeu soit égal, l'on aura $c\,\overline{(a-b)}^2+cb\overline{(2a-b)}=\overline{(2a-b)}\int y\,dx$ ou $\dfrac{\frac{1}{2}caa}{2a-b}=\int y\,dx$; mais comme nous l'avons vu ci-dessus, $\int y\,dx=bb$; donc $\dfrac{\frac{1}{2}caa}{2a-b}=bb$; ainsi le côté du carreau doit être à la longueur de la baguette, à peu-près comme $\frac{41}{31}:1$, c'est-à-dire, pas tout-à-fait double. Si l'on jouoit donc fur un damier avec une aiguille dont la longueur feroit la moitié de la longueur du côté des carrés du damier, il y auroit de l'avantage à parier que l'aiguille croisera les joints.

On trouvera par un calcul semblable, que si l'on joue avec une pièce de monnoie carrée, la somme des forts fera au fort du joueur qui parie pour le joint, comme $ac: 4abb\sqrt{\tfrac{1}{2}}-b^3-\tfrac{1}{2}Ab$, A marque ici l'excès de la superficie du cercle circonscrit au carré, & b la demi-diagonale de ce carré.

DES PROBABILITÉS. 359

causes morales; car l'action de ces causes, lorsqu'elle est répétée un grand nombre de fois, offre dans ses résultats, autant de régularité, que les causes physiques.

On peut encore déterminer par l'analyse des probabilités, comparée à un grand nombre d'expériences, l'avantage et le désavantage des joueurs, dans les cas dont la complication rend impossible leur recherche directe. Tel est l'avantage de la main, au jeu du piquet : telles sont encore les possibilités respectives d'amener les différentes faces d'un prisme droit rectangulaire, dont la longueur, la largeur et la hauteur sont inégales ; lorsque le prisme projeté en l'air, retombe sur un plan horizontal.

Enfin, on pourrait faire usage du calcul des probabilités, pour rectifier les courbes ou carrer leurs surfaces. Sans doute, les géomètres n'emploieront pas ce moyen ; mais comme il me donne lieu de parler d'un genre particulier de combinaisons du hasard, je vais l'exposer en peu de mots.

Imaginons un plan divisé par des lignes parallèles, équidistantes de la quantité a; concevons de plus un cylindre très-étroit dont $2r$ soit la longueur, supposée égale ou moindre que a. On demande la probabilité qu'en le projetant, il rencontrera une des divisions du plan.

Elevons sur un point quelconque d'une de ces divisions, une perpendiculaire prolongée jusqu'à la division suivante. Supposons que le centre du cylindre soit sur cette perpendiculaire, et à la hauteur y au-dessus de la première de ces deux divisions. En faisant tourner le cylindre autour de son centre, et nommant φ l'angle que le cylindre fait avec la perpendiculaire, au moment où il rencontre cette division; 2° sera la partie de la circonférence décrite par chaque extrémité du cylindre, dans laquelle il rencontre la division; la somme de toutes ces parties sera donc $4\int\varphi dy$, ou $4\varphi y - 4\int y d\varphi$; or on a $y = r.\cos\varphi$; cette somme est donc

$$4\varphi y - 4r.\sin\varphi + \text{constante}.$$

Pour déterminer cette constante, nous observerons que l'intégrale doit s'étendre depuis y nul jusqu'à $y = r$, et par conséquent depuis

560 THÉORIE ANALYTIQUE

$\varphi = \frac{\pi}{2}$ jusqu'à $\varphi = o$, ce qui donne

$$\text{constante} = 4r;$$

ainsi la somme dont il s'agit est $4r$. Depuis $y = a - r$ jusqu'à $y = a$, le cylindre peut rencontrer la division suivante, et il est visible que la somme de toutes les parties relatives à cette rencontre, est encore $4r$; $8r$ est donc la somme de toutes les parties relatives à la rencontre de l'une ou l'autre des divisions par le cylindre, dans le mouvement de son centre le long de la perpendiculaire. Mais le nombre de tous les arcs qu'il décrit en tournant en entier sur lui-même, à chaque point de cette perpendiculaire, est $2a\pi$; c'est le nombre de toutes les combinaisons possibles; la probabilité de la rencontre d'une des divisions du plan par le cylindre, est donc $\frac{4r}{a\pi}$. Si l'on projette un grand nombre de fois ce cylindre, le rapport du nombre de fois où le cylindre rencontrera l'une des divisions du plan, au nombre total des projections, sera par le n° 16, à très-peu près, la valeur de $\frac{4r}{a\pi}$, ce qui fera connaître la valeur de la circonférence 2π. On aura, par le même numéro, la probabilité que l'erreur de cette valeur sera comprise dans des limites données ; et il est facile de voir que le rapport $\frac{2r}{\pi}$ qui, pour un nombre donné de projections, rend l'erreur à craindre la plus petite, est l'unité; ce qui donne la longueur du cylindre égale à l'intervalle des divisions.

Concevons maintenant le plan précédent divisé encore par des lignes perpendiculaires aux précédentes, et équidistantes d'une quantité b égale ou plus grande que la longueur $2r$ du cylindre. Toutes ces lignes formeront avec les premières, une suite de rectangles dont b sera la longueur et a la hauteur. ..... la probabilité de la rencontre des divisions du plan par le cylindre, est donc

$$\frac{4.(a+b).r - 4r^2}{ab.\pi}.$$

LAPLACE
-1812-
(réf. b)

VII. *On the Theory of Local Probability, applied to Straight Lines drawn at random in a plane; the methods used being also extended to the proof of certain new Theorems in the Integral Calculus.* By MORGAN W. CROFTON, *B.A., of the Royal Military Academy, Woolwich; late Professor of Natural Philosophy in the Queen's University, Ireland. Communicated by* J. J. SYLVESTER, *F.R.S.*

Received February 5,—Read February 27, 1868.

1. THE new Theory of Local or Geometrical Probability, so far as it is known, seems to present, in a remarkable degree, the same distinguishing features which characterize those portions of the general Theory of Probability which we owe to the great philosophers of the past generation. The rigorous precision, as well as the extreme beauty of the methods and results, the extent of the demands made on our mathematical resources, even by cases apparently of the simplest kind, the subtlety and delicacy of the reasoning, which seem peculiar to that wonderful application of modern analysis—*ex extend deliret*, as it has been aptly described by LAPLACE—reappear, under new forms, in this, its latest development. The first trace which we can discover of the Theory of Local Probability seems to be the celebrated problem of BUFFON the great naturalist* — a given rod being placed at random on a space ruled with equidistant parallel lines, to find the chance of its crossing one of the lines. Although the subject was noticed so early, and though BUFFON's and one or two similar questions have been considered by LAPLACE, no real attention seems to have been bestowed upon it till within the last few years, when this new field of research has been entered upon by several English mathematicians, among whom the names of SYLVESTER and WOOLHOUSE† are particularly

* The mathematical ability evinced by BUFFON may well excite surprise; that one whose life was devoted to other branches of science should have had the sagacity to discern the true mathematical principles involved in a question of so entirely novel a character, and to reduce them correctly to calculation by means of the integral calculus, thereby opening up a new region of inquiry to his successors, must move us to admiration for a mind so rarely gifted.

† Many remarkable propositions on the subject, by these eminent mathematicians, have appeared in the mathematical columns of the 'Educational Times' and other periodicals. A very important principle has been introduced by Professor SYLVESTER, which may be termed *decomposition of probabilities*. For instance, he has shown that the probability of a group of three points, taken at random within a given triangle, fulfilling a given intrinsic condition (*i. e.* one depending solely on the internal relations of the points among each other), may be expressed as a linear function of two simpler probabilities; viz. first of the same condition being fulfilled when one of the points are fixed at a vertex of the triangle, and a second restricted to the opposite side; (2) when all three points are restricted, one to each side of the triangle. The order of the integrations required

MDCCCLXVIII.
2 D

distinguished. It is true that in a few cases differences of opinion have arisen as to the principles, and discordant results have been arrived at, as in the new celebrated *three-point* problem, by Mr. WOOLHOUSE, and the *four-point* problem of Professor SYLVESTER; but all feel that this arises, not from any inherent ambiguity in the subject matter, but from the weakness of the instrument employed; our undisciplined conceptions of a novel subject requiring to be repeatedly and patiently reviewed, tested, and corrected by the light of experience and comparison, before they are purged from all latent error.

The object of the present paper is, principally, the application of the Theory of Probability to straight lines drawn at random in a plane; a branch of the subject which has not yet been investigated. It will be necessary to begin by some remarks on the general principles of Local Probability. Some portion of what follows I have already given elsewhere*.

2. The expression "*at random*" has in common language a very clear and definite meaning; one which cannot be better conveyed than by Mr. WILSON's expression "*according to no law.*" It is thus of very wide application, being often used in cases altogether beyond the province of mathematical measurement or calculation.

In Mathematical Probability, which consists essentially in arithmetical calculation, when we speak of a thing of any kind taken at random, there is always a direct reference to the *assemblage of things* to which it belongs and from which it is taken, at random,—which here comes to the same thing as saying that any one is as likely to be taken as any other. When we have a clear conception of what the assemblage is, from which we take, and not till then, we can proceed to sum up the favourable cases.

In many problems on probability there is no difficulty in forming a clear conception of the total number of cases. Thus if balls are drawn from an urn, the number of cases is the number of balls, or of certain combinations of them; and if the number of balls be supposed infinite, no obscurity arises from this. But there are several classes of questions in which the totality of cases is not merely infinite, but of an inconceivable nature. Thus if we try to imagine how to determine completely by experiment the probability of a hemisphere thrown into the air falling on its base, we may suppose an infinite number of persons to make one trial each; afterwards we may suppose each person to make two, three, or an infinite number of trials; again, we may suppose for every trial that has taken place an infinite number of others, varying, for instance, in the substance, size, &c. of the body employed; and so on. We can thus continually suppose variations of the experiment, each variation giving a new infinity of cases. Now problems of this nature are treated by means of the following principle:—

In any question of probability regarding an infinite number of cases, all equally pro-

is thus reduced by three. The same method applies to any polygon, and also to the points taken in space within a tetrahedron. It is to be hoped that Professor SYLVESTER will give these remarkable results to the public in a detailed form: a general account of them was given to the British Association at Birmingham in 1865.

* Educational Times, May 1866.

MR. M. W. CROFTON ON THE THEORY OF LOCAL PROBABILITY. 183

lable, the result will be unaltered if we take, instead of these cases, *any lesser cases, chosen at random from among them*.

3. The case of a point or straight line taken at random in a plane or in space is a problem of the above description. Thus, if a point be taken at random in a plane, the total number of cases is of an inconceivable nature, inasmuch as a plane cannot be *filled* with mathematical points, any infinitesimal element of the plane containing an unlimited number of points. We see, however, by means of the above principle, that we may consider the assemblage we are dealing with, as *an infinity of points all taken at random in the plane*.

Let us examine the nature of this assemblage. As the points continue to be scattered at random over the plane, their density tends to become uniform. It is evident, in fact, that a random point is as likely to be in any element *dS* of the surface, as in any equal element *dS'*; and therefore by continuing to multiply points, the number in *dS* will be equal (or *sedequard*, to use a term of Professor DE MORGAN's) to that in *dS'*. Of course, though the density tends to become uniform, the disposition of the points does not tend to become symmetrical; those within any element *dS* will be dispersed in the most irregular manner over that element? However, it is important to remark that, *for all purposes of calculation*, the ultimate disposition may be supposed symmetrical; for as the position of any point is determined by that of the element *dS*, within which it falls, it matters not what arbitrary arrangement we assume for the points within the element.

* This proposition, of which, in a somewhat different form, a mathematical demonstration is given by LAPLACE (Théorie Analytique des Probabilités, chap. 3), may be regarded as almost axiomatic. Thus, suppose an urn to contain an infinite number of black and white balls, in the proportion of 2 to 3; if any lesser infinite number of balls be drawn from it, the black ones among them will be to the white as 2 to 3. For, imagine all the balls ranged in a row AGH, the black ones from A to C, the white from C to B: if we now select an infinite number at random from among them, it appears self-evident that, if the line be divided into five equal parts, the numbers of balls taken from each part will be the same, or rather, will *tend* to equality on being increased indefinitely. Hence the black balls selected will be to the white as AC to CB, or as 2 to 3. When the numbers are made infinite, this principle is approximately true, and forms, as is well known, the basis of most of the practical applications of Probability. Thus the chance of an infant living to the age of twenty is as truly found from, say, 1,000,000 of observed cases, as it would be from the total number.

In its strict mathematical form, the proposition may be thus stated:—In any unlimited number of cases, divided into favorable and unfavorable, if *p* be the ratio of the favourable to the whole number of cases, and if we select any infinite number of cases at random from among them, *the probability is infinitely small, that the same ratio, as determined from the selected cases, shall differ from p by a finite quantity*.

† Order thus results from disorder, the uniform density of the aggregate being unaffected by the disorder and irregularity of arrangement of its ultimate constituents; much as a nebula of uniform brightness is related to the stars which compose it. This remarkable law is to be traced, under one form or another, in most of the applications of the Theory of Probability.

"Elle mérite l'attention des philosophes, en faisant voir comment la régularité finit par s'établir dans les choses mêmes qui nous paraissent entièrement livrées au hasard."—*Laplace*.

A familiar illustration of the tendency to uniform density in the random points may be derived by observing the drops of rain on a pavement at the commencement of a shower: as the drops multiply, it will be evident to the eye that their density tends more and more to uniformity.

184 MR. M. W. CROFTON ON THE THEORY OF LOCAL PROBABILITY.

Hence we may, if we please, assume that, when a point is taken at random in a plane, those from which it is taken are an infinite number symmetrically disposed over the plane.

Likewise, points taken at random in a line may be supposed equidistant. And if random values be taken for any *quantity*, they may be supposed to form an arithmetical series, with an infinitesimal difference.

Let us now consider the case of a straight line drawn at random in an infinite plane: the assemblage from which we select it is, as before, *an infinity of lines drawn at random in the plane*. What is the nature of this aggregate? First, since any direction is as likely as any other, as many of the lines are parallel to any given direction as to any other. Consider one of these systems of parallels; let them be cut by any infinite perpendicular. As this infinite system of parallels is drawn at random, they are as thickly disposed along any part of the perpendicular as along any other; the intersections being in fact random points on the perpendicular. Hence it is easily seen that, for all purposes of calculation, the assemblage of lines may be thus conceived. Divide the angular space round any point into a number of equal angles *δθ*, and for every direction let the plane be ruled with an infinity of equidistant parallel lines, the common infinitesimal distance being the same for every set of parallels. Or we may suppose one such system of parallels drawn, and then turned through an angle *δθ*, then through another equal angle, and so on, till they have returned to their former direction.

If we take any fixed axes in the plane, a random line may be represented by the equation

$$x \cos\theta + y \sin\theta = p,$$

where *p* and *θ* are constants taken at random.

There is no difficulty in extending now our conceptions to points, straight lines, and planes, taken at random in *space*.

4. We may take any plane area as the *measure* of the number of random points within it: in the case of random lines, I proceed to prove the following important principle:—

The measure of the number of random lines which meet a given closed convex plane boundary, is the length of the boundary.

Draw any system of parallels meeting the boundary, their common infinitesimal distance being *δp*. If we take this distance as unity, the number of these parallels is AB, a line cutting them at right angles. Let AB=ξ, and let θ be its inclination to any fixed direction in the plane; conceive now a consecutive system of parallels inclined to the former at an angle *δθ*, then a third, and so on, till the parallels return to the direction in the figure; then the total number of lines will be

$$\frac{1}{\delta p}\int \xi d\theta;$$

or, if O be any fixed pole inside the boundary, and OV=p, the perpendicular on the

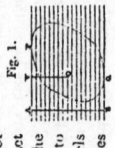

Fig. 1.

tangent to the boundary, θ its inclination to a fixed axis, the measure of the number of lines* is

$$N=\int_0^{2\pi} p\, d\theta.$$

Now the integral $\int p\, d\theta$ extended through four right angles gives the *whole length of the boundary*, whatever be its nature, provided it be convex†.

Hence if L be the length of the boundary,

$$N=L.$$

This result may be obtained also as follows. It may be shown very simply by the above principles that the measure of the number of random lines which meet any finite straight line of length a, is $2a$ (it may indeed be assumed as self-evident that the number is proportional to a). Conceiving now the boundary L as consisting of straight elements, the number of lines meeting any element ds, is $2ds$; so that the whole number which meet the boundary would be $2L$; but as each line cuts the boundary in *two points*, we should thus count each line twice over; hence the true number is L.

Hence if L be the length of any convex boundary, and l that of another, lying wholly inside the former, the probability that a line drawn at random across L shall also intersect l, is

$$p=\frac{l}{L}$$

It is important to observe that the measure of the number of lines which meet any *non-convex boundary* is *the length of a string drawn tightly round it*; as is obvious on consideration. The same is true for a boundary which is not closed.

5. Let there be any two boundaries external to each other; let X be the length of an endless band passing round both, and crossing between them, and Y the length of another endless band also enveloping both, but not crossing; then *the measure of the number of random lines which meet both boundaries is X−Y.*

It will be easily found from the principles explained above, that the number required will be the integral $\int p\, d\theta$ (referred to O as pole), taken for the left-hand curve from the position RR' of its tangent, to the position P'O of its tangent, to the position SS; then for the right-hand one, from the position P'O of its tangent, to the position SS; then for the left-hand one, from SS' to QQ; then for the right-hand one, from QO to RR. Now the values of these integrals are, drawing the perpendiculars OV, OW to RR', SS,

Fig. 2.

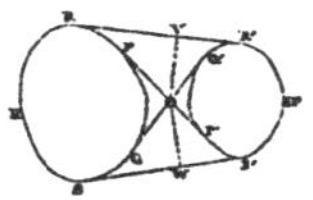

* It will be well to remember that this *measure* of the number of lines, N, means *the actual number multiplied by the constant factor* π. Our notation is thus simplified, and no confusion need arise from sometimes using "the number of lines," for shortness, instead of "the measure of the number of lines." As $d\theta$ remains constant throughout our investigation, henceforth we will denote it by π.

† As $L=\int_0^{2\pi} p\, d\theta$, we see that *the mean breadth of any convex area is equal to the diameter of a circle whose circumference equals the length of the boundary.* By *breadth* is meant the distance between two parallel tangents, whose direction is supposed to alter by uniform increments.

1. the mixed line RPO — RV,
2.   ,,   ,,   SPO — SW,
3.   ,,   ,,   SQO — SW,
4.   ,,   ,,   RQO — RV,

and the sum of these is evidently equal to X−Y.

I will add a different proof of this proposition, deduced from art. 4, as it is interesting to see our results verified.

For shortness, I will use the symbol N(S) for "the number of random lines meeting the space S," and N(S, S') for the number meeting both S and S'.

The number of lines meeting both boundaries is evidently identical with the number meeting both the mixtilinear figures OPHQ, OP'H'Q'. These two figures together form the mixtilinear reentrant figure HPP'H'QQ', and by art. 4, N(HPP'H'QQ')=Y.

Now N(OPHQ)+N(OP'H'Q')=N(HPP'H'QQ')+N(OPHQ, OP'H'Q'). But OPHQ, OP'H'Q' being convex figures, the number of lines meeting each is represented by its length; therefore

$$X=Y+N(HP'Q, HP'Q');$$

The probability that a line drawn at random across a given convex boundary of length L shall also meet a given *external* boundary is therefore

$$p=\frac{X-Y}{L}$$

6. If two convex boundaries L, L' intersect each other, in two or more points, it may be proved in a similar manner that the number of random lines which meet both is represented by L+L'−Y, where Y is the length of an endless band passing round both. Hence the probability that a line which meets L shall also meet L', is

$$p=\frac{L+L'-Y}{L}$$

7. It may easily be proved that the measure of *the number of random lines which pass between two given convex boundaries* is

$$N=PP'+QQ'-\text{arc } PQ-\text{arc } P'Q',$$

where PP', QQ' are the two common tangents which cross each other.

Thus the number of random lines which pass between the two branches of an hyperbola is represented by Δ, the difference between the whole length of the hyperbola and that of its asymptotes. This difference, as is known, is given by the definite integral

$$\Delta = 4a\int_\alpha \sqrt{1-e^2\sin^2\theta}\, d\theta,$$

where $\sin\alpha=\dfrac{1}{e}$,

CROFTON 1

SUR UN
PROBLÈME DE CALCUL INTÉGRAL,

Par M. J.-A. SERRET,
MEMBRE DE L'INSTITUT.

I.

Un savant anglais, M. Crofton, a communiqué, il y a quelque temps, à l'Académie un théorème de calcul intégral qui n'a pas manqué de fixer l'attention des géomètres, tant à cause de l'élégance du résultat obtenu que de la méthode singulière et ingénieuse dont l'auteur a fait usage pour l'établir. Voici en quels termes M. Crofton a énoncé son théorème :

« Soit un contour convexe de forme quelconque, dont la longueur totale est L, et qui renferme un espace Ω; si l'on appelle θ l'angle des deux tangentes menées d'un point extérieur (x, y) à ce contour, on aura l'intégrale

$$\iint (\theta - \sin\theta)\, dx\, dy = \tfrac{1}{2} L^2 - \pi\Omega$$

pour toute la surface du plan, extérieure au contour (*). » x et y désignent, bien entendu, des coordonnées rectangulaires.

Il est très-remarquable que ce théorème subsiste lorsque le contour convexe L, au lieu d'être une courbe continue, est formé de parties droites ou courbes faisant entre elles des angles quelconques. L'angle θ

(*) Comptes rendus des séances de l'Académie des Sciences, t. LXV, p. 994.

Annales scientifiques de l'École Normale supérieure. Tome VI.   25

---

est toujours celui sous lequel le contour L est vu du point dont les coordonnées rectangulaires sont x et y; mais les droites *circonscrivantes* qui se sont les côtés ne sont plus nécessairement des tangentes, et elles peuvent pivoter autour des divers sommets du contour.

Il est évident que, pour établir la formule de M. Crofton dans toute sa généralité, il suffit de se borner au cas où le contour L est un polygone rectiligne convexe d'un nombre quelconque n de côtés; la démonstration peut être alors présentée d'une manière très-simple, comme il suit.

L'origine des coordonnées étant placée à l'intérieur du polygone, soit ω l'angle formé par le rayon vecteur du contour L avec la direction des abscisses positives. Nous supposerons que cet angle croisse lorsque le rayon vecteur se meut en s'élevant de l'axe des x vers l'axe des y, et nous représenterons par $A_0, A_1, \ldots, A_{n-1}$ les sommets du polygone dans l'ordre où ils sont rencontrés, chaque indice pouvant être, si l'on veut, augmenté de n; nous désignerons par ω_{i-1} la valeur de ω, lorsque le rayon vecteur est perpendiculaire au côté $A_{i-1} A_i$. Posons

$$(1) \qquad v = \iint (\theta - \sin\theta)\, dx\, dy,$$

et désignons par $V_{i,j}$ la partie de l'intégrale V qui répond au cas où les droites circonscrivantes ne font que pivoter autour des sommets respectifs A_i, A_j. Soient α, β les valeurs que prend ω quand le rayon vecteur du contour est perpendiculaire à ces droites; les variables α, β seront liées aux coordonnées x, y par les équations

$$(2) \qquad \begin{cases} (x_i - x)\cos\alpha + (y_i - y)\sin\alpha = 0, \\ (x_j - x)\cos\beta + (y_j - y)\sin\beta = 0, \end{cases}$$

et si l'on fait

$$(3) \qquad \begin{cases} A_{i,j} = (x_i - x_j)\cos\alpha + (y_i - y_j)\sin\alpha, \\ B_{j,i} = (x_j - x_i)\cos\beta + (y_j - y_i)\sin\beta, \end{cases}$$

on aura, par les formules (2),

$$\frac{dx}{d\alpha}\frac{dy}{d\beta} - \frac{dz}{d\alpha}\frac{dy}{d\beta} = \frac{A_{i,j}\, B_{j,i}}{\sin^2(\beta - \alpha)};$$

ANALYSE MATHÉMATIQUE. — *Sur quelques théorèmes de calcul intégral.*
Lettre de M. CROFTON à M. J.-A. Serret.

« J'ai vu avec beaucoup d'intérêt la démonstration que vous avez donnée, dans les *Comptes rendus*, de mon théorème; je suis content de voir vérifier ce théorème par les méthodes ordinaires de la Géométrie, d'autant plus que j'avais essayé moi-même de le démontrer par quelque procédé usuel, mais sans réussir. Je vous ai envoyé de Londres un exemplaire de mon Mémoire dans les *Philosophical Transactions of the Royal Society*, où j'ai donné celui-ci avec d'autres théorèmes analogues sur des intégrales, qui se sont présentés à moi dans des recherches sur la probabilité. Si vous pensez qu'il y en ait parmi eux qui soient dignes de l'attention de l'Académie des Sciences, je serais honoré si vous vouliez bien les lui communiquer de ma part. J'ai obtenu aussi d'autres intégrales, qui ne sont pas dans le Mémoire, par des procédés semblables. La suivante me paraît assez remarquable par sa grande généralité.

» Étant donné un contour convexe quelconque, on mène une corde quelconque C; C sera une fonction de p et θ, p étant la perpendiculaire abaissée d'un pôle fixe O sur la corde, et θ étant l'inclinaison de p à une direction fixe OX. Soit I, la longueur du contour, Ω l'espace qu'il renferme, on aura

$$\int\int C^2 dp\, d\theta = 3\Omega^2,$$

où l'intégration s'étend à toutes les valeurs de p et θ qui donnent une corde réelle.

» Je n'ai pas pu vérifier cette intégrale par les moyens ordinaires; je l'ai obtenue en considérant l'espace Ω comme rempli d'une infinité de points disposés avec une densité uniforme; en considérant ensuite le système infini de lignes droites qu'on obtient en joignant chaque paire de points, le nombre de ces lignes sera proportionnel à Ω² : mais on peut les compter aussi d'une manière qui fait voir que leur nombre sera représenté par l'intégrale

$$\frac{1}{3}\int\int C^3 dp\, d\theta.$$

» D'après la théorie exposée dans mon Mémoire, il est facile de tirer la conclusion suivante de ce théorème :

» La valeur moyenne du *cube d'une corde menée au hasard dans un contour convexe quelconque* est $\frac{3\Omega^2}{L}$.

» Je n'ai pas pensé à mentionner dans le Mémoire la proposition suivante, qui découle immédiatement de cette théorie :

» La longueur moyenne d'une corde menée au hasard dans un contour est $\frac{\pi\Omega}{L}$.

» Ceci donne un résultat curieux : la longueur moyenne d'une corde menée au hasard dans un espace triangulaire est *la même que pour une corde menée au hasard dans le cercle inscrit*, savoir : le quart de la circonférence de ce cercle.

» Pour un polygone circonscrit à un cercle, il en est de même.

» Il est probable que la méthode que vous avez suivie dans les *Comptes rendus* s'appliquera à quelques autres de ces théorèmes. »

ASTRONOMIE. — *Analyse spectrale de la lumière de quelques étoiles.*
Note de M. WOLF.

« En 1867, nous avons fait connaître, M. Rayet et moi, l'existence, dans la constellation du Cygne, de trois petites étoiles (n°s 4001 et 4013, zone + 35 degrés; et 39.56, zone + 36 degrés du Catalogue de Bonn) dont le spectre présente des lignes brillantes, fait encore assez rare dans le ciel Au mois d'octobre de la même année, j'ai vérifié de nouveau l'existence et la position de ces lignes; en 1868, l'absence du télescope de $0^m,40$, emporté en Cochinchine pour l'observation de l'éclipse, ne m'a pas permis de poursuivre cette étude; mais nous avons pour cette même année les observations faites à Rome par le P. Secchi. Or, dans le Catalogue des étoiles colorées, dont le spectre a été observé au Collège romain, se trouve la Note suivante :

« *Étoiles de M. Wolf.* — On a examiné ces étoiles sans y rien voir d'extraordinaire; les faibles lignes lumineuses qu'on y voit à intervalles sont-elles de l'ordre de celles dues à la scintillation? (*Astronomische Nachrichten*, n° 1737.) »

» Dès que ces trois étoiles ont pu être observées ici, j'ai de nouveau analysé leur lumière; et, avec d'autres apparences que ceux qui m'avaient servi en 1867, j'ai retrouvé identiquement les apparences que j'ai décrites autrefois. Les lignes brillantes sont très-vives, et se voient d'une façon continue sans qu'il soit possible de trouver la aucun phénomène dû à la scintillation.

(réf. f)

The problem may be transformed by supposing two of the points of the group to range on the contour itself, according to a law which may be expressed by saying that the probability of their being found on any arc shall vary as the product of the segment included between the arc and its chord, multiplied by the time of describing the arc about any centre of force arbitrarily chosen within or upon the contour,—a theorem which, accepting the idea of negative probability, admits also of extension to the case of a centre of force exterior to the contour.

Among other problems which the author readily resolves by aid of his principle of transformation, may be mentioned that of determining the mean value of a triangle whose angles are taken at random anywhere within a given triangle, parallelogram, ellipse, or ellipsoid. In this description of questions a peculiar difficulty arises, from the fact that the figure which is to be integrated in order to determine the numerator of the fraction which gives its mean value must always be taken positive, whereas its algebraical expression will repeatedly change its sign, according to a more or less complicated law. This quality of the analytical exponent of the arithmetical value of the figure constitutes in fact, a sort of polarization which has to be got rid of; and the depolarizing process is effected with great ease by virtue of the simplified form impressed upon the data by the method set forth in the paper.

The author further took occasion briefly to allude to the form in which his own problem of four and Mr. Woolhouse's problem of three points were originally proposed, viz. in each case without a specified boundary, and to express his opinion that the principle which had been applied to them, and in which he had formerly acquiesced, was erroneous, as it could be made to lead to contradictory conclusions, and must be abandoned. He was strongly inclined to believe that, under their original form, these questions do not admit of a determinate solution.

On Professor Price's Modification of Arbogast's Method.
By Professor Sylvester, F.R.S.

SYLVESTER

-1865-

On the Calculation of the Potential of the Figure of the Earth.
By W. H. L. Russell.

The object of this paper was to simplify and under-symmetrical certain portions of Professor Ulrich's investigations on the figure of the earth. In that paper the reduction of the expression for the potential to a convenient form is effected by the introduction of a discontinuous quantity; the author of the present paper has found that the required form is obtained much more shortly by dividing the original definite integral into two parts, and then expanding separately.

On the application of D'Alembert's Principle to the Rotation of a Rigid Mass.
By Dr. Stevelly.

The author explained that the present method of applying D'Alembert's principle to the investigation of the spontaneous axis assumed by a free, rigid mass, under the action of force, in all the works he was acquainted with, led to what he showed to be a false conclusion, viz. that that axis must be a principal axis of the rigid mass. He showed how the error arose from neglecting, in applying the principle of D'Alembert, to take into account not only that part of the motion of each elementary part of the body which related to the *magnitude* of its motion, but also that part which relates to its *direction*, and from which its centrifugal endeavour at each instant arises. But if the force impressed tend to produce rotatory motion round an unstable spontaneous axis, how can the present mode of applying D'Alembert's principle lead to a true conclusion, when it proceeds on the method of bringing the body into such a state that the equations of equilibrium (that is, of no after-change) shall give the direction-courses of the axis?

On a Special Class of Questions on the Theory of Probabilities.
By Professor Sylvester, F.R.S.

After referring to the nature of geometrical or local probability in general, the author of the paper drew attention to a particular class of questions pertaking of that character in which the condition whose probability is to be ascertained is one of pure form. The chance of three points within a circle or sphere being apices of an acute or obtuse-angled triangle, or of the quadrilateral formed by joining four points, taken arbitrarily within any assigned boundary, constituting a reentrant or convex quadrilateral, will serve as types of the class of questions in view. The general problem is that of determining the chance that a system of points, each with its own specific range, shall satisfy any prescribed condition of form. For instance, we may suppose two pairs of points to be limited respectively to segments of the same indefinite straight line; the chance of their anharmonic ratio being under or over any prescribed limit will belong to this category of questions, to which, provisionally, the author proposed to attach the name of form-probability. In questions of form-probability, in which all the ranges are either collinear segments or coplanar areas, or defined portions of space, rules may be given for transforming the data, so as to make the required probability depend on one or more probabilities of a simpler kind, tending to summations of an order inferior by two degrees to those required by the methods in ordinary use. Thus Mr. Woolhouse's question relating to the chance of a triangle within a circle or sphere being acute can be made to depend upon an easy simple integration, the solutions heretofore given of this problem involving complicated triple integrals. It was shown, as a further illustration, that the form-probability of a group of points all ranging over the same triangle remains unaltered when the range of one of them is limited to any side of the triangle chosen at will, and, again, (for convenience of expression distinguishing the contour into a base and two sides) will be the mean of the two probabilities resulting from limiting one point to range over either side with uniform probability, and simultaneously therewith a second point of the group over the base, with a probability varying as its distance from that end of the base in which it is met by the side. An analogous rule can be given for transforming the form-probability of a group limited to any the same parallelogram. So again for a group of points ranging over a plane figure bounded by any curvilinear contour.

ON AN EXPERIMENTAL DETERMINATION OF π.

By Asaph Hall.

In his *Theorie analytique des Probabilités*, Chap. v., Laplace has shown that we may make use of this calculus to determine the lengths of curves and to find their surfaces; and he has pointed out very briefly how this may be done. Imagine a plane on which are drawn equidistant and parallel right lines, and let there be thrown on this plane at random a right line of given length. It is required to find the probability that the right line will intersect one of the parallel lines. This is one of the questions solved by Laplace, and by varying his solution a little, it is easy to find that its probability is expressed by the definite integral

$$\int_0^{\frac{1}{2}\pi} \frac{2l}{a\pi} \cos\phi\, d\phi = \frac{2l}{a\pi};$$

where a is the interval of the parallel lines, and l is the length of the random line. If we denote by m the whole number of times the line is thrown on the plane, and by n the number of intersections, and if m be very great and the trials be made so that there is no systematic error in the experiments, we may assume that the probability is expressed by the ratio $\frac{n}{m}$. Equating this to the rigorous value, we have

$$\pi = \frac{2ml}{an} \quad \ldots\ldots\ldots\ldots\ldots\ldots (1).$$

In this expression a and l are known, and m and n are to be found by observation.

In 1864, my friend Capt. O. C. Fox was unable to do active duty on account of a severe wound, and I proposed that he should make some experiments for determining the ratio $\frac{n}{m}$. Capt. Fox had made a plane wooden surface ruled with equidistant parallel lines, and on this he threw at random a fine steel wire. After making the first set of experiments, and in order to avoid as much as possible any constant error that might arise from his position or manner of holding the rod over the surface, the surface was given a

VOL. II.
1

slight rotatory motion before dropping the rod; the following are the results of the experiments of Capt. Fox:

m	n	l	a	
500	236	3 inches	4 inches	surface stationary.
530	253	3 "	4 "	" revolved
590	939	5 "	2 "	" revolved.

Substituting these numbers in formula (1), we have

$$\pi = \frac{2\cdot500\cdot3}{4\cdot236} = 3\cdot1780,$$

$$\pi = \frac{2\cdot530\cdot3}{4\cdot253} = 3\cdot1423,$$

$$\pi = \frac{2\cdot590\cdot5}{2\cdot939} = 3\cdot1416.$$

Washington, June 5, 1872.

PROPOSITIO 3. THEOREMA

16. *In omni solido hedris planis incluso summa omnium angulorum planorum, qui in eius hedris existunt, aequalis est quater tot angulis rectis, quot sunt anguli solidi, demtis octo; seu si numerus angulorum solidorum sit = S, summa omnium angulorum planorum aequatur 4S — 8 angulis rectis.*

DEMONSTRATIO

In solido quocunque sit numerus angulorum solidorum = S, summa autem omnium angulorum planorum aequatur R angulis rectis, ita ut demonstrari oporteat esse $R = 4S — 8$. Iam modo ante indicato abscindatur a solido unus angulus solidus, ut numerus angulorum solidorum, quos habebit, sit = $S — 1$, et summa angulorum planorum erit = $R — 4$ angulis rectis. Si denuo angulus solidus resecetur, ut reliquorum numerus sit $S — 2$, angulorum planorum summa erit = $R — 8$, atque ita pergendo patebit pro quovis angulorum solidorum numero summam omnium angulorum planorum fore, ut tabella sequens indicat.

Numerus angulorum solidorum S	Summa omnium angulorum planorum R angulis rectis
$S — 1$	$R — 4$
$S — 2$	$R — 8$
$S — 3$	$R — 12$
:	:
$S — n$	$R — 4n$

Cum igitur hac continua mutilatione pervenerimus ad $S — n$ angulos solidos, summa angulorum planorum erit = $R — 4n$ angulis rectis. At hoc modo tandem pervenietur ad id angulos solidos, quo casu corpus abibit in pyramidem triangularem, in qua constat summam omnium angulorum planorum esse aequalem 8 angulis rectis: hoc est, si sit $S — n = 4$. erit $R — 4n = 8$ seu $R = 4n + 8$ At inde est $n = S — 4$, quo valore hic substituto fiet

$$R = 4S — 16 + 8 = 4S — 8 ,$$

ita ut in quovis solido summa angulorum planorum aequetur quater tot angulis rectis, quot sunt anguli solidi, demtis octo. Q. E. D.

SCHOLION

17. Quanquam alterum Theorema ita ab hoc pendet, ut, cum hoc fuerit demonstratum, simul illius veritas sit evicta, tamen ex problemate praemisso etiam alterius Theorematis demonstratio confici potest sequenti modo.

PROPOSITIO 4. THEOREMA

18. *In omni solido hedris planis incluso numerus hedrarum una cum numero angulorum solidorum binario excedit numerum acierum.*

DEMONSTRATIO

Sit in solido quocunque proposito:

numerus angulorum solidorum $= S$,
numerus hedrarum $= H$,
numerus acierum $= A$,

atque ante vidimus, si resectione unius anguli solidi numerus S unitate minuatur, ut sit $S — 1$, tum differentiam inter numerum acierum et numerum hedrarum futuram esse $= A — H — 1$. Continuata ergo hac mutilatione,

si numerus angulorum solidorum sit	excessus numeri acierum super numerum hedrarum erit
S	$A — H$
$S — 1$	$A — H — 1$
$S — 2$	$A — H — 2$
$S — 3$	$A — H — 3$
:	:
$S — n$	$A — H — n$

Quando ergo hoc modo ad pyramidem triangularem devenietur, in qua numerus angulorum solidorum est $= 4$, numerus hedrarum $= 4$ et numerus acierum $= 6$, ita ut excessus numeri acierum supra numerum hedrarum futurus sit $= 2$, evidens est, si fiat $S — n = 4$, fore $A — H — n = 2$. Inde ergo est $n = S — 4$, hinc vero $n = A — H — 2$; sicque habetur

$$S — 4 = A — H — 2 \text{ seu } H + S = A + 2 ;$$

unde constat in omni solido hedris planis incluso numerum hedrarum H una cum numero angulorum solidorum S binario superare numerum acierum A. Q. E. D.

EULER
-1755-
(réf. h)

ANALYSE MATHÉMATIQUE. —*Note sur divers théorèmes relatifs à la rectification des courbes, et à la quadrature des surfaces;* par M. AUGUSTIN CAUCHY.

« Dans un Mémoire lithographié en 1832, j'ai donné les propositions suivantes:

» 1er *Théorème.* p désignant l'angle polaire que forme une droite OO' tracée à volonté dans un plan avec un axe fixe; S le système d'une ou de plusieurs lignes droites ou courbes, fermées ou non fermées; A la somme des projections absolues des divers éléments de S sur la droite OO', et π le rapport de la circonférence au diamètre, on aura

$$(1) \qquad S = \frac{1}{4} \int_{-\pi}^{\pi} A\,dp.$$

» 2e *Théorème.* p désignant l'angle formé par une droite quelconque OO' avec un axe fixe OP; q l'angle formé par le plan des deux droites OP, OO', avec un plan fixe qui renferme la première; S le système d'une ou de plusieurs surfaces planes ou courbes, et A la somme des projections absolues des divers éléments de S sur un plan HHK perpendiculaire à la droite OO'; on aura

$$(2) \qquad S = \frac{1}{2\pi} \int_{-\pi}^{\pi} \int_{0}^{\pi} A \sin p \, dp \, dq.$$

» Ces théorèmes entraînent évidemment les suivants.

» 3e *Théorème.* Le périmètre d'un polygone ou d'une courbe est toujours égal ou inférieur à la circonférence d'un cercle qui aurait pour rayon le quart de la plus grande somme que puissent fournir les projections des diverses parties de ce périmètre sur un axe quelconque.

» 4e *Théorème.* L'aire d'un polyèdre ou d'une surface courbe est toujours égale ou inférieur au double de la plus grande somme que puissent fournir les projections des diverses parties de cette aire sur un plan quelconque.

» D'autres théorèmes qui se rapportent aux quadratures et aux cubatures, et qui seront développés dans les *Exercices d'Analyse et de Physique mathématique,* se déduisent aisément des principes exposés dans mes leçons orales à l'École Polytechnique. Je me bornerai à énoncer le suivant.

» 5e *Théorème.* Supposons qu'une aire plane, renfermée dans le périmètre S d'un polygone convexe ou d'une courbe convexe, ait été partagée en rectangles égaux et très-petits par deux systèmes de droites parallèles à deux axes donnés. Soient h, k, les dimensions de chaque rectangle, mesurées parallèlement au premier et au second axe. Soient encore H et K les projections du contour S sur le premier ou sur le second axe. Si l'on prend pour valeur approchée de l'aire A la somme des rectangles qui sont entièrement renfermés dans cette aire, sans être traversés par le contour S, l'erreur commise sera inférieure au double de la somme

$$hK + kH.$$

» Considérons maintenant une aire plane A renfermée entre les périmètres de deux polygones construits de manière que les côtés du second polygone, parallèles à ceux du premier, en soient constamment séparés par la distance h. L'aire A, composée de trapèzes dont les hauteurs seront égales à h, aura évidemment pour mesure le produit de la distance h par la demi-somme des périmètres des deux polygones donnés. ou, ce qui revient au même, par le périmètre d'un troisième polygone dont chaque côté divisera la distance h en parties égales. Or, il suffira de transformer ce troisième polygone en une courbe plane dont le rayon de courbure surpasse constamment la distance $k = \frac{1}{2}h$, pour obtenir la proposition suivante.

» 6e *Théorème.* Supposons que le centre d'un cercle, dont le diamètre est 2k, se meuve, dans un plan donné, sur une courbe fermée, dont le rayon de courbure surpasse constamment le rayon k. L'aire comprise entre les deux enveloppes intérieure et extérieure de l'espace parcouru par le cercle, aura pour mesure le produit

$$2kS,$$

S désignant le périmètre de la courbe.

» Le théorème précédent fournit ur moyen facile de trouver la limite de l'erreur commise, quand on substitue à l'aire d'une courbe plane, l'aire d'un polygone inscrit ou circonscrit à cette courbe.

(réf. i)

STEINER

-1840-

Ueber parallele Flächen.

(Bericht über eine am 14. Mai 1840 in der Akademie der Wissenschaften zu Berlin gelesene Abhandlung.)

Unter parallelen ebenen Curven versteht man bekanntlich solche, die überall gleichweit von einander abstehen, oder die gemeinschaftliche Normalen haben, oder die Evolventen einer und derselben Curve sind. Leibniz scheint zuerst solche Curven angedeutet zu haben; Kästner und de Prasse haben sich später mit ihrer Betrachtung beschäftigt. In neuerer Zeit hat Crelle zwei wesentliche Sätze über dieselben aufgestellt und bewiesen (Annales de Mathém.). Zu diesen zwei Sätzen kann man auch auf elementarem Wege gelangen. Rollt ein constanter Kreis, dessen Radius gleich h, auf einer gegebenen Curve A, so beschreibt sein Mittelpunct eine mit A parallele Curve B. Wird nun anfänglich die Curve A als Vieleck angenommen, so ergeben sich die genannten zwei Eigenschaften unmittelbar. Nämlich es zeigt sich, dass B gleich $A \pm h\varphi$, wo φ der Winkel zwischen den gemeinschaftlichen Normalen in den Endpuncten der Bogen A, B (oder die Totalkrümmung des Bogens A) ist; und dass der von beiden Bogen und jenen Normalen eingeschlossene Flächenraum gleich $\frac{1}{2}h(A+B)$ ist. Der letzte Satz wurde bereits in der Abhandlung von 5. April 1838*) auf diese Art bewiesen.

Bei Curven von doppelter Krümmung kann der Parallelismus durch constanten Abstand im engeren oder weiteren Sinne bestimmt werden: entweder durch gerade oder durch bestimmte krumme Linien. Durch die gegebene Curve A (von doppelter Krümmung) denke man irgend eine krumme Fläche F und auf dieser alle kürzesten Linien, die zu A rechtwinklig sind, schneide von denselben (auf einerlei Seite von A) gleich lange Stücke gleich h ab, so liegen die Endpuncte in einer Curve B, die auf den nämlichen kürzesten Linien rechtwinklig ist, und welche der Curve A parallel heisst (Grass Disqu. gen. cir. supf. curv.). Ist nun die Fläche F geradlinig (d. h. durch Bewegung einer Geraden erzeugt), und ist A zu den

Geraden rechtwinklig, so sind diese das vorgenannte System von kürzesten Linien, auf denen man die constante Strecke h abzutragen hat, um die mit A parallele Curve B zu erhalten. Und ist ferner die Fläche F insbesondere eine abwickelbare, so ist ihre Knotenlinie eine gemeinsame Evolute der parallelen Curven A und B, und in diesem Falle allein haben letztere die Eigenschaft, dass auch ihre Tangenten in entsprechenden Puncten parallel sind. Für beliebige parallele Curven A und B auf einer abwickelbaren Fläche F findet der obige zweite Satz auf analoge Weise statt, was sogleich folgt, wenn die Fläche F auf einer Ebene abgewickelt wird. — Parallele sphärische Curven A, B haben die besondere Eigenschaft, dass sie zugleich in einer abwickelbaren Fläche F liegen und zu ihrem System von Geraden normal sind, so dass also sowohl ihr sphärischer Abstand h, als auch ihr geradliniger Abstand η, constant ist; jener (h) ist ein Bogen des Hauptkreises (kürzeste Linie auf der Kugel) und dieser (η) die zugehörige Sehne. Die Differenz der Curvenbogen A und B lässt sich hier auf zwei verschiedene Arten angeben, den beiden Flächen gemäss, in denen sie liegen. Noch leichter sind die Räume zu finden, welche die Bogen A und B mit ihren Grenznormalen auf beiden Flächen begrenzen; dieselben sind von einander abhängig, nämlich es verhält sich der sphärische Raum zum Raume auf der geradlinigen Fläche F, wie y zu sin h.

Zur Bestimmung paralleler, krummer Flächen kann derselbe Begriff dienen, wie bei Curven. Zwei Flächen A und B sollen parallel heissen, wenn sie gemeinschaftliche Normalen haben, oder wenn sie überall gleich weit von einander abstehen, etc. Dann folgt umgekehrt: werden von den Normalen der Fläche A auf einerlei Seite derselben gleiche Stücke gleich h, abgeschnitten, so liegen die Endpuncte in einer mit A parallelen Fläche B; oder: rollt eine constante Kugel, deren Radius gleich h, auf der gegebenen Fläche A, so beschreibt ihr Mittelpunct M eine mit A parallele Fläche B. Aus dieser Entstehungsart paralleler Flächen A, B ergeben sich leicht Ausdrücke für ihre Differenz, so wie für den zwischen ihnen liegenden Körperraum. Man denke sich, für einen Augenblick die gegebene Fläche A polyedrisch und lasse die Kugel M auf ihrer convexen Seite rollen, so sieht man, dass die Fläche B, sowie der zwischen beiden Flächen enthaltene Raum, aus folgenden Theilen bestehen:

α) Während die Kugel auf der nämlichen Seitenfläche a von A rollt, beschreibt ihr Mittelpunct ein der a gleiches ebenes Vieleck a, in der Fläche B, und der zwischen den Flächentheilen a und a, befindliche Körperraum ist ein senkrechtes Prisma, dessen Inhalt gleich ha. Die Summe aller solcher Vielecke a, ist gleich A und die Summe aller Prismen gleich hA.

β) So lange die Kugel eine und dieselbe Kante γ von B berührt, beschreibt ihr Mittelpunct M ein Stück γ, von B, welches einer geraden

*) Cf. Bd. II. S. 97—159 dieser Ausgabe.

(réf. j)

Cylinderfläche angehört, die γ zur Axe und h zum Radius hat, und der zwischen γ und γ_1 befindliche Körperraum ist ein Ausschnitt des Cylinders. Heisst der an der Kante γ liegende Nebenflächenwinkel ζ, so ist

$$\tau_1 = \gamma h \zeta, \quad \text{und} \quad c = \tfrac{1}{2} h^2 \zeta.$$

Wird die Summe aller solchen Flächenstücke τ_1 durch K und die Summe aller Cylinderausschnitte c durch C bezeichnet, so ist

$$K = h\Sigma(\gamma\rho) \quad \text{und} \quad C = \tfrac{1}{2} h K = \tfrac{1}{2} h^2 \Sigma(\gamma\rho).$$

γ) So lange die Kugel die nämliche Ecke s der polyedrischen Fläche A berührt, beschreibt ihr Mittelpunct ein sphärisches Vieleck s_1, in der Fläche B, das ebenso viele Seiten hat, als die Ecke s Kanten, welche Seiten die an diesen Kanten liegenden Nebenflächenwinkel messen. Der zwischen der Ecke s und dem Vielecke s_1 liegende Raum ist eine sogenannte Kugelpyramide P_1, deren Inhalt gleich $\tfrac{1}{3} h s_1$. Die Summe aller sphärischen Vielecke s_1, heisse E und die Summe der Pyramiden ρ sei P, so ist

$$E = \Sigma s_1 \quad \text{und} \quad P = \tfrac{1}{3} h \Sigma s_1 = \tfrac{1}{3} h E.$$

Hiernach hat man für die Fläche B und für den zwischen beiden Flächen A und B liegenden Körperraum I folgende Ausdrücke:

(1) $\quad B = A + h\Sigma(\gamma\rho) + \Sigma s_1 = A + K + E,$

(2) $\quad I = hA + \tfrac{1}{2}h^2\Sigma(\gamma\rho) + \tfrac{1}{3}h\Sigma s_1 = hA + \tfrac{1}{2}hK + \tfrac{1}{3}hE;$

oder, wird irgend eine bestimmte Länge des willkürlichen Abstandes h zur Einheit angenommen, gleich 1 gesetzt, und werden für diesen Fall die Grössen K und E durch k und e bezeichnet, wo dann für jeden anderen Fall K gleich hk und E gleich h^2e ist, so hat man:

(3) $\quad B = A + hk + h^2e,$

(4) $\quad I = hA + \tfrac{1}{2}h^2k + \tfrac{1}{3}h^3e = \tfrac{1}{2}h(A + B - \tfrac{1}{3}h^2e).$

Die Constante k ist eine Längen-Grösse, nämlich

$$k = \Sigma(\gamma\rho),$$

d. h. gleich der Summe der Producte aus den Kanten des Polyeders A in die anliegenden Nebenflächenwinkel, diese in Zahlen ausgedrückt; wogegen e gleich Σs_1 eine Zahl ist, nämlich die Summe der Zahlenwerthe der den Ecken s des Polyeders A entsprechenden Polar-Körperwinkel. Da die Grössen k und e bloss von den Krümmungen der Fläche A abhängen, so mögen sie die Krümmungs-Summen derselben heissen, und zwar „k die Summe der Kanten-Krümmung" und „e die Summe der Ecken-Krümmung".

Die obigen Formeln bleiben offenbar bestehen, wenn die polyedrische Fläche A in eine krumme Fläche übergeht. In diesem Falle gelangt man aber zu neuen Ausdrücken für die Grössen B und I, so wie für k und e. In irgend einem Puncte der gegebenen Fläche A seien die Hauptkrümmungsradien r und r_1; das Flächenelement sei a. In correspondirenden Puncte der mit A parallelen Fläche B heisse das Flächenelement b, so ist

(5) $\quad b = a\left(1 + \dfrac{h}{r}\right)\left(1 + \dfrac{h}{r_1}\right) = a + h\left(\dfrac{a}{r} + \dfrac{a}{r_1}\right) + h^2 \dfrac{a}{r r_1}.$

Für die Summe aller Elemente b, oder für die Fläche B, hat man demnach

(6) $\quad B = A + h\Sigma\left(\dfrac{a}{r} + \dfrac{a}{r_1}\right) + h^2\Sigma \dfrac{a}{r r_1},$

und für den Körperraum I:

(7) $\quad I = hA + \tfrac{1}{2}h^2\Sigma\left(\dfrac{a}{r} + \dfrac{a}{r_1}\right) + \tfrac{1}{3}h^3\Sigma \dfrac{a}{r r_1}.$

Aus den Formeln (3) und (6) folgt:

(8) $\quad k = \Sigma\left(\dfrac{a}{r} + \dfrac{a}{r_1}\right)$

und

(9) $\quad e = \Sigma \dfrac{a}{r r_1},$

woraus erkannt wird, welche Bedeutung die Grössen $\Sigma\left(\dfrac{a}{r} + \dfrac{a}{r_1}\right)$ und $\Sigma \dfrac{a}{r r_1}$ bei der krummen Fläche A haben. Sie sind zusammen die „Total-krümmung" der Fläche A. Gauss giebt diesen Namen dem Ausdrucke $\Sigma \dfrac{a}{r r_1}$ allein, welcher aber nur die Summe der Eckenkrümmung e repräsentirt.

Die Grösse e lässt sich im Allgemeinen bestimmen, die Grösse k nicht. In einigen besonderen Fällen kann jedoch k auf e zurückgeführt werden, wie z. B., wenn für alle Puncte der Fläche A die Summe der Krümmungsradien $r + r_1$ gleich s constant ist, denn alsdann ist

$$k : e = s : h.$$

Ist insbesondere A eine kleinste Fläche, so sind bekanntlich in jedem Puncte derselben die Krümmungsradien einander gleich und entgegengesetzt, also

$$r = -r_1, \quad \text{und} \quad \dfrac{a}{r} + \dfrac{a}{r_1} = 0 \quad (\text{auch } k = 0),$$

und daher

(10) $\quad B = A - h^2\Sigma \dfrac{a}{r^2},$

d. h. „jede kleinste Fläche A hat die Eigenschaft: 1) dass in jedem Puncte derselben die Kantenkrümmung $\dfrac{a}{r} + \dfrac{a}{r_1}$ Null ist; 2) dass sie unter allen mit ihr parallelen Flächen (B) ein Maximum ist, und dass von diesen Flächen (B) je zwei, welche gleichweit von jener abstehen (auf entgegengesetzten Seiten), gleich gross sind."

BARBIER

-1860-

NOTE

sur

LE PROBLÈME DE L'AIGUILLE ET LE JEU DU JOINT COUVERT;

Par M. E. BARBIER,
Élève de l'École Normale.

§ II. —. *Généralisation du problème de l'aiguille. — Divisions quelconques du plan. — Problèmes analogues au problème de l'aiguille.*

1. Dans le paragraphe précédent nous avons toujours supposé qu'un disque était jeté sur un parquet, formé de bandes égales indéfinies; on pourrait se proposer le cas où le disque serait remplacé par un fil flexible de longueur l; le seul théorème simple auquel on puisse arriver sans aucune hypothèse sur la manière de jeter le fil est celui-ci :

Le nombre des intersections qu'on peut mathématiquement espérer entre le fil et les joints du parquet est $\frac{l}{\pi a}$, *ce qui veut dire que d'après le théorème de Jacques Bernoulli la moyenne du nombre d'intersections α pour limite* $\frac{l}{\pi a}$ *quand le nombre des coups augmente indéfiniment, ou encore, la moyenne oscille indéfiniment autour de cette quantité* $\frac{l}{\pi a}$.

Ce résultat moyen $\frac{l}{\pi a}$ s'obtient par des raisonnements que je n'ai pas besoin de reproduire, ils ont été donnés au n° 3 du § I.

Je vais supposer que sur le plan où le fil est jeté, on considère non plus des lignes droites, mais un fil réparti sur la surface, de manière que chaque mètre carré en contienne L mètres, ce fil affectant une forme quelconque, d'ailleurs et variable avant chaque coup.

Pour faire voir que la forme du fil, comprise dans chaque mètre carré, n'influe pas sur la limite de la moyenne du nombre d'intersections d'un fil de longueur l qu'on jette sur le plan, nous irons du simple au composé comme nous l'avons fait au n° 3 du § I.

Nous supposerons d'abord qu'on jette une aiguille sur le plan, et qu'un dessin que nous tracerons dans un mètre carré pris sur le plan se répète fidèlement dans tous les mètres carrés juxtaposés pour former un plan indéfini.

1°. Mettons dans un mètre carré une ligne droite de longueur d, supposons la limite de la moyenne du nombre d'intersections de l'aiguille avec ces lignes d égale à m.

Changeons de place la ligne d, toutes les autres sont supposées se déplacer en même temps, la moyenne m reste constante.

2°. Considérons n droites égales à d dans un mètre carré, la moyenne sera évidemment mn, si on compte les intersections de l'aiguille qu'on jette avec toutes ces droites, etc.; bref, des raisonnements simples comme ceux du n° 3 du § I nous permettent d'énoncer le théorème suivant :

(réf. k)

2. THÉORÈME I. — *Un plan contient par mètre carré un fil flexible de longueur* L *mètres, affectant une forme variable, on y jette au hasard un fil flexible de longueur* l *mètres, la moyenne du nombre des points d'intersection oscille indéfiniment, quel que soit le nombre des épreuves, autour de* $\frac{2Ll}{\pi}$.

Nous dirons toujours, dans ce qui va suivre, *moyenne* au lieu de *limite de moyenne*, ou de *nombre autour duquel la moyenne oscille indéfiniment*. Avant de donner quelques énoncés de théorèmes de moyennes, il est utile de fixer le sens les deux expressions.

1°. Une direction quelconque dans un plan est celle du rayon d'un cercle situé dans ce plan, qui prend indifféremment toutes les orientations possibles.

2°. Une direction quelconque dans l'espace est celle du rayon mené à un point de la surface d'une sphère, qui prend indifféremment toutes les positions possibles sur cette surface [*].

La moyenne des projections d'une ligne plane l sur une direction quelconque du plan est $\frac{2l}{\pi}$, comme pour le cercle.

Il s'agit de la somme arithmétique des projections, la même remarque s'applique aux théorèmes suivants.

La moyenne des projections d'une surface, sur un plan dont l'axe a une direction quelconque, est $\frac{1}{2}$ comme pour la sphère.

La moyenne des projections d'une ligne i sur une direction quelconque est $\frac{i}{2}$.

La moyenne des projections d'une ligne l sur un plan dont l'axe est une direction quelconque, est $\frac{\pi i}{4}$.

5. THÉORÈME II. — *Supposons un espace indéfini, divisé par la pensée en cubes de 1 mètre de côté, et chacun de ces mètres cubes contenant s mètres carrés d'une étoffe (qui peut n'être pas développable sur un plan); un fil de longueur l, passé au hasard dans cet espace, traverse moyennement l'étoffe en* $\frac{sl}{2}$ *points.*

Ce théorème donne à peu près la moyenne du nombre de feuilles traversées par une flèche très-fine qui parcourt une distance connue à travers un feuillage.

THÉORÈME III. — *Chaque mètre cube d'un espace indéfini est traversé par un fil de* L *mètres de longueur; une étoffe de s mètres carrés est traversée moyennement en* $\frac{Ls}{2}$ *points par le fil.*

Ce théorème résout à peu près cette question : Un bassin renferme un acide qui peut altérer une étoffe; le liquide distille par un certain nombre de trous. Combien de points d'une surface connue de cette étoffe ont été altérés par des gouttes d'acide?

THÉORÈME IV. — *Supposons enfin que chaque mètre cube de l'espace renferme* S *mètres carrés de surfaces, la longueur moyenne de la courbe d'intersection de ces surfaces, par une surface de s mètres carrés, est* $\frac{3\pi Ss}{2}$.

---

[*] Une aiguille H horizontale, mobile autour d'un axe vertical qui la traverse en son centre de gravité, lancée de manière à faire un assez grand nombre de tours, s'arrête dans une direction, justement réputée quelconque dans le plan horizontal.

Il est bien entendu que l'aiguille n'est pas magnétique.

Au moyen de conventions faciles à imaginer, la même aiguille H nous déterminera un méridien quelconque d'une surface de révolution, un point quelconque d'un cercle ou d'une ligne de longueur donnée.

Un point d'une sphère, situé en un point quelconque d'un méridien quelconque, peut occuper toutes les positions possibles sur la sphère; mais il n'y est pas quelconque, il se portera de préférence vers les pôles.

Mais un point d'un méridien quelconque projeté en un point quelconque de la ligne des pôles d'une sphère peut être censé quelconque sur la sphère. L'aiguille H peut donc servir à déterminer un point quelconque d'une sphère par suite une direction quelconque dans l'espace.

Remarque. L'équivalence des zones de même hauteur dans la sphère fait voir qu'une surface plane prenant indifféremment toutes les directions possibles, se projette sur un plan fixe de manière que sa projection ait indifféremment toutes ses valeurs possibles.

MINKOWSKI
-1903-

XXVI.
Volumen und Oberfläche.

Herrn Rudolf Lipschitz zum fünfzigjährigen Doktorjubiläum, 9. August 1903, in herzlicher Verehrung gewidmet vom Verfasser.
(Mathematische Annalen, Band 57, S. 447—495).

Für die konvexen Körper gibt es einen elementaren Weg, um den Begriff der Oberfläche aus dem einfacheren Begriff des Volumens heraus zu entwickeln, und in Verfolg dieses Weges gelangt man zu sehr bemerkenswerten Erweiterungen der Tatsache, worach unter allen Körpern gleichen Volumens die Kugel die kleinste Oberfläche besitzt.

Liegt ein konvexer Körper $\Re$ vor und versteht man unter x, y, z rechtwinklige Koordinaten eines Punktes aus $\Re$, so nimmt ein linearer Ausdruck $ux + vy + wz$, wo u, v, w feste Größen sind, in $\Re$ immer einen bestimmten größten Wert $H(u, v, w)$ an; und diese Funktion $H(u, v, w)$ von drei beliebigen reellen Argumenten, die Stützebenenfunktion von $\Re$, charakterisiert den konvexen Körper $\Re$ vollkommen. Das Volumen des Körpers $\Re$ erscheint als ein gewisser homogener Ausdruck dritten Grades $V\beta$ in den sämtlichen Werten $H(u, v, w)$. Aus diesen Ausdrucke entspringt für drei beliebige konvexe Körper $\Re_1, \Re_2, \Re_3$ eine polare Bildung, ein symbolisches Produkt $V_{\Re_1} V_{\Re_2} V_{\Re_3}$, das gemischte Volumen der drei Körper $\Re_1, \Re_2, \Re_3$. Diese Größe ist invariant bei beliebigen Translationen der einzelnen Körper. Werden zwei der Körper mit einem bestimmten Körper $\Re$, der dritte aber mit einer Kugel vom Radius 1 identifiziert, so ist das dreifache ihres gemischten Volumens die Oberfläche von $\Re$.

Für die gemischten Volumina gilt der wichtige Satz: Für irgend drei Körper vom Volumen 1 wird das gemischte Volumen stets ≥ 1 und nur dann = 1, wenn die drei Körper miteinander homothetisch sind. Daß jeder konvexe Körper, der keine Kugel ist, eine größere Oberfläche hat als eine Kugel von demselben Volumen, ist nur ein spezieller Fall dieses Satzes.

Diese fundamentale Ungleichung läßt weiter die folgende Auslegung zu: Man bezeichne in der Mannigfaltigkeit aller möglichen Funktionen $H(u, v, w)$ von drei reellen Argumenten u, v, w eine einzelne Funktion $H(u, v, w)$ als einen „Punkt", den Inbegriff der aus zwei Funktionen H_1 und H_2 abzuleitenden Funktionen $(1-t)H_1 + tH_2$ für $0 \leq t \leq 1$ als die H_1 und H_2 verbindende „Strecke", alsdann besitzt die Gesamtheit der Stützebenenfunktionen H zu allen denjenigen konvexen Körpern, welche ein Volumen ≥ 1 haben, die Eigenschaft, mit irgend zwei Punkten stets die ganze sie verbindende Strecke zu enthalten, stellt also ein „konvexes Gebilde" in jener Mannigfaltigkeit vor.

Geht man auf die Tangentialebenen des Gebildes ein, so ist sein konvexer Charakter gleichbedeutend mit folgendem Theorem:

Auf der Kugelfläche vom Radius 1 mit dem Nullpunkt als Mittelpunkt denke man sich Masse in einer beliebigen stetigen und durchweg positiven Flächendichtigkeit ausgebreitet, doch so, daß der Schwerpunkt der ganzen Belegung in den Nullpunkt fällt; alsdann existiert eine geschlossene konvexe Fläche, bei welcher an jeder Stelle das Produkt der Krümmungsradien gleich der Flächendichtigkeit an dem Punkte der Kugel mit gleicher Normale ist; und diese Fläche ist völlig bestimmt bis auf eine beliebige Translation, durch die man sie noch variieren kann.

In diesem Theorem erkennt man eine Aussage über eine gewisse quadratische partielle Differentialgleichung zweiter Ordnung, deren Lösbarkeit unter bestimmten Bedingungen hier durch eine eigenartige, wohl noch mancher weiteren Anwendungen fähige Methode sichergestellt wird.

§ 1. Stützebenenfunktion eines konvexen Körpers.

1. Es seien x, y, z rechtwinklige Koordinaten eines Punktes im Raume, und $\mathfrak{M}$ bedeute eine *abgeschlossene* Menge von Punkten x, y, z, die ganz in einer Kugel von *endlichem* Radius enthalten ist, aber nicht völlig in eine einzige Ebene fällt. Sind u, v, w irgendwelche festen Werte, so hat der Ausdruck $ux + vy + wz$ für die Gesamtheit der Punkte x, y, z in $\mathfrak{M}$ ein bestimmtes *Maximum*, das $H(u, v, w)$ heiße.

Diese Funktion $H(u, v, w)$ von drei beliebigen reellen Argumenten erfüllt offenbar *folgende Bedingungen* (1)—(4):

$$H(0, 0, 0) = 0,$$ (1)

$$H(tu, tv, tw) = tH(u, v, w),$$ (2)

wenn $t > 0$ ist. Sind u_1, v_1, w_1 und u_2, v_2, w_2 irgend zwei Systeme der Argumente, so gibt es in $\mathfrak{M}$ immer wenigstens einen Punkt x, y, z, wofür

$(u_1 + u_2)x + (v_1 + v_2)y + (w_1 + w_2)z = H(u_1 + u_2, v_1 + v_2, w_1 + w_2)$

wird, und da für diesen Punkt sicherlich

$u_1 x + v_1 y + w_1 z \leq H(u_1, v_1, w_1), \quad u_2 x + v_2 y + w_2 z \leq H(u_4, v_2, w_2)$

(réf. 1)

ist, so gilt daher immer:

(3) $H(u_1 + u_2, v_1 + v_2, w_1 + w_2) \leqq H(u_1, v_1, w_1) + H(u_2, v_2, w_2)$.

Das Maximum von $-(ux + vy + wz)$ in $\mathfrak{M}$ ist $H(-u, -v, -w)$, also gilt in $\mathfrak{M}$ stets:

$$-H(-u, -v, -w) \leqq ux + vy + wz \leqq H(u, v, w).$$

Wenn die Werte $u, v, w \div 0, 0, 0$ sind, muß daher, da $\mathfrak{M}$ nicht ganz in einer Ebene liegen soll, stets

(4) $H(u, v, w) + H(-u, -v, -w) > 0$

sein.

2. Eine Ebene, welche wenigstens einen Punkt der Begrenzung von $\mathfrak{M}$ enthält, aber außer den Punkten, die sie mit $\mathfrak{M}$ gemein hat, $\mathfrak{M}$ ganz auf einer Seite von sich liegen läßt, nennen wir eine *Stützebene* an $\mathfrak{M}$.

Ist $H(u, v, w)$ eine beliebige reelle Funktion von drei reellen Argumenten, u, v, w, welche allen den Bedingungen (1)—(4) genügt, so bezeichnen wir den Bereich $\mathfrak{R}$ von Punkten x, y, z, welcher durch die sämtlichen Ungleichungen:

(5) $ux + vy + wz \leqq H(u, v, w)$

für alle möglichen Wertsysteme u, v, w definiert ist, als einen *konvexen Körper.*

Die Funktion H nennen wir die *Stützebenenfunktion* von $\mathfrak{R}$, da aus den Ungleichungen (5) offenbar genau die Stützebenen an $\mathfrak{R}$ zu erkennen sind. Ist $H(u, v, w)$ wie in 1. aus der Punktmenge $\mathfrak{M}$ hergeleitet, so wird der durch die Ungleichungen (5) definierte Bereich $\mathfrak{R}$ *der kleinste*, $\mathfrak{M}$ enthaltende konvexe Körper, d. h. $\mathfrak{R}$ ist ein notwendiger Bestandteil jedes konvexen Körpers, der $\mathfrak{M}$ ganz in sich aufnimmt.

Ist $\mathfrak{R}^*$ ein zweiter konvexer Körper mit der Stützebenenfunktion $H^*(u, v, w)$, so ist dann und nur dann $\mathfrak{R}$ ganz in $\mathfrak{R}^*$ enthalten, wenn stets

$$H(u, v, w) \leqq H^*(u, v, w)$$

ausfällt.

3. Ein konvexer Körper ist andererseits völlig durch die Eigenschaften zu charakterisieren, *erstens*, daß jede Gerade mit ihm sei es eine Strecke, sei es einen Punkt, sei es keinen Punkt gemein hat, *zweitens*, daß zu ihm wenigstens vier nicht in einer Ebene gelegene Punkte gehören.

4. Ist $\mathfrak{p}$ ein beliebiger Punkt, so verstehen wir unter $\mathfrak{R} + \mathfrak{p}$ den Körper, der aus $\mathfrak{R}$ durch diejenige *Translation* entsteht, durch welche der Nullpunkt nach $\mathfrak{p}$ gelangt. Sind a, b, c die Koordinaten von $\mathfrak{p}$, so wird die Stützebenenfunktion von $\mathfrak{R} + \mathfrak{p}$:

$$H(u, v, w) + au + bv + cw.$$

Unterwerfen wir $\mathfrak{R}$ einer *Dilatation* von Nullpunkt aus nach allen

Richtungen in einem festen positiven Verhältnisse $t:1$, so bezeichnen wir den entstehenden Körper mit $t\mathfrak{R}$; eine Stützebenenfunktion wird $t\,H(u, v, w)$.

5. Wir bezeichnen mit $\mathfrak{G}$ die Kugel $x^2 + y^2 + z^2 \leqq 1$, vom Radius 1 mit dem Nullpunkt o als Mittelpunkt, mit $\mathfrak{G}$ die Kugelfläche $x^2 + y^2 + z^2 = 1$, mit α, β, γ die Koordinaten eines beliebigen Punktes auf $\mathfrak{G}$, bzw. die *Richtung* vom Nullpunkte nach diesem Punkte. Infolge der Eigenschaft (2) sind alle Werte der Funktion H bereits durch deren Werte $H(\alpha, \beta, \gamma)$ für die Punkte auf $\mathfrak{G}$ bestimmt. Die Ungleichung

$$\alpha x + \beta y + \gamma z < H(\alpha, \beta, \gamma)$$

bezeichnen wir als die *Bedingung der Stützebene an $\mathfrak{R}$ mit der äußeren Normale (α, β, γ).*

Die Funktion $H(u, v, w)$ ist nach den Eigenschaften (1)—(4) eine *stetige* Funktion ihrer Argumente, und besitzt infolgedessen die Werte $H(\alpha, \beta, \gamma)$ auf $\mathfrak{G}$ ein bestimmtes *Maximum* G. Ist ein Wert $H(-\alpha, \beta, \gamma) \leqq 0$, so ist nach (4) der zugehörige Wert $H(-\alpha, -\beta, -\gamma)$ positiv und jedenfalls > 0. Mit Hilfe von (3) und (2) gewinnen wir die Ungleichung

(6) $|H(u-u_0, v-v_0, w-w_0) - H(u_0, v_0, w_0)| \leqq G\sqrt{u'^2 + v'^2 + w'^2}.$

§ 2. Annäherung an einen beliebigen konvexen Körper durch vollkommene Ovaloide.

6. Ist $\varphi = 0$ die Gleichung einer Ebene und der konvexe Körper $\mathfrak{R}$ ganz im Bereiche $\varphi \leqq 0$ enthalten, so heißt $\varphi \leqq 0$ ein *Halbraum um* $\mathfrak{R}$. Ist ein Halbraum $\tau \varphi \leqq 0$ um $\mathfrak{R}$ so beschränkt, daß man *nicht* $\varphi = t_1 \varphi_1 + t_2 \varphi_2$ setzen kann, so daß $t_1 > 0$, $t_2 > 0$ und $\tau_1 \leqq 0$, $\varphi_2 < 0$ *zwei verschiedene* Halbräume um $\mathfrak{R}$ sind, so heißt $\varphi \leqq 0$ ein *extremer Halbraum um* $\mathfrak{R}$. Die Ebene $\varphi = 0$ ist dann jedenfalls eine Stützebene an $\mathfrak{R}$ und heißt eine *extreme Stützebene* an $\mathfrak{R}$. Ein konvexer Körper mit einer endlichen Anzahl von extremen Stützebenen heißt ein *(konvexes) Polyeder.*

7. Unter einem *vollkommenen Ovaloid* wollen wir einen konvexen Körper verstehen, bei dem die Begrenzung durch eine *analytische* Gleichung in den rechtwinkligen Koordinaten x, y, z definiert wird und überdies in jedem Punkte eine *bestimmte* und immer nur eine *Berührung erster Ordnung eingehende Tangentialebene* besitzt.

8. Ist $\mathfrak{R}$ ein *beliebiger konvexer Körper mit dem Nullpunkte als innerem Punkt und ε eine beliebige positive Größe,* so läßt sich stets *ein vollkommenes Ovaloid Ω bestimmen, so daß Ω den Körper $\mathfrak{R}$ enthält und selbst in $(1 + \varepsilon)\,\mathfrak{R}$ enthalten ist.*

Es sei $H(u, v, w)$ die Stützebenenfunktion von $\mathfrak{R}$. Da der Nullpunkt im Inneren von $\mathfrak{R}$ liegt, ist jede Größe $H(\alpha, \beta, \gamma) > 0$. Es sei nun g

PROCÉDÉ MÉCANIQUE POUR DÉTERMINER LA COMPOSITION DES ROCHES.

Supposons que le volume occupé par la roche soit rapporté à un système de coordonnées, et soit p la surface occupée par l'un des minéraux constituants dans une section formée par un plan parallèle aux xy : pour obtenir exactement le volume occupé par ce minéral dans la roche, il faudrait pouvoir connaître les valeurs successives de p quand on mène une série de plans infiniment rapprochés et parallèles à xy; l'intégrale $\int p\,dz$ donnerait alors l'expression du volume cherché.

p est une fonction de z qui peut tantôt croître, tantôt décroître et même passer par plusieurs maxima et minima; mais si on désigne par m et M la plus *petite* et la plus *grande* valeur de p, l'intégrale $\int p\,dz$ sera toujours comprise entre mz et Mz, z représentant la hauteur du volume considéré : d'ailleurs les valeurs extrêmes m et M différeront d'autant moins entre elles que le minéral sera réparti et développé d'une manière plus égale dans la roche; il est facile de concevoir une distribution géométrique telle que p reste constant pour des sections d'égale largeur; alors il est évident que le volume du minéral serait représenté par pz, ou serait, en un mot, équivalent à celui d'un cylindre ayant p pour base; par conséquent, comme z est commun, les volumes des différents minéraux seraient entre eux dans le rapport des surfaces des bases.

Supposer que la section donnée dans la roche par une série de plans parallèles est à peu près constante, c'est supposer un cas qui se présente dans la nature chaque fois que la roche est *homogène*, si on donne ce nom à une roche ayant ses minéraux uniformément répartis; par conséquent, pour de pareilles roches, le rapport entre les volumes des minéraux constituants sera à peu près égal à celui des surfaces qu'ils présentent dans les sections, ou, du moins, on est certain qu'il sera compris entre les valeurs maxima et minima de ces surfaces.

Soit donc p, p', p'', etc., les surfaces présentées par les minéraux constituants d'une roche homogène dans une section; les *proportions en volume* de ces minéraux seront respectivement $\frac{p}{P}$, $\frac{p'}{P}$, $\frac{p''}{P}$, de sorte que $\frac{p}{P}+\frac{p'}{P}+\frac{p''}{P}+$etc.$=1$.

Et si on désigne par $d\ d'\ d''\ D$ les densités respectives de ces minéraux ainsi que de la roche, les *proportions en poids* seront : $\frac{p}{P}\cdot\frac{d}{D}$, $\frac{p'}{P}\cdot\frac{d'}{D}$, $\frac{p''}{P}\cdot\frac{d''}{D}$, etc., et en outre $\frac{p}{P}\frac{d}{D}+\frac{p'}{P}\frac{d'}{D}+\frac{p''}{P}\frac{d''}{D}+$etc.$=1$.

Par conséquent on aura les proportions, soit en *volume*, soit en *poids*, des minéraux constituants de la roche.

Si par d'autres analyses antérieures, faites sur ces minéraux ou sur d'autres de même nature, on peut connaître leur *composition chimique*, il sera facile de calculer la *composition chimique de la masse de la roche* sans être obligé d'en faire une analyse spéciale.

En effet, si on désigne par A la proportion d'une substance, de silice par exemple, qui entre dans la roche, on aura, en représentant par $a\ a'\ a''$ les proportions de silice des autres minéraux:

$$A=\frac{p\,d}{PD}a+\frac{p'd'}{PD}a'+\frac{p''d''}{PD}a''+\text{etc.}$$

DELESSE

-1847-

(réf. m)

BUFFON AND MATHEMATICS

JACQUES ROGER
Université de Paris I

It is a little surprising for me to be the first lecturer
in a colloquium devoted to a science I know nothing about. As a histo-
rian, I am very little acquainted with the history of your science. Actually, the reason for my being here is not stereology; it is the cele-
bration of a bicentennial, that of the Essai d'Arithmétique morale, first
published by Buffon in 1777.

Buffon never invented stereology. If, however, you may consider him one of the founding fathers of your science, and the Essay a
landmark in the prehistory of your science, it is, as you know, because
in this text Buffon for the first time published an old work of his, the
Mémoire sur le jeu de franc-carreau, which he had written in 1733, and
this mémoire is the first attempt to consider a case of geometrical probability.

Now why did Buffon delay the publication of the mémoire from
1733 to 1777 ? for merely accidental reasons. In 1733 he was a young man,
trying to enter the younger class of the Académie des Sciences, and he
respectfully submitted his work to this formidable body, which received
it very well. But, since Buffon was not yet an Academician, the Academy
did not publish the mémoire, and Buffon kept it with his papers, till the
time when, writing the Essai d'Arithmétique morale, he included the old
text in the new work. Therefore, what we are celebrating today is the
bicentennial, not of the birth, but of the first public appearance of
geometrical probability.

Buffon is essentially known as a naturalist, probably the
most important one in the 18th century with his Swedish counterpart
Carl von Linné. May I remind you that he was born in 1707, in Burgundy.
Unexpectedly, in 1739, he became the Intendant of the King's Garden, the
very place where we are today. From 1749 to his death, in 1788, he published 35 volumes of his Natural History, whose 36th and last volume

appeared posthumously in 1789, on the eve of the Revolution. The success
of the work was enormous, all over Europe and America, and many translations appeared into English, German, Italian, etc. But, precisely because
of this success, it was almost entirely forgotten that Buffon's first
scientific works had not been devoted to natural history, but to other
subject-matters, such as physics, vegetal physiology and mathematics.
This interest in different fields of scientific research was not at all
exceptional at a time when specialization was by no means as narrow as
it is today.

I hardly need remind you that mathematics had been extraordi-
narily flourishing in the 17th century, and that this golden age had cul-
minated with the discovery of the calculus by both Leibniz and Newton.
Newton, however, had made something more by creating a new mechanics,
namely dynamics, and demonstrating how mathematics could help understand
the structure and equilibrium of a physical system, namely the solar system. To that extent Newton was a true son of his century, a century
when most of the modern scientists were convinced that the human mind
could, by using his own tools, really understand the physical world
around it.

However, there was something disturbing in Newton's account
of the mechanics of the solar system, and this something was the central
concept of force. Continental scientistswere very reluctant to accept
it, because they felt that Newton was reintroducing old obscure ideas
from the dark ages, sympathy between distant bodies, natural magic and
the like, all the old creeds tha 17th century believed he had gotten
rid of by adopting a mechanical view of the universe. What is more,
nobody could explain how and why this invisible and unconceivable force
of gravitation was kind enough to obey a simple mathematical law. The
relationships between mathematics and reality, which seemed to be so
clear in the cartesian world, became very obscure again.

What is worse, that difficult question was only a part of
a more general problem, carefully scrutinized by Locke in his Treatise
concerning human understanding, the general problem of the capabilities
and limits of the human mind in its attempt to understand the outside
world. Thanks to the theory of innate ideas, Descartes had been convinced
that the human mind was able to fully understand the mechanics of the

universe. God's reason, so to speak, worked the same way as man's reason. This was no longer true for Locke, who considered that all our thoughts come from sensorial data, and are not to be trusted, even if they are, or seem to be, as clear and distinct as requested by Descartes. What, then, was the nature of scientific knowledge, and what is the role mathematics can play in it, this was to become a central question in 18th century philosophy, from Locke to Kant and 19th century positivism. Hume's answer to this question is well known; Buffon's answer is less well known, though it is quite interesting, and especially here, because his main themes are probability on the one hand, and the relations of mathematics to reality, on the other hand. Buffon's most important texts in this respect are an introductory essay published in 1749, in the first volume of the Natural History, with the title : "How to study and handle Natural History ?", and the Essai d'Arithmétique morale.

Actually, Buffon was concerned with these problems from the very beginning that is, from the time of his first mathematical essays, which are to be found in his correspondence with Gabriel Cramer, a Genevan mathematician, around 1730. By Cramer, a friend of the Bernoulli family, Buffon was introduced into the lively world of mathematical probability, and to the puzzling problem which was eventually known as "St Petersburg paradox". Buffon's reaction to the problem is typical, and here I would like to say that the most interesting feature of Buffon's mathematical works is that they demonstrate that Buffon is not a mathematician, not because of a lack of skill, but because of a lack of interest in pure mathematics. What Buffon is interested in is applied mathematics or, if I may say so, useful mathematics. Therefore, when mathematical probability says that the gain of a gambler may be intinite, provided that the game lasts long enough, Buffon flatly denies this possibility by introducing factual elements, which have nothing to do with pure mathematics, namely, the lack of time and the lack of money. This "realism", however, leads Buffon to the necessity of introducing, several years before Daniel Bernoulli, the concept of mathematical expectation and the principle of utility.

It is in this context that Buffon, in 1733, wrote his Mémoire sur le jeu de franc-carreau and invented the famous "problem of the needle", which he solved very elegantly by using integral calculus, for

the first time in the history of probability. The discussion of the
"St Petersburg paradox" and the Mémoire sur le jeu de franc-carreau did
not appear before 1777 however and the Essai d'arithmétique morale itself,
buried as it was in the Supplement to Natural History, passed almost
unnoticed till the time when Crofton, in 1869, discovered this text with
surprise, admiring "that one whose life was devoted to other branches of
science should have had the sagacity to discern the true mathematical
principles involved in a question of so entirely novel a character".
Crofton, of course, did not know that the Mémoire had be written in 1733.

Incidentally, it is somewhat ironic that Buffon, who was so
anxious to make mathematics useful, has furnished a remote starting point
for a branch of applied mathematics in a work which probably was for him
a study in pure mathematics and whose practical results could not be
foreseen by anybody at that time.

But, from that time on, probability was to get a much more
general importance for Buffon, directly related to his concern with the
theory of knowledge. In his essay of 1749 "How to study and handle Natural History ?", Buffon asks the difficult question of truth. There is
not one truth, he says, but several types of truth. The first of them
is furnished by mathematics. For us, a mathematical truth is perfectly
demonstrated and evident, because mathematics are the product of our
mind. The trouble is that this perfection is useless, because mathematics tell us nothing about physical reality. As far as physical truth
is concerned, we have no demonstration, no evidence whatsoever. A physical truth, necessarily based upon observation or experiment, is only
probable. What does it mean ? First, that no physical phenomenon, be
it as constant as the rising of the sun every day, may be considered
certain for the future. We are morally sure that the sun will rise to-
morrow morning : we have no demonstration of it. Here, Buffon is very
close to Hume's concept of belief. This first point is given further
development in the Essai d'arithmétique morale, were Buffon tries to
compute, so to speak, the degree of certainty of what he calls "physical
certitude", and compare it with the more usual "moral certitude". Of
course, it turns out that "physical certitude" is much greater than
"moral certitude".

But here a second and more interesting point appears.   If
probability is, according to Buffon, a degree of certainty, it is a degree
of belief as well and, as such, it should be able to direct human behaviour.
This, of course, is not original, but Buffon does not stop there.   Once
again, he tries to compute the probability, or the degree of belief, able
to determine our behaviour.   The difficulty is to define a suitable unit
for this computation.   Buffon starts with the idea that the most important
thing for a man is his own death.   Now, the tables of mortality show that
the probability for a man of 56 to die within the following 24 hours is
a little less than one out of 10.000.   But an average healthy man of 56
is not afraid of dying because he does not believe he will die the next
day.   Accordingly, any event whose probability is equal or inferior to
one tenthousend is of no concern for us, and we should observe this rule
in our every-day life.

I do not intend to follow Buffon in all the consequences he
draws from this computation, nor to comment at full length this attempt to
bridge the gap between "esprit de finesse" and "esprit géométrique", to
use Pascal's phrases.   I also could emphasize the fact that this curious
mixture of psychology and mathematics is somehow akin to our modern theory
of games.   I prefer to point out a last aspect of Buffon's thought on probability, which unfortunately he did not develop enough, but which is
interesting, even at an embryonic stage.

It is quite remarkable that Buffon speaks very little of causality and very often of analogy.   He seems to ignore the old formula :
"Post hoc, ergo propter hoc", and never openly objects to the causality
principle, as Hume did so energetically.   However, causality seems to be
linked, for him, with elementary mechanisms and never be as visible as in
man-made machines.   As far as Nature is concerned, Buffon generally speaks
of "secondary effects" whose causes are one or several "primary effects"
working in Nature.   Clearly, Buffon's model is Newtonian : according to
him, Newton has linked together "secondary effects", namely planetary
motions, tides and falling bodies.   By linking them by "the forces of
analogy" he was able to discover the common "primary effect", that is,
the force of gravitation.   But this is a great change in the meaning of
the word "cause".   In Descartes' transparent world, God had given the

universe a certain quantity of motion, and causality was nothing but a transmission of motion from one moving body to another, as in the simple mechanism of a clock, where a rotating gear transmits its motion to another gear. Here, the causality link is visible. In Newton's world things are different : gravitation is not visible. We do not see the sun attracting the earth, the moon attracting the sea, the earth attracting the falling bodies. The only way we have to discover, not a common "cause" for, but a unique fact in these phenomena is, according to Buffon, analogy, and the only evidence for the presence of a unique fact is, that the same mathematical law applies to all the phenomena. But the cause remains hidden and causality mysterious : we do not know how and why material bodies attract each other. Causality is a label we put on a black box : we know nothing about what happens within the box.

From this interpretation of Newton, Buffon draws two consequences. First, if the only way we have to understand Nature is by comparing phenomena and discovering analogies, we must be careful that analogy does not deceive us. We must "weigh" analogies, as Buffon says. Here, probability appears again, that is, probability understood as a degree of certainty and belief. All the more so, because in many fields of research and, for example, in biology, physical phenomena are not as simple as in celestial mechanics, and cannot be measured, in such a way that no mathematical law, that is, no final demonstration can be reached. Any general conclusion on a "primary effect" is only "probable", and the measure of that probability is the measure of analogies.

The second consequence is perhaps more interesting. If causality is such a mysterious thing, it is possible, at least in some cases, to dispose of it. The first example given by Buffon is precisely gravitation. As we already saw, gravitation is nothing but an empty word, which gives us no real knowledge of the cause of the phenomena : it is only a way to unite these phenomena into a single corpus and under the same mathematical law.

But the same phenomenological approach my be used for more complex phenomena, too complex for a cause to be discovered. Here, we may dispose of the cause and be content with establishing the law of the phenomena, and the law can be discovered by statistical studies only.

Buffon gives two examples of this approach. Dice are never perfect, but it is impossible to directly measure their irregularity in shape, weight, etc., which is the real cause of the particular results one gets with them. However, after a great number of throws, one can know the statistical result of this irregularity, that is, its law, without knowing its cause. And one is able to predict the statistical results of future throws and use the dice for an unfair game. The other example is the laws of mortality among men in a given society. There are so many causes of death, and so many of them are random events, that no causal analysis may be achieved. Statistics is the only way to deal with these phenomena and, for each case separately, that is, for each human being, the only possible knowledge is that of a probability.

The interesting thing with these two examples is, that they already oppose a microscopic and a macroscopic level though, as far as human mortality is concerned, we live at the microscopic level. On the other hand, all the difficult problems about chance and determinism are already there. Needless to say that Buffon did not see them.

Buffon is a 18th century thinker, and his problematics, his system of thought, is that of his century. I would not like to draw him towards our age and our problems. But, with Buffon as with some other 18th century scientists, science is already entering a new era and confronting new problems, almost entirely unknown to the previous century. 18th century science could not solve those problems, and sometimes hesitated between two possible solutions, one of them leading to 19th century positivism and the other to 20th century statistical and probabilistic approach. This was a very complicated situation and I never hoped to give you a full account of it. I only hope that you will read the Essai d'arithmétique morale.

THE STEREOLOGICAL ANALYSIS OF TWO-PHASE PARTICLES

RODNEY COLEMAN
Department of Mathematics, Imperial College, London SW7 2BZ

SUMMARY

Three stages in the stereological analysis of two-phase particles are identified. These are illustrated by (a) random probes and sections through a sphere with a spherical nucleus, (b) random probes through discs with a chord as phase interface, and (c) random probes and sections through a population of right cylinders with convex cross-sections and perpendicular cross-section phase interfaces.

1. INTRODUCTION

Examples of two-phase particles abound: the earth's mantle surrounds a molten core, the earth's surface is continent and ocean, people are flesh and bone. A cirrhotic liver is a two-phase structure (Takahashi, this volume). On a plum the fruit surrounds a stone. Microscopically a cell is a nucleus surrounded by cytoplasm. This paper developed from a study of two-phase mineral particles by colleagues in the Mineral Resources Engineering Department at Imperial College, London (Horton, 1976, Jones et al, 1978). They believe that the use of linear analysers, texture analysers and other sophisticated instruments, together with a sound theory, will lead to accurate assessment of the structure of mineral grains and bring about significant improvements in the economics of mineral processing.

Let us consider particles which consist of just two phases: a phase of interest, A — which might be gold — and another phase, B — base metal. For any particle the proportion by volume of phase A is called its grade or its true volumetric grade. An isotropic, uniform random (IUR) linear test probe is taken through a particle (Miles, this volume, Davy and Miles, 1977). This was called a μ-random probe in Coleman (1969, 1973). The proportion of the intercept that is in phase A is the apparent grade of the particle. Alternatively when we take an IUR planar test section through a particle the proportion of the area of the intercept that is in phase A is the apparent grade.

There are three stages of stereological analysis.

(i) Geometric probability. What is the distribution of the apparent grade for a random linear test probe or planar test section through a particle which has a known structure?

(ii) <u>Stereology</u>. How would we estimate the true volumetric grade of a particle
from a sequence of measurements of the apparent grade made by independent
random test probes or sections through a particle?

(iii) <u>Geometric statistics</u>. For a population of particles having a distribution of
grades, how would we estimate this grade distribution from a single measurement
of the apparent grade of each of a probability sample from the population?

We shall look briefly at three examples which illustrate these problems.

2. A RESULT FROM INTEGRAL GEOMETRY

We note first of all an important general result for two-phase particles (Kendall and
Moran, 1963, Sections 3.7, 3.38, 4.8). Consider a convex body K_1 having surface
area S_1 and mean caliper diameter (or integral of mean curvature) M_1. Suppose that
K_1 is wholly inside a convex body K_0 having surface area S_0 and mean caliper diameter
M_0. (The subscripts 1 for INNER and 0 for OUTER seem to be a useful mnemonic.) If
we take an IUR linear test probe Λ through K_0, then the probability that it hits K_1
is S_1/S_0, and for an IUR planar test section π through K_0 the probability that it
hits K is M_1/M_0.

For an IUR linear test probe through a convex body K_0 in the plane, where $K_1 \subset K_0$,
we have similarly that the probability the probe hits K_1 is B_1/B_0, where B_1 and B_0
are the perimeter lengths of K_1 and K_0.

When K_1 flattens to become a line of length ℓ (in the plane case) or a plate of
area a (in the three-dimensional case), the results still hold and we take $B_1 = 2\ell$,
that is, the perimeter length of a line of length ℓ, and $S_1 = 2a$, that is, the surface
area of a plate of area a.

3. PROBE THROUGH CONCENTRIC SPHERES

We take as our two-phase particle K_0 a spherical nucleus K_1 of phase A of radius r
(0<r<1), surrounded by phase B of thickness 1-r, making together a ball of unit
radius. Consider an IUR linear test probe Λ, and let the perpendicular distance
from the centre of the sphere to Λ be the random variable X. Then

$$P(X \leq x) = \frac{S'}{S_0} = \frac{4\pi x^2}{4\pi} = x^2 \qquad (0 \leq x \leq 1),$$

where S' is the surface area of a sphere K' of radius x. (This probability, x^2, is
also the ratio of the areas πx^2 and π of the projections of the spheres K' and K_0 in
a direction perpendicular to the beam containing Λ.) Thus

$$P(X^2 \le x^2) = x^2 \qquad (0 \le x^2 \le 1),$$

i.e. X^2 has a rectangular distribution on $[0,1]$.

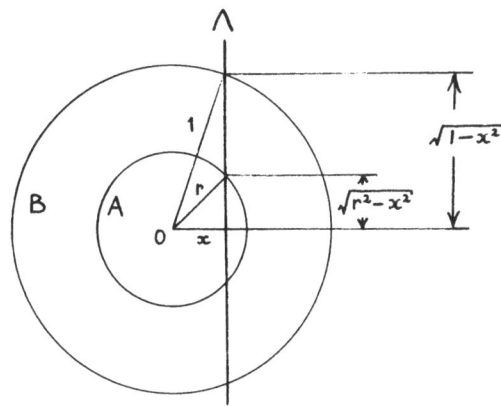

FIGURE 1. The section through concentric spheres which contains
their centre O and in which a random linear test probe Λ lies.

From simple triangle geometry (see Figure 1), the apparent grade of the particle

$$G = \begin{cases} 0 & (X \ge r) \\[2mm] \sqrt{\dfrac{r^2 - x^2}{1 - x^2}} & (X < r). \end{cases}$$

The equation

$$g = \sqrt{\frac{r^2 - x^2}{1 - x^2}} \qquad (0 \le x \le r)$$

has solution

$$x^2 = \frac{r^2 - g^2}{1 - g^2} \qquad (0 \le g \le r),$$

so

$$P(G \le g) = P(X^2 \ge \frac{r^2 - g^2}{1 - g^2})$$

$$= 1 - \frac{r^2 - g^2}{1 - g^2} = \frac{1 - r^2}{1 - g^2} \qquad (0 \le g \le r).$$

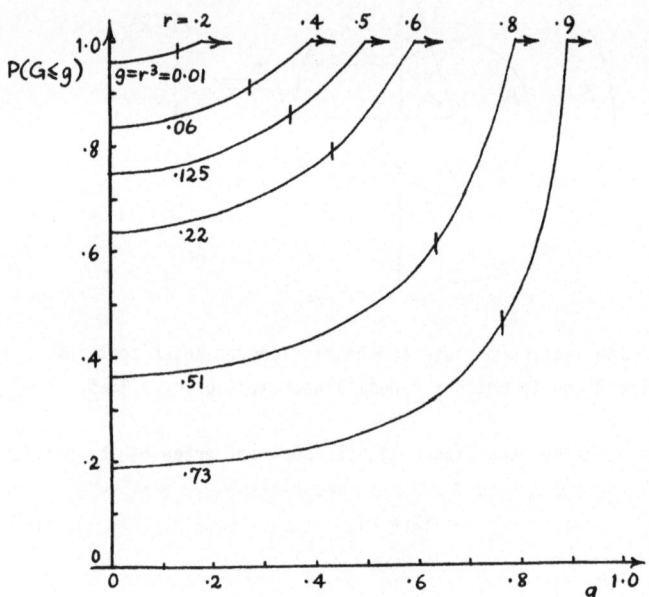

FIGURE 2. The apparent grade distribution function for a random linear test probe through a unit sphere which is a spherical nucleus of radius r of phase A surrounded by phase B. The mean of the distribution conditional on the probe hitting the nucleus is marked.

The apparent grade distribution is just as easily found for an IUR planar test section.

From an independent sequence of n observations of the apparent grade, $g_1, g_2, \ldots, g_n$, the maximum likelihood estimate for r is $\max(g_1, \ldots, g_n)$, so the maximum likelihood estimate for the true volumetric grade r^3 is $\{\max(g_1, \ldots, g_n)\}^3$. These estimates are biased too small, but are asymptotically unbiased as n becomes large. For a population of particles with r having even a very simple parametric distribution, there are in general no sufficient statistics for the parameters. Estimators are being developed but a simulation comparison to select the best will probably be necessary.

4. PROBE THROUGH A DISC SECTIONED BY A CHORD

We now consider a two-phase particle which is a unit disc and has the two phases separated by a chord C, which can be given a coordinate ϕ, the angle the chord makes to the centre at the circumference. Let the larger phase be phase A.

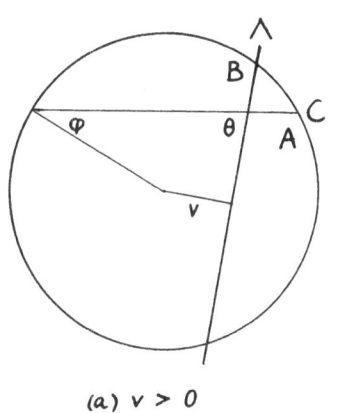 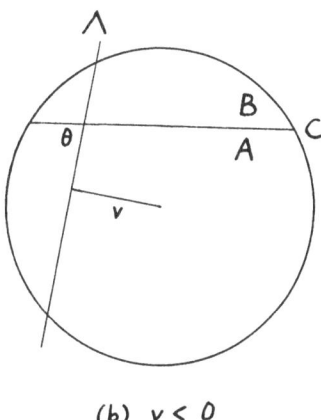

(a) v > 0   (b) v < 0

FIGURE 3. A unit disc of two phases separated by a chord C.
A test probe Λ is given coordinates θ ($0 < \theta < \frac{1}{2}\pi$) and
v ($-1 < v < 1$) (a) v > 0, (b) v < 0.

An IUR linear test probe Λ can be given coordinates θ and v, where θ is the angle made by the probe Λ with the chord C, and can by symmetry be taken to lie between 0 and $\frac{1}{2}\pi$. For an IUR probe θ has a rectangular distribution. For any fixed θ, an IUR probe is chosen uniformly from the beam making angle θ to the chord C. The coordinate v is the perpendicular distance from the centre of the disc to the probe Λ, and is measured both positively and negatively. Thus the distribution of v given θ is rectangular on $(-1,1)$. The joint density of θ and v is therefore $1/\pi$ ($0 < \theta < \frac{1}{2}\pi$, $-1 < v < 1$). Let G be the apparent grade. Then, from Section 2, the probability Λ hits C, so $0 < G < 1$, is

$$P(\uparrow) = \frac{2 \times \text{length of C}}{\text{perimeter length of disc}} = \frac{2 \cos \phi}{\pi}$$

The probes which hit C have their (θ,v) coordinates in set

$$\mathcal{A}(\phi) = \{0 < \theta < \tfrac{1}{2}\pi, \quad -\sin(\theta + \phi) < v < \sin(\theta - \phi)\}$$

The probes which are entirely in the smaller phase, phase B, have their (θ,v) coordinates in set

$$\mathcal{B}(\phi) = \{0 < \theta < \tfrac{1}{2}\pi - \phi, \quad -1 < v < -\sin(\theta + \phi)\}.$$

Therefore

$$P(G = 0) = \iint_{\mathcal{B}(\phi)} \frac{1}{\pi}\,d\theta dv = \tfrac{1}{2} - \frac{\phi}{\pi} - \frac{\cos\phi}{\pi}$$

so

$$P(G = 1) = 1 - P(\uparrow) - P(G = 0) = \tfrac{1}{2} + \frac{\phi}{\pi} - \frac{\cos\phi}{\pi} \quad .$$

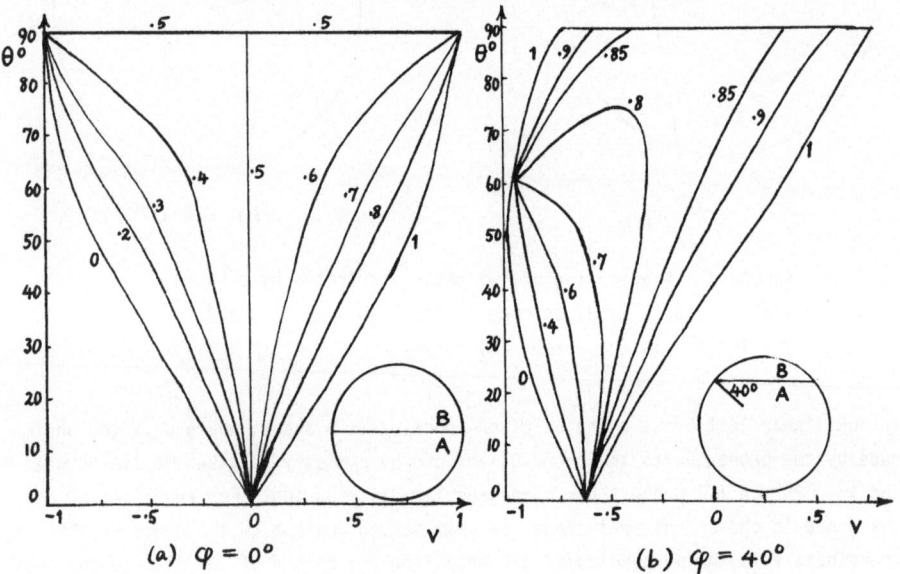

FIGURE 4. Contours of constant apparent grade g* for random probes which cut the phase-separating chord, C, of a unit disc. (a) $\phi = 0°$ (b) $\phi = 40°$.

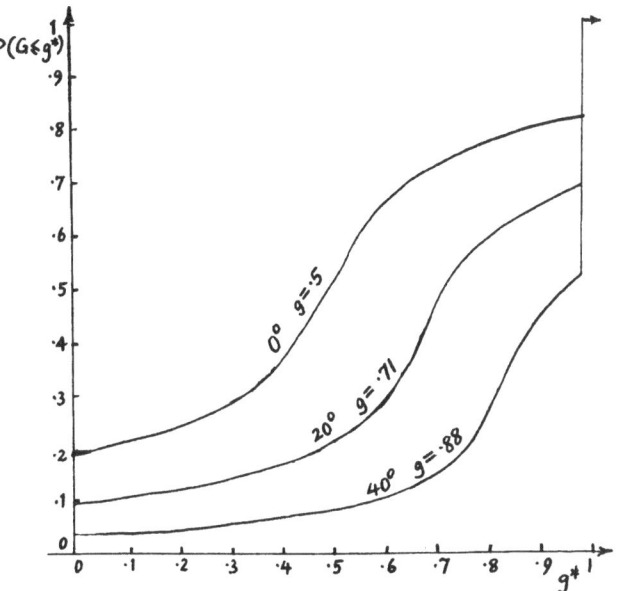

FIGURE 5. The distribution function P(G ≤ g*) for the apparent
grade G of a random probe through a two-phase disc with a chord
as phase interface.

Let us now consider only those probes which hit the chord C. The joint density of
θ and v is a constant, $(2 \cos \phi)^{-1}$ over $\mathcal{A}(\phi)$. The apparent grade of a probe (θ, v)
in $\mathcal{A}(\phi)$ is

$$g = \tfrac{1}{2} + \tfrac{1}{2} \frac{\sin \phi + v \cos \theta}{\sin \theta \sqrt{1 - v^2}}$$

so

$$P(G \leq g^* | \uparrow) = \frac{1}{2 \cos \phi} \iint_{\{\mathcal{A}(\phi) : g \leq g^*\}} d\theta dv .$$

In Figure 4 the contours in the (θ, v) plane are plotted; $P(G \leq g^* | \uparrow)$ is then just
the proportion of area 'below' the contour g*. These areas can be calculated
exactly only in the case $\phi = 0$, when the chord C is a diameter. Monte-Carlo
integration was used for $\phi \neq 0$.

Since we have an explicit formula for the apparent grade density only when C is a diameter — otherwise the formula is an unpleasant integral — these results do not lead to the efficient estimation of ϕ from a random sample of probes. However if we ignore the actual apparent grade measurements, and note only the numbers of independent probes entirely in phase A, entirely in phase B, and cutting C, we can obtain a maximum likelihood estimate of ϕ from the trinomial distribution of these numbers.

5. PROBES THROUGH CYLINDERS

Some general results for probes through single phase cylinders have been obtained by Kellerer (1971). We now consider IUR linear test probes and planar test sections through two-phase right cylinders of unit height with convex cross-sections, with the two phases separated by a perpendicular cross-section. We shall consider a population of cylinders. Some will be entirely of phase A, some entirely of phase B, and the rest will be composite. When a probe is made through a composite cylinder, it will pass entirely through phase A, entirely through phase B, or it will hit the phase interface.

Consider a cylinder chosen uniformly from the population, and an IUR linear test probe through it. Let G be the grade of the cylinder, i.e. the height of phase A is G and of phase B is 1 - G. Let G* be the apparent grade of the probe, i.e. G* is the proportion of the intercept that is in phase A. Define events:

$$A \ : \ G = 1, \qquad B \ : \ G = 0, \qquad C \ : \ 0 < G < 1;$$
$$A^* \ : \ G^* = 1, \qquad B^* \ : \ G^* = 0, \qquad C^* \ : \ 0 < G^* < 1.$$

The probability that the probe through a composite cylinder hits the phase interface is, from Section 2,

$$P(C^*|C) \ = \ \frac{2 \times \text{area of cross-section}}{\text{surface area of the cylinder}} \ = \ \frac{1}{1 + \gamma}$$

where

$$\gamma \ = \ \frac{\text{perimeter length of cross-section}}{2 \times \text{area of cross-section}} \ .$$

For example, when the cylinder is a unit cube or a tin can of diameter 1, $\gamma = 2$. This does not depend on the value of G.

Since the probe can be composite only if the cylinder is, i.e. $C^* \subseteq C$,

$$P(C^*) \ = \ P(C^* \cap C) \ = \ P(C^*|C)P(C)$$

i.e.

$$P(C) \ = \ (1 + \gamma)P(C^*).$$

An IUR probe is taken uniformly from a beam having a rectangularly distributed orientation. For any beam, with probability one, the projection of the cylinder in the plane perpendicular to the beam will be a rectangle. For a particular beam making angle θ to the axis of a cylinder suppose this rectangle has width ν. If the cylinder is composite of grade G, the probability that a probe chosen uniformly from this beam is entirely in phase A relative to the probability it is entirely in phase B is νG sin θ to ν(1 - G)sin θ, i.e. G to 1 - G, regardless of the direction of the beam.

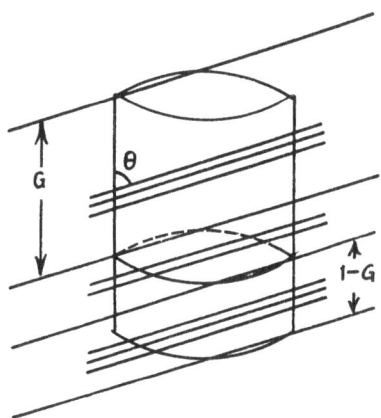

FIGURE 6. A beam passing through a cylinder of grade G.

If in Figure 6 we visualize the beam to be a beam of planar sections into the page, we see that this result holds for IUR planar sections also. Therefore

$$E(G^* | C \wedge \overline{C^*}) = \frac{G}{G + (1 - G)} = G \qquad \text{(if C).}$$

But $G^* = G$ also when G = 0 or 1 ⟶ i.e. on $\overline{C}$ ⟶ so

$$E(G^* | \overline{C^*}) = G.$$

We thus have that G^* is unbiased for G over the entire set $\overline{C^*}$, so any bias in estimating the true grade from the apparent grade comes only from those probes which cut the phase interface (and clearly only then if they have at least one end in the wall of the cylinder). Although MILES and DAVY (1976) have shown that G^* is in general a biased estimator of G, we have seen that for most probes there is no bias, and in a simulation study with cubic partieles no bias could be detected.

We shall therefore use the <u>approximate</u> result

$$EG^* = EG.$$

Then approximately

$$EG = P(A^*) + E(G^*|C^*)P(C^*).$$

Since the chance that a probe hits the phase interface of a composite cylinder does not depend on the grade of the cylinder, we have that

$$E(G|C^*) = E(G|C).$$

Then approximately

$$P(A) = P(A^*) - \gamma E(G^*|C^*)P(C^*).$$

Thus suppose we observe for a random sample of cylinders that the proportion of probes entirely in phase A is p_A, the proportion entirely in phase B is p_B, and that the average apparent grade of those probes which cut the phase interface is $\bar{g}$. Then by these formulae we would obtain estimates, if $p_C = 1 - p_A - p_B$,

$$\widehat{EG} = p_A + \bar{g}p_C$$
$$\widehat{P(C)} = (1 + \gamma)p_C$$
$$\widehat{P(A)} = p_A - \gamma\bar{g}p_C$$
$$\widehat{P(B)} = 1 - \widehat{P(A)} - \widehat{P(C)}.$$

As a numerical example, suppose for unit cubes ($\gamma = 2$), we observed $p_A = 0.7$, $p_B = 0.2$, $\bar{g} = 0.6$, then $p_C = 0.1$ and

$$\widehat{EG} = 0.76, \quad \widehat{P(C)} = 0.30, \quad \widehat{P(A)} = 0.58, \quad \widehat{P(B)} = 0.12.$$

For random sections, γ is a function of the mean caliper diameters of a disc and a cylinder. DeHoff (this volume) shows how the integrals of mean curvature, which are proportional to the mean caliper diameters for convex bodies, can be calculated. Alternatively if we take a composite cylinder, so $P(C) = 1$, and independent IUR sections, then

$$1 = P(C) = (1 + \gamma)P(C^*),$$

so γ can be estimated from the proportion of composite test sections.

We must now   examine the estimates   for particular shapes of cylinder, and
seek formulae for estimating the higher moments. In Figure 7 are plotted the
apparent grade distribution functions from a simulation study by R. Horton (1976) for
linear probes through unit cubes of constant grade g (g = 0.1,0.2,...,0.9), the phase
interface being parallel to a face. We note that a weighted average of the apparent
grades of the probes cutting the interface, weighted by their corresponding total intercept lengths, wight be expected to give an estimate of mean apparent grade with
smaller variance   than if we use $\bar{g}$.

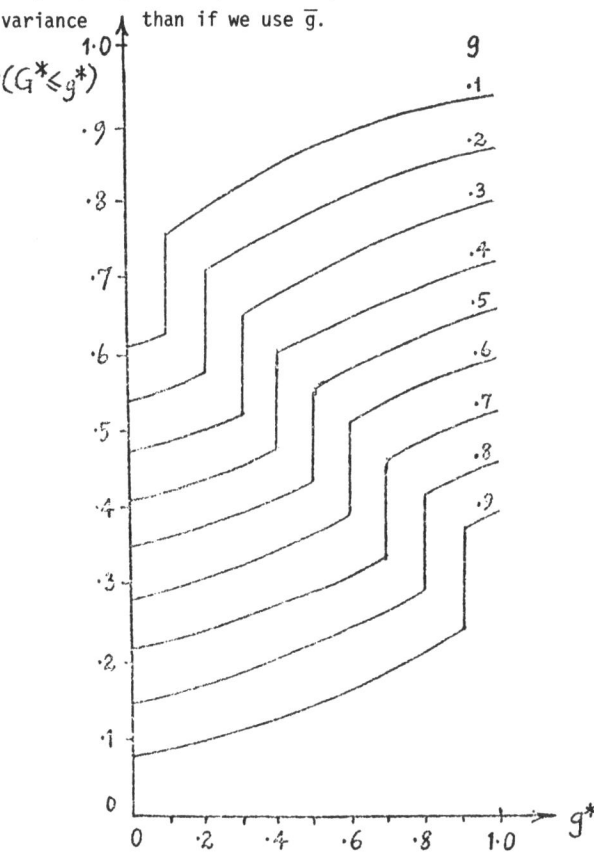

FIGURE 7. The apparent grade distribution function for IUR probes
through cubes of constant grade g, the phase interface being
parallel to a face.

ACKNOWLEDGEMENT. I am grateful for the interest shown by Mr M.P. Jones, Mr R. Horton,
and Mr C.H.J. Beavan of the Mineral Resources Engineering Department at Imperial
College, and for their generosity in making their results available to me.

REFERENCES

COLEMAN R. (1969) Random paths through convex bodies. J. Appl. Prob. 6, 430-441.

COLEMAN R. (1973) Random paths through rectangles and cubes. Metallography 6, 103-114.

DAVY P. and MILES R.E. (1977) Sampling theory for opaque spatial specimens. J.R. Statist. Soc. B 39, 56-65.

DEHOFF R.T. (1978) Stereological uses of the area tangent count. This volume.

HORTON R. (1976) Automatic linear measurement and assessment of composite particles. Unpublished report.

JONES M.P., COLEMAN R. and HORTON R. (1978) The assessment of composite particles from linear measurements. Proceedings of the Second European Symposium on Quantitative Analysis of Microstructures in Biology, Materials Science and Medicine. Caen, France, October 1977. To appear in Practical Metallography, special issue Vol.8.

KELLERER A.M. (1971) Considerations on the random traversal of convex bodies and solutions for general cylinders. Radiation Research 47, 359-376.

KENDALL M.G. and MORAN P.A.P. (1963) Geometrical probability. Griffin, London.

MILES R.E. (1978) The importance of the proper specification of underlying models in stereology. This volume.

TAKAHASHI T. (1978) Stereological and topological analysis of cirrhotic livers as a linkage of regenerative nodules multiply connected in the form of three-dimensional network. This volume.

MILES R.E. and DAVY P. (1976) Precise and general conditions for the validity of a comprehensive set of stereological fundamental formulae. J. Microsc. 107, 211-226.

ON THE DISTANCE BETWEEN POINTS IN POLYGONS

HAROLD RUBEN, *McGill University*

1. *Introduction and summary*

The need to evaluate the probability distribution, or at least the first few moments, of the distance between two independent uniform points, each in a planar and bounded polygonal domain, appears to surface periodically in numerous diverse areas of investigation. For some early isolated results, with applications in bombing, particle counting, and forestry sample surveys [or, more generally, problems of topographic variation -- see Matérn (1947) and Ghosh (1949)] etc., refer to Borel (1925), Ghosh (1943a,b; 1951), Matérn (1947) and Armitage (1949). Evaluation by direct integration, or by some such device as Crofton's 'second' theorem [more precisely, a slight extension of this theorem -- see Kendall and Moran (1963), pp.65-66] is at best intolerably tedious and time-consuming[1], and at worst intractable. In this paper, I present a method which should enable future investigators to determine systematically, and with relatively little effort, the probability distribution and moments of the distance, once the coordinates of the vertices of the polygonal domains have been specified.

To achieve this objective, it will be helpful to generalize the problem by considering more general sets (not necessarily polygons) in Euclidean space of arbitrary dimensionality (not necessarily 2). Specifically, let R and S denote sets in E^n, and $g \equiv g(|r-s|)$ a real or complex-valued function on R×S. Double[2] space integrals μ of the type

$$\mu[g;R,S] = \int_{R \times S} g \qquad (1)$$

have been discussed by Ruben (1970) [henceforth referred to as DSI], in which the basic result is the following [(*) and (**) below]. Let $g \equiv g(|r-s|)$ denote a real or complex-valued function on $\partial'R \times \partial'S$, where $\partial'R$ is the set of points on ∂R for which an outward normal is well defined (with a similar meaning for $\partial'S$), and let $\theta \equiv \theta(r,s)$ denote the angle between the outward normals at r and s. Define double[3] surface integrals ν by

$$\nu[g;\partial'R,\partial'S] = \int_{\partial'R \times \partial'S} g \cos \theta . \qquad (2)$$

---

[1] A partial exception occurs when the domains are similarly oriented (parallel) rectangles. Here the displacements between the two random points along directions parallel to two orthogonal sides of a rectangle are independent, and the squared distance between the points is the convolution of the squares of the two displacements. Note also that each of the displacements is distributed as a (in general, non-homogeneous) linear combination of two independent standard uniform variates.

[2] In the present context, the double space integrals in (1) can be viewed also (by Fubini) as iterated space integrals.

[3] In the present context, the double surface integrals in (2) can be viewed also (by Fubini) as iterated surface integrals.

Then [(*) of DSI, p.214]

$$\mu[g;R,S] = \nu[g^*;\partial'R,\partial'S] , \qquad (*)$$

where g^* is an arbitrary solution of the differential equation

$$\frac{d}{dy}\left\{y^{n-1}\frac{dg^*(y)}{dy}\right\} = -y^{n-1}g(y) \qquad (\ell < y < L) \qquad (**)$$

with $\ell = \inf\limits_{r\in R, s\in S}|r-s|$, $L = \sup\limits_{r\in R, s\in S}|r-s|$. Sufficient conditions for the
validity of (*) (in the sense of the existence and equality of the two sides of (*))
are as follows:

Conditions on R and S. Each of the sets R, S is a finite disjoint union of
standard domains in E^n, in the sense of Whitney (1957)[4], such that $|\partial'D| < \infty$
for each standard domain D which is a component set of R or S.

Conditions on g. (i) g has a finite set of discontinuities on (ℓ,L) ,

(ii) $\int_\ell^L y^{n-1}|g(y)|dy < \infty$,

(iii) $\int_0^u y^{n-1}g(y)dy = O(u^{n-1})$ for $u \to 0+$, if $\ell = 0$.

An application of (*) and (**) to geometrical probability[5] is immediate. If ξ and
η are independent uniform points in R and S, respectively, and $X = |\xi-\eta|$,
then, since

$$Eg(X) = \mu[g;R,S]/|R\times S| , \qquad (3)$$

we have, by (*) ,

$$Eg(X) = \nu[g^*;\partial'R,\partial'S]/|R\times S| . \qquad (4)$$

In particular, if $g \equiv g_x$ is the indicator function defined by

$$g_x(y) = \begin{cases} 1 & (y \le x) , \\ 0 & (y > x) , \end{cases} \qquad (5)$$

then, by (**), a suitable choice of $g^* \equiv g_x^*$ is

---

[4]See also DSI (pp.212-213). The notion of a standard domain D in E^n imposes
restrictions on the local 'convolutedness' of $\partial'D$ and on the 'size' of $\partial D-\partial'D$
(the latter set having zero $(n-1)$-extent, in the sense of Whitney).

[5]For an application to mensuration, specifically, a formula for the content of a
domain (not necessarily convex, and not necessarily in E^2) which is radically
different from that supplied by Crofton's second theorem for area, see Ruben (1967)
and DSI (p.219), and for an application to statistical mechanics, see Ruben (1964)
and DSI (p.220). [There are two transcription errors on p.219 of DSI. In Eq.(3.1),
replace $\|x_2-x_1\|^2$ by $\|x_2-x_1\|^2\cos\theta$, and in Eq. (3.1'), replace $\|s-r\|^2$ by
$\|s-r\|^2\cos\theta$. Also, in $\ell.4$ of Section 3(A), replace $g(\rho)$ by $g^*(\rho)$.]

$$
g_x^*(y) = \begin{cases} -\dfrac{y^2}{2n} & (y \le x;\ n \ge 1), \\[2ex] -\dfrac{x^2}{4} - \dfrac{x^2}{2}\log\dfrac{y}{x} & (y > x;\ n=2), \\[2ex] -\dfrac{x^2}{n-2}\left[\dfrac{1}{2} - \dfrac{1}{n}\dfrac{x^{n-2}}{y^{n-2}}\right] & (y > x;\ n \ne 2), \end{cases} \tag{6}
$$

and (3) becomes, writing F_X for the distribution function of X,

$$
F_X(x) = \nu[g_x^*;\ \partial'R, \partial'S]/|R \times S| . \tag{7}
$$

Hence also the density function F_X' of X is given by

$$
F_X'(x) = \nu[\tfrac{\partial}{\partial x} g_x^*;\ \partial'R, \partial'S]/|R \times S| , \tag{8}
$$

where

$$
\frac{\partial}{\partial x} g_x^*(y) = \begin{cases} 0 & (y \le x;\ n \ge 1), \\[2ex] -x\cdot\log\dfrac{y}{x} & (y > x;\ n=2), \\[2ex] \dfrac{x}{n-2}\left(\dfrac{x^{n-2}}{y^{n-2}} - 1\right) & (y > x;\ n \ne 2). \end{cases} \tag{9}
$$

For the moments Ex^k of X, k integral, with[6] $k \ge -1$ if $n > 1$ and $k > -1$ if $n = 1$, take $g(y) = y^k$. Then

$$
g^*(y) = -\frac{y^{k+2}}{(k+2)(k+n)} . \tag{10}
$$

Equations (4) and (10) jointly give the moments of X.

We now specialize our results to the case $n = 2$ with R and S polygonal domains in E^2. Denote the sides of R by $V_1, \ldots, V_p$ and the sides of S by $W_1, \ldots, W_q$. Let the angle between the outward normals of V_i and W_j be θ_{ij}. Then the angle between the outward normal, with respect to R, at a point r on V_i and the outward normal, with respect to S, at a point s on W_j is independent of r and s, and is θ_{ij} for all $r \in V_i$, $s \in W_j$. For a real or complex-valued function $h \equiv h(|r - s|)$ on $\Gamma \times \Lambda$, where Γ and Λ denote curves in E^2, define the iterated line integral

$$
\chi[h;\Gamma,\Lambda] = \int_{r \in \Gamma} dr \int_{s \in \Lambda} h(|r-s|)ds . \tag{11}
$$

With this notation, and in view of the preceding remark, we obtain[7] the following results for the distribution and density functions (refer to Eqns. (6)-(9)), as well as the moments (refer to Eqns. (4) and (10)), of X.

---

[6] $E(1/X)$ for $n = 2$ has arisen in the past as a practical problem.

[7] The previous conditions on R and S are here automatically satisfied.

$$\begin{cases} |R||S|F_X(x) = \sum_{i=1}^{p} \sum_{j=1}^{q} \chi[g_x^*; V_i, W_j] \cos\theta_{ij} , \\\\ g_x^*(y) = \begin{cases} -\dfrac{y^2}{4} & (y \le x) , \\\\ -\dfrac{x^2}{4} - \dfrac{x^2}{2}\log\dfrac{y}{x} & (y > x) . \end{cases} \end{cases} \qquad \text{(A)}$$

$$\begin{cases} |R||S|F_X'(x) = \sum_{i=1}^{p} \sum_{j=1}^{q} \chi[\dfrac{\partial}{\partial x} g_x^*; V_i, W_j] \cos\theta_{ij} , \\\\ \dfrac{\partial}{\partial x} g_x^*(y) = \begin{cases} 0 & (y \le x) , \\\\ -x\log\dfrac{y}{x} & (y > x) . \end{cases} \end{cases} \qquad \text{(B)}$$

$$\begin{cases} |R||S|EX^k = \sum_{i=1}^{p} \sum_{j=1}^{q} \chi[g^*; V_i, W_j] \cos\theta_{ij} , \\\\ g^*(y) = -\dfrac{y^{k+2}}{(k+2)^2} & (k \ge -1) . \end{cases} \qquad \text{(C)}$$

It will be noted that the pair V_i, W_j does not contribute to the series in (A), (B) and (C) if V_i, W_j are orthogonal.

Formulae (A), (B) and (C) show that the distribution and density functions, as well as the moments, of X can be expressed as linear combinations of χ-integrals on the product sets $V_i \times W_j$. A further considerable simplification and reduction, which makes possible a systematic evaluation of the probability distribution and moments of X, can be effected on noting that the general χ-functional [see (11)] is additive with respect to each of the set components Γ, Λ. This implies that if $\theta_{ij} \neq 0, \pi$ (V_i, W_j non-parallel) then $\chi[\cdot; V_i, W_j]$ can be expressed as a linear combination, with coefficients $+1$ or -1, of four[e] χ-functionals, each of the type $\chi[\cdot; OA, OB]$, where the line segments OA, OB have a common endpoint O, while if $\theta_{ij} = 0$ or π (V_i, W_j parallel) then $\chi[\cdot; V_i, W_j]$ can be expressed as the difference of two χ-functionals, each of the type $\chi[\cdot; CD, EF]$, where the line segments CD, EF are parallel and CE is orthogonal to CD and EF. [See Section 2(a).] Call these two types of χ-functionals *canonical χ-functionals*, and, correspondingly, when the function argument in χ is specified, *canonical χ-integrals*. Then it follows that, for specified angles θ_{ij}, F_X, F_X', *and* EX^k *in* (A), (B), *and* (C) *can be expressed as linear combinations of canonical χ-integrals,* in which the coefficients depend on the θ_{ij} (cf. DSI, p.219). Explicit formulae for the two types of canonical χ-integrals needed in (A), (B), and (C), i.e. with the function argument

---

[e] In 'degenerate' cases, two or one (instead of four).

in χ equal successively to g_x^*, $(\partial/\partial x)g_x^*$, g^*, as defined in (A), (B), and (C), are derived in Section 2(b).

The basic results in this paper are provided by Eqns. (A), (B), (C), (D), (E), (A'), (B'), (C'), and Tables 1 and 2. Eqns. (A), (B), and (C) represent the distribution function, density function, and moments, respectively, of X as linear combinations of χ-integrals, while Eqns. (D) and (E) represent each such χ-integral as a linear combination of the two types of canonical χ-integrals, called L and M integrals, so that Eqns. (A), (B), (C), (D), and (E), taken jointly, represent the distribution function, density function, and moments of X as linear combinations of L and M integrals. Table 1 gives the L-integrals, and Table 2 the M-integrals for the distribution function, density function, and moments of X in terms of certain auxiliary functions. Eq. (A') gives these auxiliary functions for (A), i.e., when the distribution function of X is required; Eq. (B') gives the auxiliary functions for (B), i.e., when the density function of X is required; Eq. (C') gives the auxiliary functions for (C), i.e., when the moments of X are required.

2. *Systematic evaluation of the probability distribution and moments*

(a) *Reduction of iterated line integrals to canonical form*

The two 'canonical' configurations of a pair of line segments Γ, Λ in the plane, of lengths a, b, are depicted in Fig. 1. For a Type I configuration, the mutual angle of inclination between Γ and Λ is denoted by $\varepsilon(\neq 0,\pi)$, and for a Type II configuration the mutual separation between Γ and Λ is denoted by d. In both cases, $\chi[h;\Gamma,\Lambda]$ is, for given h, a three-parameter function, and we write $\chi[h;\Gamma,\Lambda] = L(a,b;\varepsilon)$ for the Type I configurations, and $\chi[h;\Gamma,\Lambda] = M(a,b;d)$ for the Type II configurations (the function h being suppressed in both cases and supplied by the context). L and M are the canonical χ-functionals of Section 1. Observe that these two functionals are not independent (the M-functional being expressible as the limit of a linear combination of four L-functionals -- see Eqn. (12)). However, M is best computed independently in each instance.

Type I	Type II

$|\Gamma| = a$, $|\Lambda| = b$;
angle between Γ and $\Lambda = \varepsilon(\neq 0,\pi)$

$|\Gamma| = a$, $|\Lambda| = b$, Γ,Λ parallel and line joining left endpoints of Γ and Λ orthogonal to Γ and Λ; distance between Γ and $\Lambda = d$

Fig. 1. The two canonical configurations of a pair of line segments Γ, Λ

54

We now represent $\chi[\cdot;V,W]$ for an *arbitrary* pair of line segments V,W (i.e., V,W not necessarily either of Type I or Type II) as a linear combination of χ-functionals. If V, W are non-parallel, then there are three possible configurations of the pair, according as the point of intersection of the two lines in which V and W lie below to none, one, or two of the line segments. These three configurations are depicted in Figs. 2(i), 2(ii), and 2(iii). If V,W are parallel (call them 'horizontal'), then there are two possible configurations (these are, however, not fundamentally distinct) of the pair according as the left endpoint of V lies to the left or right of the left endpoint of W. These two configurations are depicted in Figs. 3(i) and 3(ii).

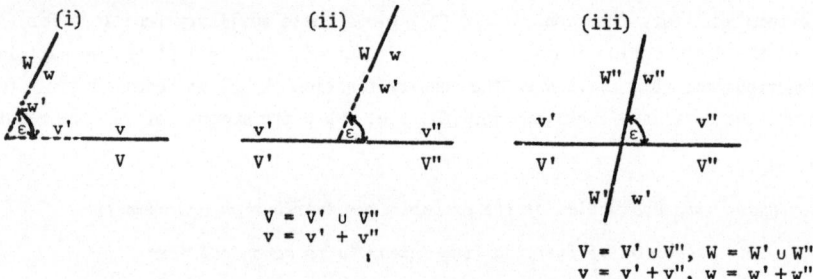

$$V = V' \cup V''$$
$$v = v' + v''$$

$$V = V' \cup V'', \quad W = W' \cup W''$$
$$v = v' + v'', \quad w = w' + w''$$

Fig. 2. The three possible configurations of non-parallel line segments V, W
$[|V| = v, \ |W| = w, \ \text{angle between} \ V \ \text{and} \ W = \varepsilon(\neq 0,\pi)]$

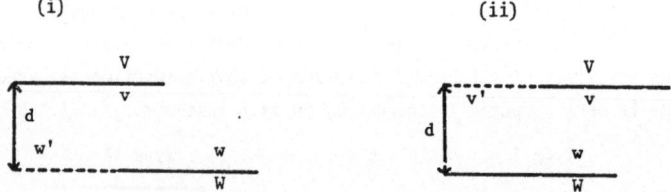

Left endpoint of V to left of left endpoint of W

Left endpoint of V to right of left endpoint of W

Fig. 3. The two possible (essentially non-distinct) configurations of parallel line segments V, W
$[|V| = v, \ |W| = w, \ \text{separation between} \ V \ \text{and} \ W = d]$

A glance at Figs. 2 and 3 give the following reductions for $\chi \equiv \chi[h;V,W]$.

Fig. 2(i): $\chi = L(v+v',w+w';\varepsilon) - L(v',w+w';\varepsilon) - L(v+v',w';\varepsilon) + L(v',w';\varepsilon)$. ⎫

Fig. 2(ii): $\chi = L(v',w+w';\pi-\varepsilon) - L(v',w';\pi-\varepsilon) - L(v'',w';\varepsilon) + L(v'',w+w';\varepsilon)$. ⎬   (D)

Fig. 2(iii): $\chi = L(v',w';\varepsilon) + L(v',w'';\pi-\varepsilon) + L(v'',w';\pi-\varepsilon) + L(v'',w'';\varepsilon)$. ⎭

Fig. 3(i): $\chi = M(v,w+w';d) - M(v,w';d)$. ⎫

Fig. 3(ii): $\chi = M(v+v',w;d) - M(v',w;d)$. ⎬   (E)
                                                                     ⎭

The first of the three equations in (D) shows incidentally that the L and M-functionals are not independent. In fact, in the second diagram of Fig. 1, Λ may be regarded as the limiting position of a line segment of fixed length b with fixed left endpoint and inclined at an angle ε to Γ, so that

$$M(a,b;d) = \lim \{L(a+d\cot\varepsilon, b+d\,\mathrm{cosec}\,\varepsilon;\varepsilon) - L(d\cot\varepsilon, b+d\cos\varepsilon;\varepsilon)$$
$$- L(a+d\cot\varepsilon, d\,\mathrm{cosec}\,\varepsilon;\varepsilon) + L(d\cot\varepsilon, d\,\mathrm{cosec}\,\varepsilon;\varepsilon)\}. \quad (12)$$

We illustrate the use of the three fundamental formulae (A), (B) and (C) for F_χ, F'_χ, and Ex^k, together with the two ancillary sets of reduction formulae (D) and (E), by four examples [see diagrams below]. This will result in F_χ, F'_χ and Ex^k being represented as a linear combination of L and M-integrals.

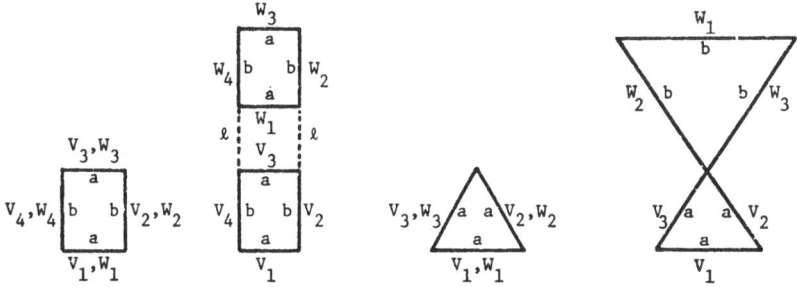

In the first example, R = rectangle of sides a and b = S. Here $V_i = W_i$ for i = 1, 2, 3, 4, and $\theta_{11} = \theta_{22} = \theta_{33} = \theta_{44} = 0$, $\theta_{13} = \theta_{24} = \theta_{31} = \theta_{42} = \pi$, while all the other θ_{ij} are $\pi/2$. Pairs of sides V_i, W_j for which $\theta_{ij} = \pi/2$ do not contribute to the series on the right of (A), (B) and (C). Hence these series are

$$\chi[\cdot;V_1,V_1]\cos 0 + \chi[\cdot;V_1,V_3]\cos\pi + \chi[\cdot;V_2,V_2]\cos 0 + \chi[\cdot;V_2,V_4]\cos\pi$$
$$+ \chi[\cdot;V_3,V_1]\cos\pi + \chi[\cdot;V_3,V_3]\cos 0 + \chi[\cdot;V_4,V_2]\cos\pi + \chi[\cdot;V_4,V_4]\cos 0$$
$$= 2\{\chi[\cdot;V_1,V_1] - \chi[\cdot;V_1,V_3] + \chi[\cdot;V_2,V_2] - \chi[\cdot;V_2,V_4]\}$$
$$= 2\{M(a,a;0) - M(a,a;b) + M(b,b;0) - M(b,b;a)\}.$$

In the second example, R, S are congruent rectangles of sides a and b, and are aligned so that V_2 is collinear with W_2 and V_4 with W_4. The separation between R and S is ℓ. As in the previous example, mutually orthogonal sides do not contribute to the series in (A), (B) and (C). Hence these series are

$$\chi[\cdot;V_1,W_1] \cos\theta_{11} + \chi[\cdot;V_1,W_3]\cos\theta_{13} + \chi[\cdot;V_2,W_2]\cos\theta_{22} + \chi[\cdot;V_2,W_4]\cos\theta_{24}$$

$$+ \chi[\cdot;V_3,W_1]\cos\theta_{31} + \chi[\cdot;V_3,W_3]\cos\theta_{33} + \chi[\cdot;V_4,W_2]\cos\theta_{42} + \chi[\cdot;V_4,W_4]\cos\theta_{44}$$

$$= M(a,a;b+\ell) - M(a,a;2b+\ell) + \chi[\cdot;V_2,W_2] - \chi[\cdot;V_2,W_4]$$

$$- M(a,a;\ell) + M(a,a;b+\ell) - \chi[\cdot;V_4,W_2] + \chi[\cdot;V_4,W_4] ,$$

and since (by (E))

$$\chi[\cdot;V_2,W_2] = M(b,2b+\ell;0) - M(b,b+\ell;0) = \chi[\cdot;V_4,W_4] ,$$

$$\chi[\cdot;V_2,W_4] = M(b,2b+\ell;a) - M(b,b+\ell;a) = \chi[\cdot;V_4,W_2] ,$$

the series reduce to

$$2M(a,a;b+\ell) - M(a,a;\ell) - M(a,a;2b+\ell) - 2M(b,b+\ell;0) + 2M(b,2b+\ell;0)$$

$$+ 2M(b,b+\ell;a) - 2M(b,2b+\ell;a) .$$

In the third example, R is an equilateral triangle of side $a = S$. Here $V_i = W_i$, and $\theta_{ij} = 0$ or $2\pi/3$ according as $i = j$ or $i \neq j$ for $i,j = 1,2,3$, and $\varepsilon_{ij} = \pi/3$ for $j \neq i$ (ε_{ij} = angle between V_i and W_j). By symmetry, the series in (A), (B) and (C) are

$$3\sum_{j=1}^{3} \chi[\cdot;V_1,W_j] \cos\theta_{1j} = 3\{\chi[\cdot;V_1,V_1] + 2\chi[\cdot;V_1,V_2]\cos\theta_{12}\}$$

$$= 3\{M(a,a;0) - L(a,a;\pi/3)\} .$$

In the fourth example, R and S are equilateral triangles of sides a and b ($\geq a$) with a common vertex, V_2 is collinear with W_2 and V_3 with W_3 . Here $\theta_{ij} = \pi$ or $\pi/3$ according as $i = j$ or $i \neq j$ for $i,j = 1,2,3$. We have

$$\chi[\cdot;V_1,W_1] = M\left(\frac{b+a}{2}, b; \frac{\sqrt{3}(b+a)}{2}\right) - M\left(\frac{b-a}{2}, b; \frac{\sqrt{3}(b+a)}{2}\right)$$

(from (E), after noting that the distance between V_1 and W_1 is $\sqrt{3}(b+a)/2$) ,

$$\chi[\cdot;V_2,W_2] = M(a,a+b;0) - M(a,a;0) = \chi[\cdot;V_3,W_3]$$

(by (E)),

$$\chi[\cdot;V_1,W_2] = L(a,a+b;\pi/3) - L(a,a;\pi/3) = \chi[\cdot;V_1,W_3] ,$$

$$\chi[\cdot;V_2,W_1] = L(b,a+b;\pi/3) - L(b,b;\pi/3) = \chi[\cdot;V_3,W_1] ,$$

(by the first equation in (D), and noting that $\varepsilon_{12} = \varepsilon_{13} = \varepsilon_{21} = \varepsilon_{31} = \pi/3$) ,

$$\chi[\cdot;V_2,W_3] = L(a,b;2\pi/3) = \chi[\cdot;V_3,W_2]$$

(since $\varepsilon_{23} = \varepsilon_{32} = 2\pi/3$), giving for the series in (A), (B), and (C),

$$-M\left(\frac{b+a}{2},b;\frac{\sqrt{3}(b+a)}{2}\right) + M\left(\frac{b-a}{2},b;\frac{\sqrt{3}(b+a)}{2}\right) - 2M(a,a+b;0) + 2M(a,a;0)$$

$$+ L(a,a+b;\pi/3) - L(a,a;\pi/3) + L(b,a+b;\pi/3) - L(b,b;\pi/3) + L(a,b;2\pi/3) .$$

(b) Evaluation of the canonical L *and* M *functions for the probability distribution and moments*

Since $L(a,b;\varepsilon) = L(b,a;\varepsilon)$ and $M(a,b;d) = M(b,a;d)$, *we shall assume from this point on* (for the sake of definiteness and without loss of generality) *that* $a \leq b$ *in both* L *and* M.

To evaluate the L-function, refer to the first diagram of Fig. 1, and let t and u be the respective distances of a point r on Γ and a point s on Λ from the point of intersection of Γ and Λ. Let ρ denote the distance between r and s. Then

$$L(a,b;\varepsilon) = \iint\limits_{0 \leq t \leq a, 0 \leq u \leq b} h(\rho)dtdu \, ,$$
$$\rho = (t^2 + u^2 - 2tu \cos \varepsilon)^{\frac{1}{2}} \, . \right\}} \tag{13}$$

Setting

$$t = \frac{1}{2} \left(\frac{t'}{\sin \frac{1}{2} \varepsilon} + \frac{u'}{\cos \frac{1}{2} \varepsilon} \right) , \quad u = \frac{1}{2} \left(\frac{t'}{\sin \frac{1}{2} \varepsilon} - \frac{u'}{\cos \frac{1}{2} \varepsilon} \right),$$

we obtain

$$(\sin \varepsilon)L(a,b;\varepsilon) = \iint\limits_{OACB} h(\rho)dt'du' \, , \tag{14}$$

where $\rho = (t'^2 + u'^2)^{\frac{1}{2}}$ and OACB is the parallelogram [see Fig.4]

$0 \leq \frac{1}{2}(t'/\sin\frac{1}{2}\varepsilon + u'/\cos\frac{1}{2}\varepsilon) \leq a$, $0 \leq \frac{1}{2}(t'/\sin\frac{1}{2}\varepsilon - u'/\cos\frac{1}{2}\varepsilon) \leq b$ with vertices $0(0,0)$, $A(a\sin\frac{1}{2}\varepsilon, a\cos\frac{1}{2}\varepsilon)$, $B(b\sin\frac{1}{2}\varepsilon, -b\cos\frac{1}{2}\varepsilon)$, $C((a+b)\sin\frac{1}{2}\varepsilon, (a-b)\cos\frac{1}{2}\varepsilon)$. Here $\angle OAC = \varepsilon$, and $OA = a$, $OB = b$, $OC = c = (a^2 + b^2 - 2ab\cos\varepsilon)^{\frac{1}{2}}$. Further, writing $\angle AOC = \alpha$, $\angle BOC = \beta$, we have

$$\alpha = \cos^{-1}\{(a-b \cos \varepsilon)/c\} , \quad \beta = \cos^{-1}\{(b-a \cos \varepsilon)/c\} . \tag{15}$$

We note also (these will be useful in the sequel) that

$$\sin(\alpha+\varepsilon) = (a \sin \varepsilon)/c , \qquad \sin(\beta+\varepsilon) = b \sin \varepsilon/c ,$$
$$\cot(\alpha+\varepsilon) = (a \cot \varepsilon - b \operatorname{cosec} \varepsilon)/a, \quad \cot(\beta+\varepsilon) = (b \cot \varepsilon - a \operatorname{cosec} \varepsilon)/b, \right\}} \tag{15'}$$

and

$$\cot \varepsilon - \cot(\alpha+\varepsilon) = b \operatorname{cosec} \varepsilon/a , \quad \cot \varepsilon - \cot(\beta+\varepsilon) = a \operatorname{cosec} \varepsilon/b . \tag{15''}$$

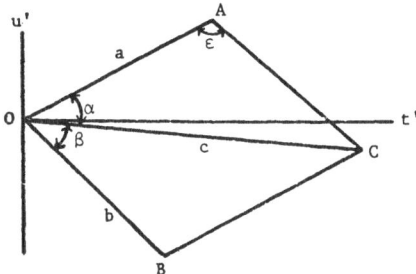

Fig. 4. The parallelogram OACB (the domain of integration in the evaluation of L)

In terms of polar coordinates (ρ,ϕ) for the triangle OAC and (ρ,ψ) for the triangle OBC (ϕ measured from OA and ψ from OB), (14) becomes

$$(\sin\epsilon)L(a,b;\epsilon) = \int_0^\alpha d\phi \int_0^{a\sin\epsilon/\sin(\phi+\epsilon)} h(\rho)\rho d\rho + \int_0^\beta d\psi \int_0^{b\sin\epsilon/\sin(\psi+\epsilon)} h(\rho)\rho d\rho,$$

so that, defining

$$H(f,\gamma) = \int_0^\gamma d\omega \int_0^{f\sin\epsilon/\sin(\omega+\epsilon)} h(\rho)\rho d\rho , \qquad (16)$$

we arrive at the result

$$(\sin\epsilon)L(a,b;\epsilon) = H(a,\alpha) + H(b,\beta) . \qquad (17)$$

Eq. (17) is suitable when $h(\rho)$ has the same functional form on its domain, as in (C) for the moments of X. However, if $h(\rho)$ assumes different functional forms for $\rho \le x$ and $\rho > x$, say

$$h(\rho) = \begin{cases} h_0(\rho) & (\rho \le x) , \\ h_1(\rho) & (\rho > x) , \end{cases} \qquad (18)$$

as in (A) and (B), for the distribution and density functions of X, it becomes necessary, in evaluating the H-functions in (17), to decompose the triangles OAC and OBC (the respective domains of integration for $H(a,\alpha)$ and $H(b,\beta)$) correspondingly. The (perpendicular) distances of O from the lines containing AC and BC are $a\sin\epsilon$ and $b\sin\epsilon$. Consider a circle G_x with centre O and radius x, and for $x \ge a\sin\epsilon$ define angles $\alpha' \equiv \alpha'_x$, $\alpha'' \equiv \alpha''_x$ with $\alpha' \le \alpha''$ (so that $\sin^{-1}(a\sin\epsilon/x)$ in (19) is interpreted as an angle $\le \pi/2$), while for $x \ge b\sin\epsilon$ define angles $\beta' \equiv \beta'_x$, $\beta'' \equiv \beta''_x$ with $\beta' \le \beta''$ (so that $\sin^{-1}(b\sin\epsilon/x)$ in (19) is interpreted as an angle $\le \pi/2$) by

$$\left.\begin{array}{ll} \alpha' + \epsilon = \sin^{-1}(a\sin\epsilon/x) , & \beta' + \epsilon = \sin^{-1}(b\sin\epsilon/x) , \\ \alpha'' + \epsilon = \pi - \sin^{-1}(a\sin\epsilon/x) , & \beta'' + \epsilon = \pi - \sin^{-1}(b\sin\epsilon/x). \end{array}\right\} \quad (19)$$

We note also incidentally (these will be useful in the sequel) that

$$\begin{array}{ll} \sin(\alpha'+\epsilon) = a\sin\epsilon/x, & \sin(\beta'+\epsilon) = b\sin\epsilon/x, \\ \cot(\alpha'+\epsilon) = (x^2-a^2\sin^2\epsilon)^{\frac{1}{2}}/a\sin\epsilon, & \cot(\beta'+\epsilon) = (x^2-b^2\sin^2\epsilon)^{\frac{1}{2}}/b\sin\epsilon, \\ \sin(\alpha''+\epsilon) = a\sin\epsilon/x, & \sin(\beta''+\epsilon) = b\sin\epsilon/x, \\ \cot(\alpha''+\epsilon) = -(x^2-a^2\sin^2\epsilon)^{\frac{1}{2}}/a\sin\epsilon, & \cot(\beta''+\epsilon) = -(x^2-b^2\sin^2\epsilon)^{\frac{1}{2}}/b\sin\epsilon. \end{array} \quad (19')$$

If G_x cuts AC in two points P', P'', then $\angle AOP' = \alpha'$ and $\angle AOP'' = \alpha''$, while if G_x cuts AC in one point, then this point is either P' or P''. Similarly, if G_x cuts BC in two points Q', Q'', then $\angle BOQ' = \beta'$ and $\angle BOQ'' = \beta''$, while if G_x cuts BC in one point, then this point is either Q' or Q''. We obtain the following scheme for the points of intersection of G_x with AC and BC.

Scheme 1. Points of intersection of G_x with (i) AC, (ii) BC .

(i)		(ii)	
	Intersection points		Intersection points
$a < c, \; \epsilon > \pi/2$		$b < c, \; \epsilon > \pi/2$	
$0 \leq x < a$	None	$0 \leq x < b$	None
$a \leq x < c$	P''	$b \leq x < c$	Q''
$x \geq c$	None	$x \geq c$	None
$a < c, \; \epsilon \leq \pi/2$		$b < c, \; \epsilon \leq \pi/2$	
$0 \leq x < a \sin \epsilon$	None	$0 \leq x < b \sin \epsilon$	None
$a \sin \epsilon \leq x < a$	P', P''	$b \sin \epsilon \leq x < b$	Q',Q''
$a \leq x < c$	P''	$b \leq x < c$	Q''
$x \geq c$	None	$x \geq c$	None
$a \geq c$		$b \geq c, \; \cos \epsilon < a/b$	
$0 \leq x < a \sin \epsilon$	None	$0 \leq x < b \sin \epsilon$	None
$a \sin \epsilon \leq x < c$	P', P''	$b \sin \epsilon \leq x < c$	Q', Q''
$c \leq x < a$	P'	$c \leq x < b$	Q'
$x \geq a$	None	$x \geq b$	None
		$b \geq c, \; \cos \epsilon \geq a/b$	
		$0 \leq x < c$	None
		$c \leq x < b$	Q'
		$x \geq b$	None

The values of $H(a,\alpha)$ and $H(b,\beta)$ follow from Scheme 1. We use the abbreviated notation

$$(\gamma,\delta) = H(a,\delta) - H(a,\gamma) , \quad ((\gamma,\delta)) = H(b,\delta) - H(b,\gamma),$$
$$(\gamma,\delta)_0 = H_0(a,\delta) - H_0(a,\gamma) , \quad ((\gamma,\delta))_0 = H_0(b,\delta) - H_0(b,\gamma), \qquad (20)$$

where

$$H_0(f,\gamma) = \int_0^\gamma d\omega \int_0^{f \sin \epsilon / \sin(\omega+\epsilon)} h_0(\rho)\rho d\rho . \qquad (21)$$

As an example, let $a \geq c$ in the triangle OAC , and let $b \geq c$, $\cos \epsilon \geq a/b$ in the triangle OBC . The condition $a \geq c$ is equivalent to $\cos \epsilon \geq b/2a$, and the condition $b \geq c$ to $\cos \epsilon \geq a/2b$, so that $\cos \epsilon \geq b/2a$ for the triangle OAC and $\cos \epsilon \geq a/b$ for the triangle OBC . Note in passing that $\cos \epsilon \geq b/2a$, $\cos \epsilon \geq a/b$ are jointly equivalent to $\cos \epsilon \geq a/b$ if $b/a < \sqrt{2}$, and to $\cos \epsilon \geq b/2a$ if $b/a \geq \sqrt{2}$. To evaluate $H(a,\alpha)$ when $\cos \epsilon \geq b/2a$ in terms of the H and H_0 –functions , refer to the third part of Scheme 1(i). This yields

$$0 \leq x < a \sin \epsilon : \quad H(a,\alpha) = (0,\alpha)$$
$$a \sin \epsilon \leq x < c : \quad H(a,\alpha) = (0,\alpha') + (\alpha',\alpha'')_0 + (\alpha'',\alpha)$$
$$c \leq x < a : \quad H(a,\alpha) = (0,\alpha') + (\alpha',\alpha)_0$$
$$x \geq a : \quad H(a,\alpha) = (0,\alpha)_0 ,$$

where (for example) the result for $a \sin \epsilon \leq x < c$ is obtained from the decomposition

$$H(a,\alpha) = \iint_{\Delta OAC} h(\rho)d\rho d\phi$$

$$= \iint_{\Delta OAP'} h(\rho)\rho d\rho d\phi + \int\int_{\Delta OP'P''} h(\rho)\rho d\rho d\phi + \iint_{\Delta OP''C} h(\rho)\rho d\rho d\phi$$

$$= \iint_{\Delta OAP'} h(\rho)\rho d\rho d\phi + \int\int_{\Delta OP'P''} h_0(\rho)\rho d\rho d\phi + \iint_{\Delta OP''C} h(\rho)\rho d\rho d\phi$$

$$= (0,\alpha') + (\alpha',\alpha'')_0 + (\alpha'',\alpha) \ .$$

Similarly, to evaluate $H(b,\beta)$ when $\cos\epsilon \geq a/b$ in terms of the H and H_0-functions, refer to the last part of Scheme 1(ii). This yields

$$0 \leq x < c : \quad H(b,\beta) = ((0,\beta))$$
$$c \leq x < b : \quad H(b,\beta) = ((0,\beta')) + ((\beta',\beta))_0$$
$$x \geq b : \quad H(b,\beta) = ((0,\beta))_0 \ .$$

Combining the two sets of values, we arrive at the following set of values of $H(a,\alpha) + H(b,\beta) = (\sin\epsilon)L(a,b;\epsilon)$ for $a \geq c$, $b \geq c$, $\cos\epsilon \geq a/b$.

$$0 \leq x < a\sin\epsilon : \quad (0,\alpha)$$
$$+ ((0,\beta))$$

$$a\sin\epsilon \leq x < c : \quad (0,\alpha') + (\alpha',\alpha'')_0 + (\alpha'',\alpha)$$
$$+ ((0,\beta))$$

$$c \leq x < a : \quad (0,\alpha') + (\alpha',\alpha)_0$$
$$+ ((0,\beta')) + ((\beta',\beta))_0$$

$$a \leq x < b : \quad (0,\alpha)_0$$
$$+ ((0,\beta')) + ((\beta',\beta))_0$$

$$x \geq b : \quad (0,\alpha)_0$$
$$+ ((0,\beta))_0 \ .$$

Analogous considerations show the existence of six distinct formulae for $(\sin\epsilon)L(a,b;\epsilon)$, corresponding to six distinct ranges of $\cos\epsilon$ when $a/b < \sqrt{2}$ and six distinct ranges of $\cos\epsilon$ when $b/a \geq \sqrt{2}$, as shown below.

	Equivalent condition on $\cos\epsilon$ when $b/a < \sqrt{2}$	Equivalent condition on $\cos\epsilon$ when $b/a \geq \sqrt{2}$
$a < c$, $\epsilon > \pi/2$; $b < c$, $\epsilon > \pi/2$	$\cos\epsilon < 0$	$\cos\epsilon < 0$
$a < c$, $\epsilon \leq \pi/2$; $b < c$, $\epsilon \leq \pi/2$	$0 \leq \cos\epsilon < a/2b$	$0 \leq \cos\epsilon < a/2b$
$a < c$, $\epsilon \leq \pi/2$; $b \geq c$, $\cos\epsilon < a/b$	$a/2b \leq \cos\epsilon < b/2a$	$a/2b \leq \cos\epsilon < a/b$
$a \geq c$; $b \geq c$, $\cos\epsilon < a/b$	$b/2a \leq \cos\epsilon < a/b$	Empty
$a \geq c$; $b \geq c$, $\cos\epsilon \geq a/b$	$\cos\epsilon \geq a/b$	$\cos\epsilon \geq b/2a$
$a < c$, $\epsilon \leq \pi/2$; $b \geq c$, $\cos\epsilon \geq a/b$	Empty	$a/b \leq \cos\epsilon < b/2a$

The values of $(\sin\epsilon)L(a,b;\epsilon)$ for the six cases just listed are given in Table 1. Of course, if $h_0(\rho) = h_1(\rho) = h(\rho)$ for all ρ, i.e. if h does not change its functional form, then the whole of Table 1 collapses to Eq. (17) (as it should). It is important to note in using Table 1 that ϵ is either θ or $\pi-\theta$, where θ denotes the angle between the outward normals to the corresponding sides V, W in Fig.2. This implies that the term $\chi[\cdot;V_i,W_j]\cos\theta_{ij}$ in the series on the right of Eqns. (A), (B) and (C), this term being evaluated via Eqns. (D) as a linear combination of (at most) four L-functions, is, for $\theta_{ij} \neq 0, \pi$, equal to $\cot\theta_{ij}$ times a linear combination of (at most) four entries in Table 1.

To evaluate the M-function, refer to the second diagram of Fig.1, and let t and u be the respective distances of a point r on Γ and a point s on Λ from the left endpoints of Γ and Λ. Let ρ denote the distance between r and s. Then

$$\left.\begin{aligned} M(a,b;d) &= \iint\limits_{0\leq t\leq a, 0\leq u\leq b} h(\rho)dtds\,, \\ \rho &= \{d^2 + (t-u)^2\}^{\frac{1}{2}}. \end{aligned}\right\} \tag{22}$$

Hence, if $\eta(z)\sqrt{2} \equiv \eta(z;a,b)\sqrt{2}$ is the length of the line segment formed by the intersection of the line $u-t = z$ with the rectangle $\{(t,u): 0 \leq t \leq a, 0 \leq u \leq b\}$, i.e.

$$\eta(z) = \begin{cases} 0 & (z < -a)\,, \\ a + z & (-a \leq z < 0)\,, \\ a & (0 \leq z < b-a)\,, \\ b - z & (b-a \leq z < b)\,, \\ 0 & (z \geq b)\,, \end{cases} \tag{23}$$

we have

$$\begin{aligned} M(a,b;d) &= \int_{-\infty}^{\infty} h(\rho)\eta(z)dz \\ &= \int_{-a}^{0} h(\rho)(a+z)dz + \int_{0}^{b-a} h(\rho)adz + \int_{b-a}^{b} h(\rho)(b-z)dz \end{aligned} \tag{24}$$

with $\rho = (d^2+z^2)^{\frac{1}{2}}$. Eq. (24) can be written as

$$M(a,b;d) = -A(a) - A(b) + A(b-a) + aB(a) + bB(b) - (b-a)B(b-a), \tag{25}$$

where

$$A(z) = \int_0^z yh(\rho)dy\,, \quad B(z) = \int_0^z h(\rho)dy \tag{25.1}$$

with $\rho = (d^2+y^2)^{\frac{1}{2}}$. Eq. (25) is suitable if h has the same functional form on its domain, as in (C) for the moments of X. However, if h assumes different functional forms for $\rho \leq x$ and $\rho > x$, i.e. h given by (18), as for the distribution and density functions of X in (A) and (B), then a different representation of M is needed (cf. the argument after Eq. (17) in relation to L). We note that if $x < d$, then $\rho \leq x$ and $\rho > x$ are respectively equivalent to

$z \in \phi$ (the empty set) and to $z \in Z$ (the space of z), while if $x \geq d$, then $\rho \leq x$ and $\rho > x$ are respectively equivalent to $|z| \leq (x^2 - d^2)^{\frac{1}{2}}$ and to $|z| > (x^2 - d^2)^{\frac{1}{2}}$, so that (18) becomes

$$\left.\begin{array}{ll} h(\rho) = \begin{cases} h_0(\rho) & (z \in \phi) \\ h_1(\rho) & (z \in Z) \end{cases} & \text{if } x < d, \\[4mm] h(\rho) = \begin{cases} h_0(\rho) & (|z| \leq (x^2-d^2)^{\frac{1}{2}}) \\ h_1(\rho) & (|z| > (x^2-d^2)^{\frac{1}{2}}) \end{cases} & \text{if } x \geq d. \end{array}\right\} \quad (26)$$

The first half of (26) applied to (24) gives

$$M(a,b;d) = \int_{-a}^{0} h_1(\rho)(a+z)dz + \int_{0}^{b-a} h_1(\rho)adz + \int_{b-a}^{b} h_1(\rho)(b-z)dz \quad (x < d) \quad (27)$$

with $\rho = (d^2 + z^2)^{\frac{1}{2}}$. For $x \geq d$, it is necessary to distinguish two cases, according as $b < 2a$ or $b \geq 2a$, and, further, within each of these two cases, to consider four sub-cases according as to the location of the points $-(x^2 - d^2)^{\frac{1}{2}}$, $(x^2 - d^2)^{\frac{1}{2}}$ with respect to the points $-b$, $-a$, $-(b-a)$, 0, $b-a$, a, b on the z-axis. This gives Scheme 2 for the computation of M, in which the 'noughts' (o) are located at $-(x^2 - d^2)^{\frac{1}{2}}$ and $(x^2 - d^2)^{\frac{1}{2}}$, the values where $h(\rho)$ changes functional form ($\rho = (d^2 + z^2)^{\frac{1}{2}}$), and the 'crosses' (×) are located at $-a$, 0, $b-a$, b, the values where $\eta(z)$ changes functional form.

Scheme 2. Computation of M when $x \geq d$.

$b < 2a$

$$\zeta \equiv \zeta_x = (x^2 - d^2)^{\frac{1}{2}}$$

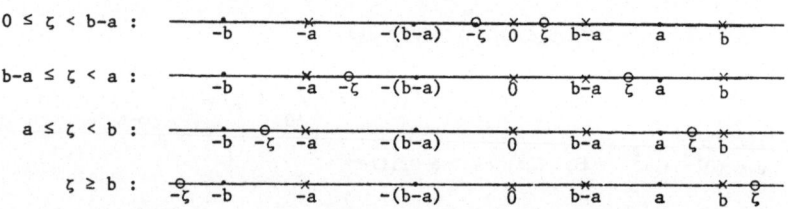

$b \geq 2a$

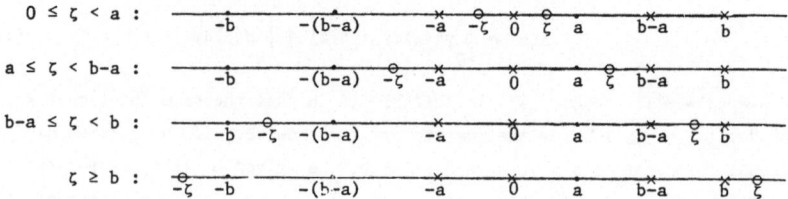

We write

$$
\begin{aligned}
I_0(z,d) &= \int h_0(\rho)(a+z)dz\,, & I_1(z,d) &= \int h_1(\rho)(a+z)dz\,, \\
J_0(z,d) &= \int h_0(\rho)adz\,, & J_1(z,d) &= \int h_1(\rho)adz\,, \\
K_0(z,d) &= \int h_0(\rho)(b-z)dz\,, & K_1(z,d) &= \int h_1(\rho)(b-z)dz\,,
\end{aligned}
\tag{28}
$$

and use the abbreviated notation

$$
\begin{aligned}
{[z',z'']}_0 &= I_0(z'',d) - I_0(z',d)\,, & [z',z'']_1 &= I_1(z'',d) - I_1(z',d)\,, \\
\{z',z''\}_0 &= J_0(z'',d) - J_0(z',d)\,, & \{z',z''\}_1 &= J_1(z'',d) - J_1(z',d)\,, \\
<z',z''>_0 &= K_0(z'',d) - K_0(z',d)\,, & <z',z''>_1 &= K_1(z'',d) - K_1(z',d)\,.
\end{aligned}
\tag{29}
$$

As an example, let $x \geq d$ and consider $b < 2a$ with $0 \leq \zeta < b-a$, i.e., $d \leq x < \{d^2 + (b-a)^2\}^{\frac{1}{2}}$. Then (24) gives with the aid of the first diagram in Scheme 2,

$$
\begin{aligned}
M(a,b;d) = & \int_{-a}^{-\zeta} h_1(\rho)(a+z)dz + \int_{-\zeta}^{0} h_0(\rho)(a+z)dz + \int_{0}^{\zeta} h_0(\rho)adz \\
& + \int_{\zeta}^{b-a} h_1(\rho)adz + \int_{b-a}^{b} h_1(\rho)(b-z)dz \\
= & [-a,-\zeta]_1 + [-\zeta,0]_0 + \{0,\zeta\}_0 + \{\zeta,b-a\}_1 + <b-a,b>_1\,.
\end{aligned}
$$

Consideration of all such cases in Scheme 2, and rewriting (27) in terms of the abbreviated notation, yields Table 2 for $M(a,b;d)$. Note that if $h_0(\rho) = h(\rho) = h_1(\rho)$ then Table 2 collapses to Eq. (24), and therefore to Eq. (25).

To complete the determination of (a) the distribution function $F_X(x)$, (b) the density function $F_X'(x)$, and (c) the moments $EX^k (k \geq -1)$ of X, it remains only to evaluate the corresponding H and H_0-functions (these are needed in Eq. (17) for the computation of L in (c) and in Table 1 for the computation of L in (a) and (b)), the I_0, I_1, J_0, J_1, K_0, K_1-functions (these are needed in Table 2 for the computation of M in (a) and (b)), as well as the A and B-functions (these are needed in Eq. (25) for the computation of M in (c)). The evaluation of the I_0, I_1, J_0, J_1, K_0, K_1 and A-functions is straightforward. We proceed to outline the derivation of the H, H_0 and B-functions (omitting computational details). For (a), set $h(\rho) = g_X^*(\rho)$ in (16), where $g_X^*(\rho)$ is defined in (A). We note that $g_X^*(\rho)$ is continuous at x, and that

$$
\frac{\partial}{\partial \rho} g_X^*(\rho) = \begin{cases} -\dfrac{\rho}{2} & (\rho \leq x)\,, \\[2mm] -\dfrac{x^2}{2\rho} & (\rho > x)\,. \end{cases}
$$

Integration by parts then gives $(-x^4/16) - (x^2\rho*^2/4)\log(\rho*/x)$, where $\rho* = f\sin\varepsilon/\sin(\omega+\varepsilon)$, for the inner integral in (16). Also, since here $h_0(\rho) = -\rho^2/4$, the inner integral in (21) is $-\rho*^4/16$. Evaluation of the outer integrals in (16) and (21) gives H and H_0, respectively. For (b), set $h(\rho) = (\partial/\partial x)g_X^*(\rho)$, where

$(\partial/\partial x)g^*_x(\rho)$ is given in (B), and proceed as for (a). For (c), set $h(\rho) = -\rho^{k+2}/(k+2)^2$, as in (C), and (16) gives

$$H(f,\gamma) = -\frac{(f \sin \varepsilon)^{k+4}}{(k+2)^2(k+4)} \int_0^\gamma \mathrm{cosec}^{k+4}(\omega+\varepsilon)\,d\omega$$

$$= \frac{(f \sin \varepsilon)^{k+4}}{(k+2)^2(k+4)} \left\{ \Psi_{k+4}(\varepsilon) - \Psi_{k+4}(\gamma+\varepsilon) \right\}, \qquad (30)$$

where

$$\Psi_m(\theta) = \int_{\pi/2}^\theta \mathrm{cosec}^m\theta'\,d\theta',$$

so that [9]

$$\Psi_m(\theta) = \begin{cases} -\displaystyle\sum_{j=0}^{(m-2)/2} \binom{(m-2)/2}{j} \frac{\cot^{2j+1}\theta}{2j+1} & (m = 2,4,\ldots) \\[4mm] -\cot\theta \displaystyle\sum_{j=0}^{(m-3)/2} \frac{(m-2)_j}{(m-1)_j} \frac{\mathrm{cosec}^{m-2-2j}\theta}{m-1-2j} - \frac{(m-2)(m-4)\ldots 1}{(m-1)(m-3)\ldots 2} \log(\mathrm{cosec}\theta + \cot\theta) \\ \hfill (m = 3,5,\ldots) \end{cases} \qquad (30.1)$$

with $(a)_0 = 1$, $(a)_j = a(a-2)\ldots(a-2j+2)$ for $j = 1,2,\ldots$. Finally, the B-function is clearly not essentially distinct from the Ψ-function. In fact,

$$B(z) = \frac{d^{k+3}}{(k+2)^2} \Psi_{k+4}(\cot^{-1}z/d).$$

The remarks are summarized in Eqns. (A'), (B') and (C') below.

For the distribution function, F_X, of X:

$$H(f,\gamma) = \left\{ (f \sin \varepsilon)^2 \frac{x^2}{4} - \frac{x^4}{16} \right\}\gamma - (f \sin \varepsilon)^2 \frac{x^2}{4} (\log \frac{f \sin \varepsilon}{x} + 1)\{\cot \varepsilon - \cot(\gamma+\varepsilon)\}$$

$$+ (f \sin \varepsilon)^2 \frac{x^2}{4} \{\cot(\gamma+\varepsilon) \log \sin(\gamma+\varepsilon) - \cot \varepsilon \log \sin \varepsilon\},$$

$$H_0(f,\gamma) = -\frac{(f \sin \varepsilon)^4}{16} \left[\cot \varepsilon - \cot(\gamma+\varepsilon) + \frac{1}{3}\{\cot^3\varepsilon - \cot^3(\gamma+\varepsilon)\} \right];$$

$$I_0(z,d) = -\frac{1}{4}(ad^2z + \frac{1}{2}d^2z^2 + \frac{1}{3}az^3 + \frac{1}{4}z^4), \qquad I_1(z,d) = -\frac{x^2}{4}(az + \frac{z^2}{2}) + \frac{x}{2}\{a\lambda(z,d) + \mu(z,d)\},$$

$$J_0(z,d) = -\frac{a}{4}(d^2z + \frac{z^3}{3}), \qquad J_1(z,d) = -\frac{x^2}{4}az + \frac{x}{2}a\lambda(z,d),$$

$$K_0(z,d) = -\frac{1}{4}(bd^2z - \frac{1}{2}d^2z^2 + \frac{1}{3}bz^3 - \frac{1}{4}z^4), \qquad K_1(z,d) = -\frac{x^2}{4}(bz - \frac{z^2}{2}) + \frac{x}{2}\{b\lambda(z,d) - \mu(z,d)\},$$

where

---

[9] From a recursion for $\int_{\pi/2}^\theta \mathrm{cosec}^m\theta'\,d\theta'$ with $\int_{\pi/2}^\theta \mathrm{cosec}\theta'\,d\theta' = -\log(\mathrm{cosec}\theta + \cot\theta)$, $\int_{\pi/2}^\theta \mathrm{cosec}^2\theta'\,d\theta' = -\cot\theta$. (The first half of (30.1) can also be derived directly.)

$$\lambda(z,d) = -xz \log \frac{(d^2 + z^2)^{\frac{1}{2}}}{x} + xz - dx \tan^{-1} \frac{x}{d} ,$$

$$\mu(z,d) = -\frac{x}{2}(d^2 + z^2) \log (d^2 + z^2)^{\frac{1}{2}} + \frac{xz^2}{4} + (\frac{x}{2} \log x)z^2 .$$

For the density fucntion, F'_X, *of X:*

$$H(f,\gamma) = \left\{ (f \sin \varepsilon)^2 \frac{x}{2} - \frac{x^3}{4} \right\} \gamma - (f \sin \varepsilon)^2 (\frac{x}{4} + \frac{x}{2} \log \frac{f \sin \varepsilon}{x}) \left\{ \cot \varepsilon - \cot(\gamma + \varepsilon) \right\}$$

$$+ (f \sin \varepsilon)^2 \frac{x}{2} \left\{ \cot(\gamma + \varepsilon) \log \sin(\gamma + \varepsilon) - \cot \varepsilon \log \sin \varepsilon \right\} ,$$

$$H_0(f,\gamma) = 0 ;$$

$$I_0(z,d) = 0 , \quad I_1(z,d) = a\lambda(z,d) + \mu(z,d) ,$$
(B')

$$J_0(z,d) = 0 , \quad J_1(z,d) = a\lambda(z,d) ,$$

$$K_0(z,d) = 0 , \quad K_1(z,d) = b\lambda(z,d) - \mu(z,d) ,$$

where $\lambda(z,d)$, $\mu(z,d)$ are as in (A').

For the moments, EX^k, *of X:*

$$(\sin \varepsilon)L(a,b;\varepsilon) = \frac{(a \sin \varepsilon)^{k+4}}{(k+2)^2(k+4)} \left\{ \Psi_{k+4}(\varepsilon) - \Psi_{k+4}(\alpha + \varepsilon) \right\} + \frac{(b \sin \varepsilon)^{k+4}}{(k+2)^2(k+4)} \left\{ \Psi_{k+4}(\varepsilon) - \Psi_{k+4}(\beta + \varepsilon) \right\},$$

$$M(a,b;d) = -\frac{1}{(k+2)^2(k+4)} \left\{ (d^2)^{\frac{k+4}{2}} - (d^2 + a^2)^{\frac{k+4}{2}} - (d^2 + b^2)^{\frac{k+4}{2}} + (d^2 + (b-a)^2)^{\frac{k+4}{2}} \right\}$$

$$+ \frac{d^{k+3}}{(k+2)^2} \left\{ a\Psi_{k+4}(\cot^{-1} \frac{a}{d}) + b\Psi_{k+4}(\cot^{-1} \frac{b}{a}) - (b-a)\Psi_{k+4}(\cot^{-1} \frac{b-a}{d}) \right\}$$

for $k = -1,1,2,\ldots$, and, in particular,[10]

$$M(a,b;0) = -\frac{1}{(k+2)^2(k+3)(k+4)} \left\{ a^{k+4} + b^{k+4} - (b-a)^{k+4} \right\}$$

for $k = -1,1,2,\ldots$, where Ψ_{k+4} is given in (30.1).

We recall that α and β are defined in (15) [these are needed in the first term on the right of $H(a,\alpha)$ and $H(b,\beta)$ in (A')], that the sines and cotangents of $\alpha + \varepsilon$, $\beta + \varepsilon$, $\alpha' + \varepsilon$, $\beta' + \varepsilon$, $\alpha'' + \varepsilon$, $\beta'' + \varepsilon$ are given in (15') and (19') [these are needed when substituting for $H(a,\alpha)$, $H(b,\beta)$, $H_0(a,\alpha)$, $H_0(b,\beta)$, as given in (A') and (B'), in Table 1, and when substituting for $\Psi_{k+4}(\alpha + \varepsilon)$, $\Psi_{k+4}(\beta + \varepsilon)$ in (C')], and that $\cot \varepsilon - \cot(\alpha + \varepsilon)$ and $\cot \varepsilon - \cot(\beta + \varepsilon)$ are given in (15'') [these are needed when substituting for $H(a,\alpha)$, $H(b,\beta)$, $H_0(a,\alpha)$, $H_0(b,\beta)$, as given in (A') and (B'), in Table 1].

---

[10] From (30.1), $\Psi_m(\theta) = -\frac{1}{m-1} \frac{1}{\theta^{m-1}} + o(\frac{1}{\theta^{m-1}})$ as $\theta \to 0+$ for $m = 2,3,\ldots$.

TABLE 1: Values of $(\sin\varepsilon)L(a,b;\varepsilon)$

α, β defined in Eqn. (15); α', β', α'', β'' defined in Eqn. (19) [$\alpha' \le \alpha''$, $\beta' \le \beta''$]; (γ,δ), $(\gamma,\delta)_0$, $((\gamma,\delta))$, $((\gamma,\delta))_0$ defined in Eqns. (20), (16) and (21).

(1) $\underline{\cos\varepsilon < 0}$

$0 \le x < a$: $(0,\alpha)$
$+ ((0,\beta))$

$a \le x < b$: $(0,\alpha'')_0 + (\alpha'',\alpha)$
$+ ((0,\beta))$

$b \le x < c$: $(0,\alpha'')_0 + (\alpha'',\alpha)$
$+ ((0,\beta''))_0 + ((\beta'',\beta))$

$x \ge c$: $(0,\alpha)_0$
$+ ((0,\beta))_0$

(2) $\underline{0 \le \cos\varepsilon < a/2b}$

$\underline{\sin\varepsilon > a/b}$

$0 \le x < a\sin\varepsilon$: $(0,\alpha)$
$+ ((0,\beta))$

$a\sin\varepsilon \le x < a$: $(0,\alpha') + (\alpha',\alpha'')_0 + (\alpha'',\alpha)$
$+ ((0,\beta))$

$a \le x < b\sin\varepsilon$: $(0,\alpha'')_0 + (\alpha'',\alpha)$
$+ ((0,\beta))$

$b\sin\varepsilon \le x < b$: $(0,\alpha'')_0 + (\alpha'',\alpha)$
$+ ((0,\beta')) + ((\beta',\beta''))_0 + ((\beta'',\beta))$

$b \le x < c$: $(0,\alpha'')_0 + (\alpha'',\alpha)$
$+ ((0,\beta''))_0 + ((\beta'',\beta))$

$x \ge c$: $(0,\alpha)_0$
$+ ((0,\beta))_0$

$\underline{\sin\varepsilon \le a/b}$

$0 \le x < a\sin\varepsilon$: $(0,\alpha)$
$+ ((0,\beta))$

$a\sin\varepsilon \le x < b\sin\varepsilon$: $(0,\alpha') + (\alpha',\alpha'')_0 + (\alpha'',\alpha)$
$+ ((0,\beta))$

$b\sin\varepsilon \le x < a$: $(0,\alpha') + (\alpha',\alpha'')_0 + (\alpha'',\alpha)$
$+ ((0,\beta')) + ((\beta',\beta''))_0 + ((\beta'',\beta))$

$a \le x < b$: $(0,\alpha'')_0 + (\alpha'',\alpha)$
$+ ((0,\beta')) + ((\beta',\beta''))_0 + ((\beta'',\beta))$

$b \le x < c$: $(0,\alpha'')_0 + (\alpha'',\alpha)$
$+ ((0,\beta''))_0 + ((\beta'',\beta))$

$x \ge c$: $(0,\alpha)_0$
$+ ((0,\beta))_0$

<u>TABLE 1 (cont'd)</u>

(3) $a/2b \leq \cos \epsilon < b/2a$ if $b/a < \sqrt{2}$ or $a/2b \leq \cos \epsilon < a/b$ if $b/a \geq \sqrt{2}$

<u>$\sin \epsilon > a/b$</u>

$0 \leq x < a \sin \epsilon$: $(0,\alpha)$
$+ ((0,\beta))$

$a \sin \epsilon \leq x < a$: $(0,\alpha') + (\alpha',\alpha'')_0 + (\alpha'',\alpha)$
$+ ((0,\beta))$

$a \leq x < b \sin \epsilon$: $(0,\alpha'')_0 + (\alpha'',\alpha)$
$+ ((0,\beta))$

$b \sin \epsilon \leq x < c$: $(0,\alpha'')_0 + (\alpha'',\alpha)$
$+ ((0,\beta')) + ((\beta',\beta''))_0 + ((\beta'',\beta))$

$c \leq x < b$: $(0,\alpha)_0$
$+ ((0,\beta')) + ((\beta',\beta))_0$

$x \geq b$: $(0,\alpha)_0$
$+ ((0,\beta))_0$

<u>$\sin \epsilon \leq a/b$</u>

$0 \leq x < a \sin \epsilon$: $(0,\alpha)$
$\div ((0,\beta))$

$a \sin \epsilon \leq x < b \sin \epsilon$: $(0,\alpha') \div (\alpha',\alpha'')_0 + (\alpha'',\alpha)$
$+ ((0,\beta))$

$b \sin \epsilon \leq x < a$: $(0,\alpha') + (\alpha',\alpha'')_0 + (\alpha'',\alpha)$
$+ ((0,\beta')) + ((\beta',\beta''))_0 + ((\beta'',\beta))$

$a \leq x < c$: $(0,\alpha'')_0 + (\alpha'',\alpha)$
$+ ((0,\beta')) + ((\beta',\beta''))_0 + ((\beta'',\beta))$

$c \leq x < b$: $(0,\alpha)_0$
$+ ((0,\beta')) + ((\beta',\beta))_0$

$x \geq b$: $(0,\alpha)_0$
$+ ((0,\beta))_0$

(4) <u>$b/2a \leq \cos \epsilon < a/b$ if $b/a < \sqrt{2}$</u>

$0 \leq x < a \sin \epsilon$: $(0,\alpha)$
$+ ((0,\beta))$

$a \sin \epsilon \leq x < b \sin \epsilon$: $(0,\alpha') + (\alpha',\alpha'')_0 + (\alpha'',\alpha)$
$+ ((0,\beta))$

$b \sin \epsilon \leq x < c$: $(0,\alpha') + (\alpha',\alpha'')_0 + (\alpha'',\alpha)$
$+ ((0,\beta')) + (\beta',\beta'')_0 + ((\beta'',\beta))$

$c \leq x < a$: $(0,\alpha') + (\alpha',\alpha)_0$
$+ ((0,\beta')) + ((\beta',\beta))_0$

$a \leq c < b$: $(0,\alpha)_0$
$+ ((0,\beta')) + ((\beta',\beta))_0$

$x \geq b$: $(0,\alpha)_0$
$+ ((0,\beta))_0$

TABLE 1 (cont'd)

(5) $\underline{\cos \epsilon \geq a/b \text{ if } b/a < \sqrt{2} \text{ or } \cos \epsilon \geq b/2a \text{ if } b/a \geq \sqrt{2}}$

$0 \leq x < a \sin \epsilon$: $(0,\alpha)$
                  $+ ((0,\beta))$

$a \sin \epsilon \leq x < c$: $(0,\alpha') + (\alpha',\alpha'')_0 + (\alpha'',\alpha)$
                  $+ ((0,\beta))$

$c \leq x < a$: $(0,\alpha') + (\alpha',\alpha)_0$
                  $+ ((0,\beta')) + ((\beta',\beta))_0$

$a \leq x < b$: $(0,\alpha)_0$
                  $+ ((0,\beta')) + ((\beta',\beta))_0$

$x \geq b$: $(0,\alpha)_0$
                  $+ ((0,\beta))_0$

(6) $\underline{a/b \leq \cos \epsilon < b/2a \text{ if } b/a \geq \sqrt{2}}$

$0 \leq x < a \sin \epsilon$: $(0,\alpha)$
                  $+ ((0,\beta))$

$a \sin \epsilon \leq x < a$: $(0,\alpha') + (\alpha',\alpha'')_0 + (\alpha'',\alpha)$
                  $+ ((0,\beta))$

$a \leq x < c$: $(0,\alpha'')_0 + (\alpha'',\alpha)$
                  $+ ((0,\beta))$

$c \leq x < b$: $(0,\alpha)_0$
                  $+ ((0,\beta')) + ((\beta',\beta))_0$

$x \geq b$: $(0,\alpha)_0$
                  $+ ((0,\beta))_0$

TABLE 2: Values of $M(a,b;d)$

$[\cdot,\cdot]_0$, $[\cdot,\cdot]_1$, $\{\cdot,\cdot\}_0$, $\{\cdot,\cdot\}_1$, $<\cdot,\cdot>_0$, $<\cdot,\cdot>_1$ defined in Eqns. (28) and (29)

$\underline{x < d}$ $[-a,0]_1 + \{0,b-a\}_1 + <b-a,b>_1$

$\underline{x \geq d, \ b < 2a}$

$$\zeta \equiv \zeta_x = (x^2 - d^2)^{\frac{1}{2}}$$

$0 \leq \zeta < b-a$: $[-a,-\zeta]_1 + [-\zeta,0]_0 + \{0,\zeta\}_0 + \{\zeta,b-a\}_1 + <b-a,b>_1$

$b-a \leq \zeta < a$: $[-a,-\zeta]_1 + [-\zeta,0]_0 + \{0,b-a\}_0 + <b-a,\zeta>_0 + <\zeta,b>_1$

$a \leq \zeta < b$: $[-a,0]_0 + \{0,b-a\}_0 + <b-a,\zeta>_0 + <\zeta,b>_1$

$\zeta \geq b$: $[-a,0]_0 + \{0,b-a\}_0 + <b-a,b>_0$

$\underline{x \geq d, \ b \geq 2a}$

$0 \leq \zeta < a$: $[-a,-\zeta]_1 + [-\zeta,0]_0 + \{0,\zeta\}_0 + \{\zeta,b-a\}_1 + <b-a,b>_1$

$a \leq \zeta < b-a$: $[-a,0]_0 + \{0,\zeta\}_0 + \{\zeta,b-a\}_1 + <b-a,b>_1$

$b-a \leq \zeta < b$: $[-a,0]_0 + \{0,b-a\}_0 + <b-a,\zeta>_0 + <\zeta,b>_1$

$\zeta \geq b$: $[-a,0]_0 + \{0,b-a\}_0 + <b-a,b>_0$.

REFERENCES

[1] ARMITAGE, P. (1949). An overlap problem arising in particle counting. *Biometrika* 36, 257-266.

[2] BOREL, E. (1925). *Principes et formules classiques du Calcul des Probabilités. Traité du Calcul des Probabilités et ses Applications*, Gauthier-Villars, Paris.

[3] GARWOOD, F. (1947). The variance of the overlap of geometrical figures with reference to a bombing problem. *Biometrika* 34, 1-17.

[4] GHOSH, B. (1943a). On the distribution of random distances in a rectangle. *Science and Culture* 8, 388.

[5] ----- (1943b). On random distances between two rectangles. *Science and Culture* 8, 464.

[6] ----- (1949). Topographic variation in statistical fields. *Calcutta Statist. Assoc. Bull.* 2(5), 11-28.

[7] ----- (1951). Random distances within a rectangle and between two rectangles. *Bull. Calcutta Math. Soc.* 43, 17-24.

[8] KENDALL, M.G., and MORAN, P.A.P. (1963). *Geometrical Probability*, Griffin, London.

[9] MATÉRN, B. (1947). Metoder att uppskatta noggraunheten vid linjeoch provy-tetaxering. (Methods of estimating the accuracy of line and sample splot surveys.) [In Swedish. English summary.] *Meddelanden från Statens skogsforskningsinstitut,* 36(1), 1-138.

[10] RUBEN, H. (1964). Generalized concentration fluctuations under diffusion equilibrium. *J. Appl. Prob.* 1, 47-68.

[11] ----- (1967). An intrinsic formula for volume. *J. Reine Angew. Math.* 226, 116-119.

[12] ----- (1970). On a class of double space integrals with applications in mensuration, statistical physics, and geometrical probability, *Proc. 12th Biennial Seminar of Canadian Mathematical Congress*, 209-230.

[13] WHITNEY, H. (1957). *Geometric Integration Theory*, Princeton Univ. Press, Princeton, N.J.

ANALYSE D'UNE POPULATION DE CELLULES SPHERIQUES
-=-

Ph. D'ATHIS - M. RAPHAEL - J.L. BINET - PITIE SALPETRIERE 91 Bld de l'Hôpital
75013 PARIS

The first part of this paper is devoted to the analysis and verification of methods derived from that of WICKSELL which allows deduction of the cell volume histogram from the histogram of cell section diameters. The second part emphasiezs the difficulties involved in the interpretation of the cell volume histogram : if it is multimodal can one conclude that it is composed of sub-histograms defined by some biological reality ? And if it is unimodal does this mean that it represents a single population or a superimposition of gaussian populations. In the last case, how does one determine the statistical law attributed to a subpopulation ?

Nous abordons ici essentiellement le problème de la reconnaissance de sous-populations au sein d'une population cellulaire donnée.

FORMULATION DU PROBLEME :

Considérons une population de cellules sphériques dont le diamètre x n'est pas nécessairement constant mais peut prendre au plus la valeur d_1.

Le problème étudié par S.D. WICKSELL puis repris par BACH, SALTYKOV et de nombreux autres auteurs considère l'expérience consistant à couper par un plan aléatoire un bloc contenant ces cellules.

Ici nous considérons l'expérience consistant à observer une cellule prise au hasard, dans la population et à y effectuer, au hasard encore, une section plane dont on mesure le diamètre z.

Cette expérience amène à considérer le diamètre cellulaire x comme une variable aléatoire de densité de probabilité f(x) et le diamètre de section z comme une variable aléatoire de densité g(z).

En exprimant qu'une cellule donnée est coupée aléatoirement, de façon uniforme, par rapport à son diamètre x, on montre que la densité conditionn-lle $g_x(z)$ du diamètre de section z vaut

$$(1) \qquad g_x(Z) = \frac{Z}{x\sqrt{x^2 - z^2}} \qquad \text{pour } 0 < Z < x$$

$$= 0 \qquad \text{pour } Z > x$$

ce qui permet de déduire la densité g(z) correspondant à une section aléatoire d'une sphère aléatoire

$$(2) \qquad g(Z) = \int_Z^{d_1} \frac{Z}{x\sqrt{x^2 - z^2}} \, f(x)dx$$

La mesure d'un grand nombre de sections permettant de connaître la fonction $g(z)$, le problème est d'en déduire $f(x)$. Nous disposons pour cela de 2 méthodes : l'une imposant à la densité $f(x)$ un type donné, l'autre étant plus générale en ne posant pas d'hypothèse de loi ; toutes deux font appel à l'informatique.

METHODE LIEE A UNE HYPOTHESE DE LOI :

Si nous pouvons admettre que le diamètre cellulaire observé résulte d'un mélange de r lois de probabilités, connues, $f_i(x)$ (pour i = 1, 2, ... r) sous la forme

$$(3) \qquad f(x) = p_1\, f_1(x) + p_2\, f_2(x) + \ldots + p_r\, f_r(x)$$

où chaque terme p_i représente la proportion relative à la loi $f_i(x)$ de telle sorte que

$$(4) \qquad p_1 + p_2 + \ldots + p_r = 1$$

et si, de plus, chaque fonction $f_i(x)$ est déterminée par la donnée de paramètres $t_{i,1}$, $t_{i,2}$, ... $t_{i,m}$ on peut obtenir une décomposition de $f(x)$ en utilisant la méthode d'estimation des moindres carrés ou, mieux encore si l'on dispose d'un outil de calcul suffisamment important, la méthode du maximum de vraisemblance.

Dans le cas d'une population cellulaire constituée par un mélange de sous-populations gaussiennes, par exemple, on aurait :

$$(5) \qquad f_i(x) = \frac{1}{s_i\sqrt{2\pi}} \exp\left(-\frac{1}{2}\left(\frac{x-m_i}{s_i}\right)^2\right)$$

et chaque fonction $f_i(x)$ serait déterminée par la donnée du diamètre moyen m_i et de l'écart-type s_i.

On chercherait alors les valeurs $\hat{p}_i$, $\hat{m}_i$, $\hat{s}_i$ des variables p_i, m_i, s_i (pour i = 1, 2, ... r) réalisant le meilleur ajustement de la fonction théorique $p_1\, f_1(x) + p_2\, f_2(x) + \ldots + p_r\, f_r(x)$ à la fonction expérimentale $f(x)$.

Méthode libre de toute hypothèse de loi : En posant

(6) $$P(Z) = \frac{g(Z)}{Z}$$

et

(7) $$Q(x) = \frac{f(x)}{2x^2}$$

on peut récrire l'équation intégrale (2) sous la forme

(8) $$P(Z) = 2 \int_{Z}^{d_1} \frac{x}{\sqrt{x^2 - Z^2}} Q(x) dx$$

et montrer que, sous de légères hypothèses de régularité à formuler pour $P(z)$, elle admet la solution

(9) $$Q(x) = -\frac{1}{\pi} \int_{x}^{d_1} \frac{P'(Z)}{\sqrt{Z^2 - x^2}} dz \qquad (\text{où } P' : \frac{dP}{dz})$$

Le principe de la méthode est alors de substituer à la fonction $P(z)$ un polynôme dont la valuation est au moins égale à 2, ce qui est pratiquement toujours possible (en utilisant, par exemple, le système orthogonal de TCHEBYCHEV) ; on arrive alors à un développement de la forme

(10) $$P'(Z) \simeq \sum_{k=0}^{n-2} (k+1) \frac{b_{k+2}}{d_1^{k+3}} Z^k$$

qui lui-même conduit à la solution

(11) $$f(x) \simeq -\left(\frac{x}{d_1}\right)^2 \sum_{k=0}^{n-2} b_{k+2} \, T_k \left(\frac{x}{d_1}\right)$$

où chaque fonction $I_k(s)$ est déterminée par

(12) $$I_k(s) : (k+1) \int_{s}^{1} \frac{t^k}{\sqrt{t^2 - s^2}} dt \qquad \text{pour } 0 < s < 1$$

COMPARAISON DES DEUX METHODES :

La méthode liée à une hypothèse possède 2 avantages :

- elle peut décomposer l'histogramme du diamètre cellulaire en sous-histogrammes d'un type donné (gaussien ou log-gaussien, par exemple) ;

- elle fournit des paramètres tels que moyenne et écart-type de sous-population.

mais elle est affectée de 3 inconvénients :

- le choix du type statistique imposé à la loi du diamètre cellulaire est souvent subjectif ;

- la programmation de la méthode est lourde ;

- son application est lente.

La méthode libre de toute hypothèse de loi a de son côté 3 avantages :

- elle repose sur la seule hypothèse de sphéricité des cellules ;

- elle est simple à programmer ;

- son exécution est rapide.

mais elle est affectée de 2 inconvénients non négligeables :

- l'interprétation de l'histogramme qu'elle fournit est difficile et nécessite souvent de nouveaux calculs (décomposition en sous-histogrammes d'un type donné, par exemple) dont l'imprécision s'ajoute à celle déjà dûe à la méthode ;

- l'estimation correcte des paramètres tels que moyenne et écart-type est compliquée (approximation numérique d'intégrale).

CONCLUSION

La généralité et la rapidité font souvent préférer la méthode libre de toute hypothèse à sa rivale, mais ces qualités ne compensent pas la richesse de l'information que cette dernière peut apporter.

En particulier, la première méthode ne montre pas toujours qu'une population peut être considérée comme un mélange de sous-populations, car la composition de plusieurs histogrammes peut donner un histogramme unimodal.

Notre problème actuel est donc la détermination d'un critère objectif permettant de préférer une méthode à l'autre ; cela fait, on pourra chercher à améliorer la méthode libre de toute hypothèse pour mieux répondre à la question de l'existence de sous-populations.

-=-=-=-=-=-=-=-=-=-=-=-

REFERENCES

d'ATHIS P. (1974) : Etude du volume de particules sphériques à partir d'une section aléatoire. C.R. Acad. Sc. Paris, t278, série D, p1249,1252.

d'ATHIS P. (1976) : Analysis of a set of spherical cells relative to their volume. Proceedings of the 4-th International Congress for stereology held at NBS, Gaithersburg, Md., September 1975 (issued 1976).

De HOFF R.T., THINES F.N. (1968 : Quantitative microscopy. Mac Graw-Hill book cy. New-New-York.

WICKSELL S.D. (1925) : The corpuscle problem : case of spherical particles. Biometrika, 17, 84.

-=-=-=-=-=-=-

(1) Département d'Informatique Médicale (Prof. agr. L. DUSSERRE), Hôpital du Bocage, 21034 DIJON CEDEX, FRANCE.

(2) Département d'Hématologie (Prof. J.L. BINET), C.H.U. Pitié-Salpétrière, 91 Bld de l'Hôpital, 75013 PARIS, FRANCE.

-=-=-=-=-=-=-

ETUDE MORPHOLOGIQUE DES FIBRES CONJONCTIVES

SUR BIOPSIES OU PIECES OPERATOIRES.

Pr. J. MIGNOT

Hopital Ambroise-Paré. 9 Avenue du Général de Gaulle   92 I00 Boulogne.

SUMMARY

  The morphological study of tissues under th light microscope is
confined basically to two types of structures : the cells, which are
usually well-defined convex structures characterized by their cytoplasm
and nucleus,and the intercellular structures, consisting of connective
tissue fibers which usually form complex interconnected network.

  Alterations in these fibrillar structures may provide important
information on abnormal tissues,but such structures are very difficult
to define and classify by visual examination.

  We have attempted to apply the simplest possible morphometric methods
as an objective aid to study of these structures. The most basic is
measurement of connective tissue volume fraction (v), but this method
is inadequate when it comes to defining alterations in patterns
characteristic of certain disease states.

Information must be obtained on alterations in distribution of fibrillar components for hostological structures of varying siez-scale. To
achieve this,we are presently attempting to take account of the
variance ^{2}(v), and,generally speaking, the entire histogram f(v) of
the volume fraction. These depend of the size of the support s (of
the small elementary zone) vhich defines this volume fraction.

  This leads to my question : are there any methods or models capable
of interpreting variations in ^2_s(V) et f_s(v) when the support s
incrases or decreases ?

INTRODUCTION

L'étude morphologique des tissus pathologiques en microscopie
optique comporte schématiquement deux types d' éléments :
1) les cellules, 2) les fibres conjonctives intercellulaires. Ces
fibres, qui constituent le "squelette mou", servent de support aux
éléments cellulaires, contribuent à maintenir l'architecture des
tissu, et interviennent dans la circulation des liquides intercellulaires. La quantité,la structure et la répartition de ces fibres
sont étroitement liées à l'activité physiologique des populations
cellulaires et à leur pathologie.L'étude de leurs modifications est
très utile dans le diagnostic d'une lésion, mais la complexité et la
variété des images de ces réseaux fibrillaires rend leur description
et leur classification très difficiles. L'application des méthodes
morphométriques peut être un moyen utile pour obtenir des paramètres
objectifs susceptibles de préciser ces structures.

C'est uniquement l'étude des fibres conjonctives que nous envisa-
gerons pour le moment.

L'emploi,en pratique courante,des méthodes morphométriques comme
aide au diagnostic anatomo-pathologique,ne peut se concevoir que si
les techniques utilisées sont relativement simples et surtout rapides.
Dans ce but,nous avons entrepris,depuis 1973,avec les moyens du
Laboratoire, la réalisation d'un appareil automatique spécialement
conçu pour essayer de répondre aux exigences particulières de ce
type d'études.

TECHNIQUES HISTOLOGIQUES

Le matériel utilisé est constitué par des biopsies fixées au
liquide de Bouin, incluses à la paraffine et coupées au microtome à une
épaisseur de 5 à 6 micromètres. Pour permettre la détection automatique des fibres conjonctives, il était nécéssaire de choisir les
colorants les plus électifs et présentant le meilleur contraste dans
le domaine spectral leplus favorable pour les photo-détecteurs au
silicium utilisés dans l'appareil. Le maximum de sensibilité de ces
éléments se trouvant dans le rouge et le proche infra-rouge, nous
avons d'abord utilisé le bleu d'aniline après mordançage à l'acide
phosphotungstique. Ce colorant se fixe assez électivement sur les
fibres,collagènes et assez bien sur les fibres dites de réticuline,
les cytoplasmes et les noyaux étant très peu colorés. Ultérieurement,
nous avons adopté la méthode d'argentation de Gordon-Sweet qui teinte

en noir intense les fibres de réticuline et en brun les fibres collagènes. Cette méthode présente l'avantage d'un contraste plus élevé entre les éléments fibrillaires et le bruit de fond. Quelque soit le soin apporté à la réalisation des coupes il est impossible d'éliminer de petites différences d'épaisseur et de coloration d'une coupe à l'autre et parfois dans la même coupe.Il peut exister également des déchirures ou des plis. Il est nécéssaire de tenir compte de la présence de ces artefacts lors de l'analyse automatique des préparations.

Les fibres conjonctives les plus fines que l'on puisse observer facilement en microscopie optique ont un diamètre d'un micromètre environ. Ces éléments,souvent désignés sous le nom de fibres de réti-culine, sont disposés autour de groupes cellulaires,et autour des capillaires. Elles se groupent pour former des structures plus épaisses notamment autour des vaisseaux. Des éléments fibrillaires plus volu-mineux sont constitués par des fibres collagènes, souvent groupées en faisceaux et dont le diamètre peut atteindre 20 micromètres. Sur les coupes histologiques,les fibres de réticuline forment des mailles de formes et de dimensions très variables, et les fibres collagènes ont tendance à se grouper en zônes denses,autour des vaisseaux, dans les cloisons qui marquent la périphérie des lobules, ou en surface de l' organe pour en former la capsule. Il en résulte une disposition hiérarchisée directement liée à la structure du réseau vasculaire et à la subdivision de l'organe en lobules et en "unités tissulaires"[1]. On est de ce fait amené à étudier des structures présentant une certaine périodicité,mais appartenant à des échelles de grandeur très différentes. Les unités tissulaires élémentaires plus ou moins bien individualisées peuvent varier de 0,2mm à 0,8 mm environ, les mailles du réseau de fibres de réticuline sont de l'ordre de grandeur de 20 à 40 micromètre , et les fibres elles mêmes ont des épaisseurs de l'ordre du micromètre.

APPAREIL UTILISE

Pour obtenir un échantillon représentatif de ces différentes structures,il est nécéssaire d'effectuer des mesures sur plusieurs millimètres avec un grossissement d'environ x400 pour détecter facilement les éléments de l'ordre du micromètre. Dans ce but, nous avons adopté un double système de balayage, l'un mécanique et l'autre électronique. Le système mécanique est constitué par le déplacement de la préparation à l'aide d'une platine commandée par un moteur pas à pas. Le pas correspond à une avance de 0,78 micromètre. Le balayage électronique est constitué par une barrette de 64 photo-diodes

disposées dans le plan de l'image microscopique,perpendiculairement
au sens du déplacement mécanique.

Cette barrette explore sur la préparation compte tenu du grossissement du microscope, une ligne de 50 micromètres.Les 64 cellules
sont adressées séquentiellement par une commande électronique, et le
courant photo-électrique de chacune de ces cellules est, après
amplification appliqué à un comparateur à seuil réglable à l'aide
d'un potentiomètre. L'information optique reçue par les cellules est
ainsi digitalisée en niveaux "O" ou "I".

La zône explorée par le déplacement de l'image sur la barrette
est une bande de plusieurs millimètres de long sur 50 micromètres
de large,dont les points testés sont disposés sur plusieurs milliers
de lignes de 64 points. L'avance de la platine du microscope est d'
environ 50 micromètres par seconde (64 pas de 0,78 micromètres).
Elle est synchronisée avec le déclenchement de l'exploration de la
barrette de photo-diodes.

Le controle de la focalisation de l'image,du niveau du seuil de
détection et les corrections nécéssitées par la présence des artefacts
sont exécutés grâce à une double visualisation simultanée de l'image
optique et du signal électronique. L'image optique est projetée sur
un écran translucide, et un oscilloscope placé au dessus de cet
écran permet d'observer l' image électronique correspondante.Cet
oscilloscope trace les variations dusignal photoélectrique direct
de la barrette des photodiodes, ainsi que le signal obtenu après digi-
talisation. On peut ainsi,au cours du déplacement de la préparation,
ajuster s'il y a lieu le seuil de la détection aux petites variations
lentes du niveau général de la coloration. Si une déchirure ou un
pli sont observés sur l'écran optique,il est possible d'actionner
une commande d'arrêt de l'enregistrement des données pendant que la
zône défectueuse défile devant les photo-diodes.

La dimension moyenne des coupes à étudier a conduit à adopter une

longueur de bande d'exploration de 6,4 mm dans la plupart des cas.
Deux ou trois bandes différemment orientées sont généralement
nécéssaires pour obtenir un échantillonnage suffisant. L'exploration
d'une bande dure environ 2 minutes, et, compte tenu du temps nécéssaire pour mettre en place la préparation et effectuer les différents
réglages on peut estimer de IO à I2 minutes le temps total de
l'opération.

TRAITEMENT DES DONNEES

Lorsqu'un Anatomo-pathologiste examine une coupe histologique,il commence toujours par l'observation de celle-ci à l'oeil nu et au très faible grossissement (x2,5) pour apprécier les structures à grande échelle. Après avoir ainsi repéré certaines zônes qui se différencient par la valeur moyenne de leur coloration, il va regarder chacune d'elles avec un grossissemet plus important (x I60) pour savoir si ces zônes sont vraimet homogènes ou s'il existe déjà une structuration à plus petite échelle. Finalement,il utilise un grossissement encore plus fort (x400 ou plus) pour obtenir une information sur les structures les plus fines discernables avec un microscope optique. Cette manière d'opérer est destinée à reconnaitre les différentes échelles de structure qui ont été décrites plus haut.Le traitement des données enregistrées par l'appareil doit être conçu pour fournir des paramètres quantitatifs pour ces différents types de structures.
L'examen au très faible grossissement constitue un filtrage optique qui fait disparaitre les éléments fibrillaires les plus fins et donne un contraste appréciable pour les accumulations denses de fibres collagènes épaisses. C'est ainsi que,dans une biopsie hépatique seul le tissu conjonctif situé en périphérie des lobules (espaces portes) peut être observé à l'état normal, alors que, au cours de cirrhoses,on verra de larges travées de fibrose (fig.1). On pourrait mesurer directement à ce faible grossissement,la fraction d'aire occupée par cette "fibrose dense" à l'aide d'un petit nombre de points-tests grâce à un réglage approprié du niveau de gris détecté, mais il faudrait alors répéter des mesures sur la même préparation avec des grossissements plus élevés pour étudier les structures plus fines, d'où une perte considérable de temps et un gaspillage des données recueillies.

La bande explorée par l'appareil utilise des points-tests de 0,78 micromètre de côté et s'étend sur 6,4 mm de long.C'est à dire qu'elle comporte tous les éléments nécéssaires pour étudier les structures fines comme les structures à grande échelle.

Pour mesurer ces zônes de "fibrose dense", la bande de 6,4 mm qui n'a que 50 micromètres de large peut être assimilée à une ligne que l'on découpe en I28 "points". Ces "points" sont en réalité de petits champs de 50 micromètres de côté,dimension négligeable vis à vis de la longueur de 6,4mm de la bande.

Pratiquement,pour réaliser cette mesure,, les informations digita-
lisées fournies par les cellules photo-électriques sont mises en
mémoire sous forme d'une matrice de 4096 données correspondant à
64 lignes de 64 points. Le nombre de points de niveau "1" de cette
matrice-mémoire est divisé par 4096 pour obtenir la fraction d'
aire [A_A] occupée par les fibres dans un champ de 50 micromètres
de côté. Les valeurs de cette fraction d'aire [A_A] des I28 champs de
la bands explorée sont mises en mémoire pour fournir un histogramme
réparti en 8 classes.

Pour obtenir unemesure des structures à grande échelle,nous avons vu
que l'on pouvait assimiler ces champs à des points-tests, mais ces
points sont affectés d'une valeur entre 0 et 1. On pourrait résou-
dre cette difficulté en fixant la valeur de A_A que l'on considère
comme une zône de fibrose dense. Il faut souligner que ce problème
se retrouve en pratique dans toutes les applications stéréologiques utilisant des points-tests. Quand on effectue des mesures de
fraction d'aire avec une grille,il arrive que le point, supposé
idéal,soit juste en bordure de la structure étudiée. Dans ce cas,
on adopte souvent un artifice qui consiste à lui donner la valeur1/2.
Cette situation est due à ce que le point matériel a toujours une
surfaceet l'on est conduit déjà à lui affecter non plus 2 mais 3
valeurs.

En pratique,plûtot que de fixer un seuil arbitraire à la valeur A_A
il est apparu beaucoup plus intéressant de garder l'ensemble des
données exprimées sous forme d'une courbe des fréquences cumulées
des I28 mesures de fractions d'aire des champs de 50 micromètres
de côté. La forme de cette fonction de répartition donne des préci-
sions à la fois sur les structures à grande et à moyenne échelle.

L'information fournie par cette courbe est par contre insuffisante
pour décrire les structures à petite échelle. Dans les champs
examinés, les valeurs nulles ou très faibles correspondent à des
plages sans éléments fibrillaires où aucune structure ne peut exister.
Dans les plages de densité voisine de I, il s'agit aussi de zônes de
structure nécéssairement homogène. Par contre, pour les valeurs
intermédiaire,une même valeur A_A du champ peut correspondre à des
structures variées. Si l'on considère,comme c'est le cas dans les
ganglions lymphatiques,que la structure fine est constituée par un
réseau de fibres,formant des mailles de taille variable avec des
cloisons souvent incomplètes, d'épaisseur également variable , la

même valeur de la fraction d'aire d'un champ peut être donnée par des
mailles larges à paroi épaisses ou par des mailles fines à parois
également fines.Il est dans ce cas utile d'avoir une information au
moins sur l'un des deux facteurs, par exemple sur la surface moyenne
des "trous" des mailles du réseau. Ce problème est assez difficile,
car,comme nous l'avons déjà souligné,le contour de ces mailles est
discontinu. Une méthode assez simple peut cependant donner une
estimation de cette surface moyenne.

Le champ élémentaire de 50 micromètres de côté est stocké, nous l'
avons déjà vu, sous la forme d'une matrice de 64 x64 points. Il est
possible,par un procédé électronique, de diviser cette matrice en 4
 carrés de 32x32,puis en 16 carrés de 16x16,en 64 carrés de 8x8 et
en 1024 carrés de 2x2. Ce système donne,pour chacune de ces dimensions
de champ, la fréquence des valeurs de densité nulle. Ce test peut
être considéré comme la probabilité pour un carré de taille donnée de
tomber entièrement dans un trou du réseau. Il est bien évident que
la forme des mailles intervient également dans ce test, mais en
première approximation,la décroissance des fréquences des cas
positifs en fonction de la taille du carré est une estimation de la
taille des mailles du réseau.

CONCLUSIONS

Cette méthode d'étude morphométrique a déjà été utilisée sur plusieurs
milliers de coupes pour évaluer son efficacité.L'étude de l'évolution
de la fibrose hépatique sur des biopsies successives au cours des
hépatites chroniques a été traitée dans une thèse récente [2,3].
Dans des biopsies de foie prélevées chirurgicalement, au cours de
laparotomies exploratrices chez des sujets atteints de maladie de
Hodgkin,il a été possible de démontrer d'une manière évidente la
présence d'une fibrose assez discrète dont l'existence était
jusqu'alors très discutée [ ].
Des mesures sur la fibrose des disques intervertébraux au cours du
vieillissement ont mis en évidence l'épaississement assez brutal des
fibres collagènes au delà de 50 ans [,].
Enfin une étude a été réalisée sur la structure des fibres conjonctives
dans les ganglions au cours de la maladie de Hodgkin [ ], et les
premiers résultats semblent indiquer que certains paramètres
morphométriques permettront de préciser la classification des différents types histologiques de cette maladie.
L'étude morphométrique des fibres conjonctives ne se limite pas
à ces quelque exemples, et l'on peut envisager son application à
d'autres organes et d'autres lésions.

REFERENCES

1. J.Delarue,J.Mignot,T.Caulet, J.Diebold et G.Daddi
   Ann. Anat Path. IO [I965] 2-20
2. P. Cassan Thèse Paris-Ouest [I974]
3. J.Mignot,A.Bergue ,E.Baviera, VERH.OTSH.GES.PATH. 59 [1975]546.
4 A.Malod Panisset. Thèse Paris-Ouest [I975]
5. L.Auquier,J.Mignot,J.B.Paolaggi,A.Bergue et E.Baviera
   Rev. du Rhum. 4I [I974] 509-5I5/
6. L.Auquier,J.Mignot,J.B.Paolaggi,A.Bergue et E.Baviera
   I.N.S.E.R.M. Rapport de synthèse,n°6, [1975]
7. O.Esperandieu, Thèse Paris-Ouest, [I977].

STEREOLOGICAL AND TOPOLOGICAL ANALYSIS OF CIRRHOTIC LIVERS AS A LINKAGE OF REGENERATIVE NODULES MULTIPLY CONNECTED IN THE FORM OF THREE-DIMENSIONAL NETWORK

Tohru Takahashi and Norio Suwa

Department of Pathology, Tohoku Univ., School of Medicine, SENDAI, Japan

Summary

The microstructure of cirrhotic liver was assimilated by a dispersion system of spherical regenerative nodules of different raii r. By assuming some distribution function for r, a stereological method was established to characterize the different patterns of cirrhosis in three-dimensional metric terms. On the other hand, reconstruction from serial sections disclosed that the nodules were united with one another to form a 'nodular network' in the space. The internodular connectivity was expressed by the estimated total connectivity number p_1 of the liver, which was widely different according to different pattern of cirrhosis. It revealed the importance not only of metric but of non-metric analysis in dealing with a biological structure like liver cirrhosis.

Introduction

Cirrhotic liver of man is a typical two-phase structure that consists of parenchymal nodules and internodular scar areas(Fig. 1). The scar originates from some injury to the parenchymal cells for instance due to viral hepatitis, while the surviving parenchymal tissues proliferate after the injury to form round nodular masses of different sizes. From a morphological viewpoint, the structure of cirrhosis is interesting not only because it is feasible to geometrical analysis, but because there are a large diversity of its patterns(Fig. 1A, 1B). Characterization of different patterns requires a pertinent combination of quantitative expressions such as the mean and deviation of nodular sizes and the volumetric ratio of the two phases.

Granulometry of Nodules as Dispersed Spheres

Our first approach was granulometry of cirrhotic nodules.[1] We used a model in which the nodules were replaced by spheres of different radii r, randomly dispersed in the space. Line sampling was performed on microscopic sections of cirrhosis to measure the length λ of chords delivered by intersection of the sampling line with nodular sections(Fig. 2). Assuming some distribution function N(r) for the sphere radius, we developed a mathematical method to estimate the parameters

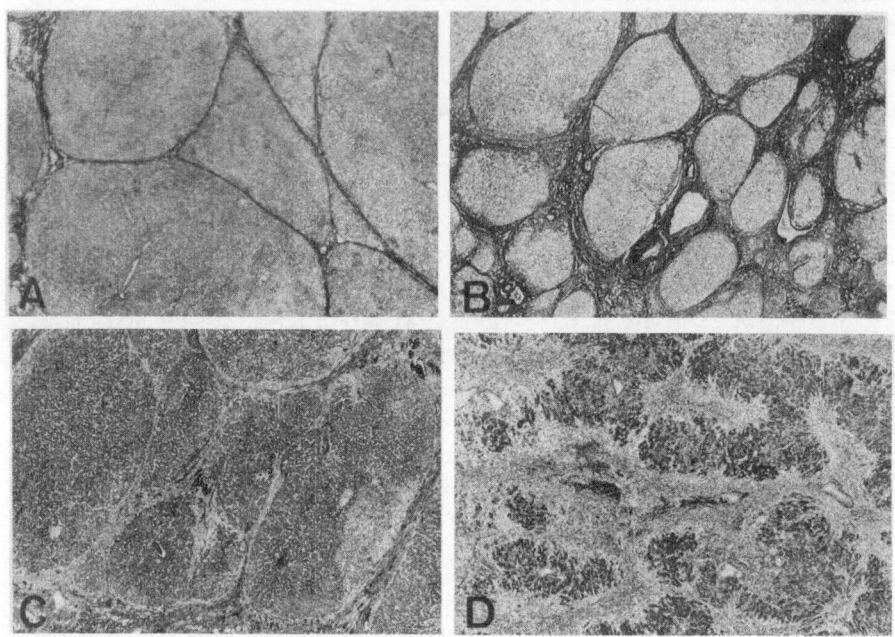

Fig. 1. Different microscopic patterns of cirrhosis and precirrhotic lesions. A: Posthepatitic cirrhosis, coarse-nodular(10×). B: Post-necrotic cirrhosis, small nodular(10×). C: Internodular coalescence in a case of postnecrotic cirrhosis(15×). D: Subacute hepatitis, 60 days after the onset(25×). Silver impregnation(A, B) and Goldner stain(C,D).

of $N(r)$ from chord measurement. Recently, the method was further extended into a general form.[2] The distribution function of chord length $F(\lambda)$ expected from assumed $N(r)$ was theoretically derived as

$$F(\lambda) = \frac{\pi}{2} \int_{\frac{\lambda}{2}}^{\infty} \lambda\, N(r)\, dr.$$

To cirrhotic nodules we applied logarithmic normal distribution or

$$N(r) = \frac{N_{vo}}{\sqrt{2\pi}\, mr} \exp\left[-\frac{(\log r - \log r_0)^2}{2m^2}\right]$$

as $N(r)$, where N_{vo} denotes the number of spheres in a unit volume, r_0 the geometric mean of r and m the standard deviation of $\log r$. Besides logarithmic normal, Weibull distribution

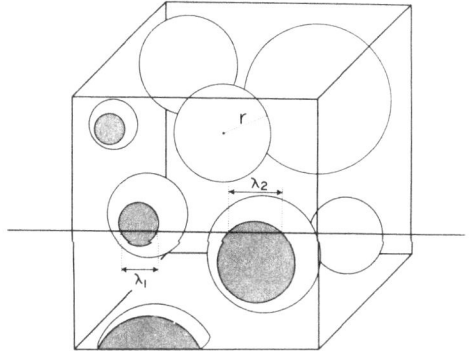

Fig. 2. A model illustrating the chord measurement with a system of dispersed spheres.

$$N(r) = N_{vo} \, m\alpha(\alpha r)^{m-1} \exp[-(\alpha r)^m]$$

as well as Gamma distribution

$$N(r) = \frac{N_{vo}}{\Gamma(m)} \, (\alpha m)^{m-1} \exp[-\alpha r]$$

may also ensure satisfactory fitting for radius distribution of cirrhotic nodules[2].

The parameters of $N(r)$ were determined as follows. If $I_n(\lambda)$ is defined by

$$I_n(\lambda) = \int_0^\infty \lambda^n \, F(\lambda) \, d\lambda,$$

n being 0 or an positive integer, $I_0(\lambda)$, $I_1(\lambda)$ and $I_2(\lambda)$ correspond respectively to the expectation of chord number, that of $\Sigma\lambda$ and of $\Sigma\lambda^2$ per unit length of sampling line. The parameters of $N(r)$ are calculated by replacing these expectations with the measurements, and by transforming the above equation into

$$I_n(\lambda) = \frac{\pi}{2} \int_0^\infty \int_{\frac{\lambda}{2}}^\infty \lambda^{n+1} N(r) \, dr \, d\lambda$$

$$= \frac{2^{n+1} \pi}{n+2} \int_0^\infty r^{n+2} N(r) \, dr.$$

When logarithmic normal distribution is used for $N(r)$, we obtain

$$I_0(\lambda) = \pi N_{vo} \, r_0^2 \, e^{2m^2} = N_\lambda/L$$

$$I_1(\lambda) = \frac{4}{3} \pi N_{vo} \, r_0^3 \, e^{\frac{9}{2}m^2} = \Sigma\lambda/L$$

$$I_2(\lambda) = 2\pi N_{vo} \, r_0^4 \, e^{8m^2} = \Sigma\lambda^2/L$$

where L is the length of sampling line. The histogram of chord length obtained from a cirrhotic liver is shown in Fig. 3, on which the theoretical curve of $F(\lambda)$ is superimposed. The curve agrees quite well with the histogram, suggesting a good adaptability of logarithmic normal to the actual distribution of nodular radius.

Cirrhotic livers of different morphologic patterns from 83 autopsy cases were submitted to the granulometric analysis. Fig. 4 shows the result of r_o. The unimodal histogram seems to make a morphological classification of cirrhosis rather inappropriate, so far as it is based upon the size of nodules. We could not find multimodal distribution with other metric indices such as the standard deviation m of log r or the volumetric ratio of nodules V_o.

Topological Analysis of Internodular Connectivity

After the granulometric analysis we found an important structural principle of cirrhotic liver which had not been properly taken into account. Careful examination on microscopic sections demonstrates that there are connections between the nodules even in the cases of seemingly complete separation of nodules(Fig. 1C). In extreme cases the nodular sections are so deeply coalesced as to give a continuous figure, making an attempt at granulometry impossible.

The three-dimensional structure of cirrhotic nodules was analyzed by reconstruction of serial histological sections. The result discloses that the nodules are by no means discrete structures, but they are connected with one another(Fig. 5A). The nodular chains form a three-dimensional network which can be better visualized if nodules are replaced by nodes and internodular connections by branches(Fig. 5A, right). The network is demonstrated in all the morphological types of cirrhosis and, consequently, it can be regarded as the common structural framework of this

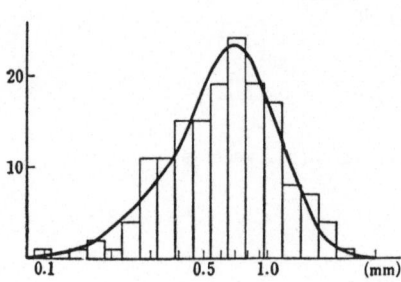

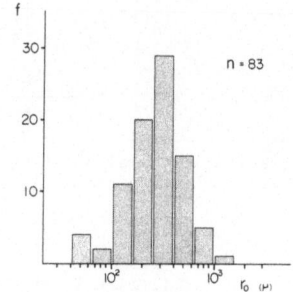

Fig. 3. Histogram of λ superimposed with its theoretical curve.

Fig. 4. r_o determined in 83 cases of cirrhosis.

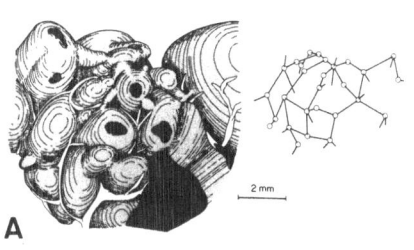

Fig. 5. A: (left) Reconstruction of nodules and blood vessels in the case of Fig. 1A. (right) A node-branch model illustrating internodular connectivity. B: Reconstruction in a case of subacute hepatitis(Fig. 1D). H: Hepatic vein. P: Portal vein.

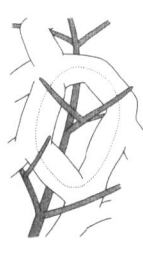

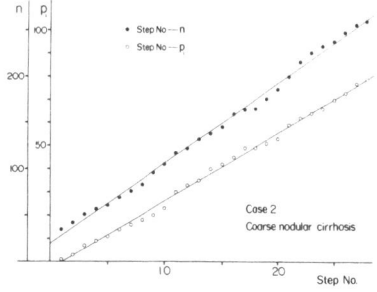

Fig. 6. A schematic illustration of regenerative hyperplasia continuously transforming the parenchymal network into nodular network.

Fig. 7. n as well as p_1 increases with advancing steps of serial sections.

disorder. On the other hand, it is closely penetrated by the blood vessels as a 'conjugate' branch system.

A similar network is also found in precirrhotic lesions such as subacute or chronic hepatitis. Fig. 5B shows the reconstruction from a case of subacute hepatitis (Fig. 1D). The parenchymal masses that survived recent injury are already united to form a network intertwined with the branches of hepatic vein. Thus, it appears likely that the distribution of injury either along the efferent or along the afferent vascular tree molds the surviving parenchyma into the form of a network.

Parenchymal tissues which survived injury undergo regenerative hyperplasia, and the parenchymal network in subacute hepatitis may be transformed into the nodular network of cirrhosis(Fig. 6). Because the process is a kind of continuous transformation, connecting relation of the network must remain unchanged, as revealed by the preservation of loop structure. On the other hand, if there is an evident heteromorphism of network between two diseased livers, the one cannot develop into the other through continuous transformation alone. In this case, a radical reformation of network must take place. The morphogenetic relation between the cirrhotic and precirrhotic lesions largely depends upon the topological properties of the structural framework.

Connectivity of the framework of chronic liver diseases was parametrized by the total connectivity number(the 1st Betti number) p_1, after the network was simplified into a node-branch system. In practice, a test space of certain volume was taken in the organ, and the number of nodules (n) and that of multiple connections (p_1) in the space were determined by cumulative counting from serial sections(Fig. 7). When a definite linear regression of n as well as p_1 was established upon the volume examined, the total number of each in the liver was easily calculated from the volume ratio of the organ to the test space. The total p_1 thus determined corresponds to the 'minimum genus', and the procedure of network simplification to the construction of 'deformation retract', each having been proposed and described by DeHoff et al.[3]

Six livers with different types of cirrhosis or precirrhotic disorders were selected from the autopsy materials and were submitted to the topological analysis. As shown in Fig. 8, the estimated total p_1 is widely different among the disease types examined. For example, p_1 in an advanced cirrhosis of posthepatitic type (Case 2) is estimated at 100,000, whereas it attains to 6,100,000 in Case 5 of chronic hepatitis which has generally been regarded as the very precursor of posthepatitic cirrhosis. The steep fall of p_1 demonstrates that a radical re-formation of the framework proceeds in the process of cirrhogenesis. Such a re-formation can be explained by assuming a destruction of nodules due to episodic parenchymal injuries(Fig. 9) which frequently involve cirrhotic livers.[4] On the other hand, cirrhosis of portal type(Case 4) retains almost as much p_1 as chronic hepatitis does (Fig. 8). It appears quite probable that this type of cirrhosis may be produced from chronic hepatitis only through continuous transformation, i.e., through nodular parenchymal hyperplasia or 'piecemeal' extension of fibrous septa. Topological analysis of the framework thus gives an insight into the morphogenetic relation of different types of chronic liver diseases. In addition, it is worthy of notice that the ratio of n : p_1 is about 3 : 1 in all the cases examined(Fig. 8). The reason for the constancy is not clear, but it seems to be concerned with the way of space division of nodular chains that are arranged in close intertwinement with the arborescent system of blood vessels.

		Case 1	Case 2	Case 3	Case 4	Case 5	Case 6
		Cirrhosis post- necrotic	Cirrhosis post- hepatitic (A)	Cirrhosis post- hepatitic (B)	Cirrhosis portal	Chronic Hepatitis	Subacute Hepatitis
Hepatic weight (g)		1030	780	590	1970	1660	580
Total n (×10⁴)		118	30	68	1859	1728	1421
Total p_1 (×10⁴)		37	10	23	635	610	501

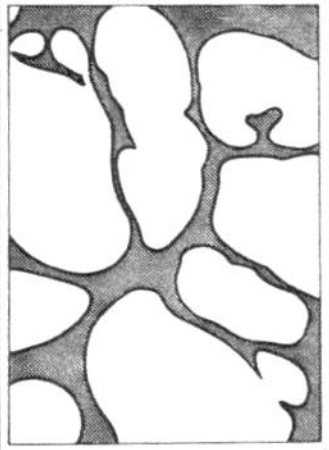

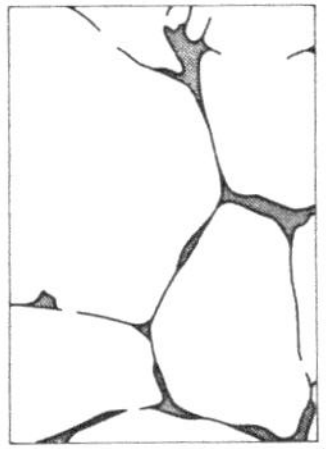

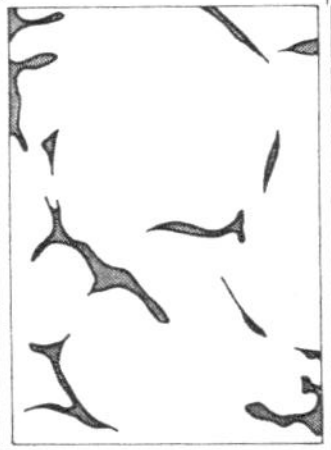

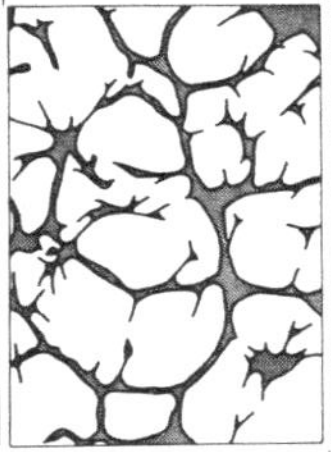

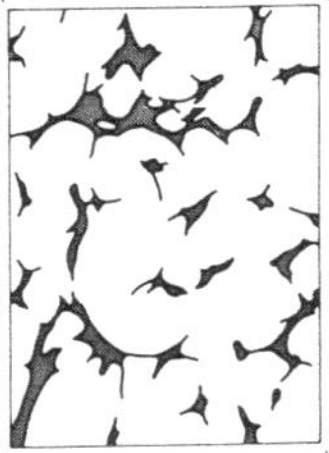

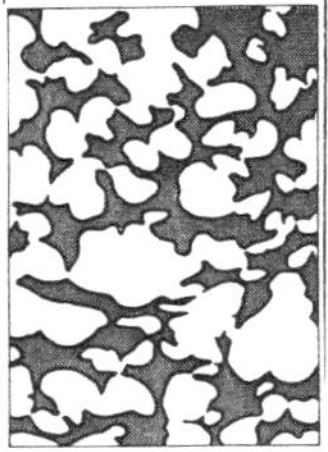

Fig. 8. The results of topological analysis tabulated with the sketches of liver histology. The sketches were prepared under an equal magnification of 15×.

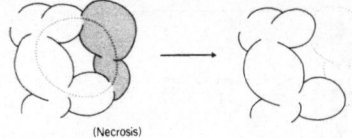

(Necrosis)

Fig. 9. Schema illustrating the split of nodular loop following destruction of nodules.

Conclusions

The method of granulometry stated above is rather effective in the analysis of discrete spherical bodies such as the islets of pancreas[5] or the nuclei of some kind of cells. Cirrhotic nodules form a system of connected spheres, of which geometry is determined not only by metric concepts but by non-metric ones such as connectivity. There are various kinds of biological structures requiring a morphological treatment from a similar viewpoint. Because the analysis of connectivity in the space is beyond the reach of stereology in its present state, scanning of the space from serial sections is unavoidable. We are looking forward to more practical way of approach to this aspect of morphology.

References

1. Suwa, N., Takahashi, T., Sasaki, Y.: Tohoku J.exp.Med.,84:1-36, 1964; Addendum and Correction: Tohoku J.exp.Med.,84:199-200, 1964

2. Suwa, N., Takahashi, T., Saito, K., Sawai, T.: Tohoku J.exp.Med.,118:101-111, 1976

3. DeHoff, R.T., Aigeltinger, E.H., Craig, K.R.: J.Microsc.,95:69-91, 1972

4. Baggenstoss, A.H., Soloway, R.D., Summerskill, W.H.J., Elveback, L.R., Schoenfield, L.J.: Human Path.,3:183-198, 1972

5. Saito, K., Iwama, N., Takahashi, T.: Tohoku J.exp.Med., in press

METHODE D'ETUDE DES ESPACES INTERCELLULAIRES DE LA GRANULOSA DES FOLLICULES OVARIENS DE RATTE ADULTE CYCLIQUE.

J.C. MARIANA & J. PERAY[+]

I.N.R.A. - Station de Physiologie de la Reproduction - 37380 Nouzilly (F).
+ Centre de Morphologie Mathématique. Ecole Nationale Supérieure des Mines de Paris. 35 Rue St-Honoré, 77305 Fontainebleau (F).

RESUME

During the growth of the folliculum, small enclavec develop in the inside of the granulosa, and progressily cluster together into a larger lacuna, called the antrum.

For a biologist, it turns out to be difficult to appreciate the begining of this process. However one must do it if one wants to study the role of hormous in folliculum growth.

We used the size distribution by bi-dimensionnal openings (notion taken from the theory of mathematical morphology), in order to accurately quantify the coming out of the antrum.

INTRODUCTION

Dans une première étude (MARIANA & MACHADO, 1976), nous avions tenté d'analyser la formation des espaces intercellulaires de la granulosa du follicule au cours de sa croissance et, plus particulièrement, quand le follicule amorce la dernière phase de son développement qui, normalement, aboutit à l'ovulation.

Nous avions également examiné l'influence des hormones sur la croissance des follicules et montré qu'après injection de 10 U.I. de PMSG au métoestrus, l'antrum apparaît plus tôt au cours de la croissance du follicule que chez des animaux témoins.

Les premières études et descriptions de l'antrum ont été faites par BRAMBELL (1928) et PARKES (1930). Dans une description plus fine, MOTTA (1942) et BRAMBELL (1956) précisent qu'aux phénomènes de vacualisation intéressant les cellules folliculaires s'ajoutent des phénomènes de secrétion de la part de celles-ci.

Il manquait à l'ensemble de ces observations des critères objectifs et quantitatifs permettant de caractériser les espaces intercellulaires et définir leur évolution.

Nous appelions début d'antrum une enclave intercellulaire d'une surface égale ou supérieure à 500 u2 sur coupe, ce qui est, évidemment, arbitraire.

METHODE D'ETUDE

La méthode proposée ici, pour l'étude des espaces intercellulaires, est celle des "granulométries" ; son application consiste en une classification quantifiée des enclaves ou pores par ordre de "taille". Si on peut déceler à l'oeil la présence d'enclaves d'une surface donnée en les comparant à une figure de référence, il est, en revanche, beaucoup plus difficile de préciser l'importance, le pourcentage en surface de ces enclaves par rapport à la totalité des vides d'un follicule.

Or, dans le processus de maturation du follicule, la formation de l'antrum corres-pond à des regroupements d'enclaves de plus en plus importantes. On conçoit que la granulométrie soit, dans ces conditions, un outil privilégié pour dater physiolo-giquement le processus et le comparer éventuellement à des processus accélérés par injection d'hormones.

Nous ne tenterons pas ici de présenter la théorie mathématique qui soustend la notion de granulométrie. Nous renvoyons le lecteur à l'ouvrage de SERRA (1968).

APPLICATION AUX ENCLAVES INTERCELLULAIRES DES FOLLICULES OVARIENS

L'ensemble de l'étude a été réalisée sur 10 follicules d'un ovaire de ratte adulte. Après fixation et inclusion dè l'ovaire dans la paraffine, nous avons pratiqué une coloration nucléaire au Feulgen sur des coupes de 5 u d'épaisseur.

Ce type de coloration réalise en effet le meilleur contraste entre les noyaux et le restant du cytoplasme et des espaces intercellulaires pour l'analyseur de texture Leitz.

Les follicules sont classés en fonction de leur taille dans la section où est présent le nucléole de l'ovocyte. Les études granulométriques ont été réalisées dans cette section seulement.

Nous avons retenu, pour illustrer l'efficacité des histogrammes, une série de 6 follicules dont la taille est comprise entre 12 350 u2 et 15 700 u2 ; pour ces follicules, l'antrum est théoriquement en cours de formation sans pouvoir être caractérisé par le critère traditionnel de l'enclave supérieure à 500 u2. Nous y avons adjoint un follicule de 16 900 u2 dans lequel l'antrum est nettement visible et répond au critère des 500 u2. Les granulométries et histogrammes correspondant ont été mesurés et calculés directement sur les photos de la planche (2) à partir de l'analyseur de textures (SERRA, 1967).

Les 6 follicules étudiés ont été numérotés de 1 à 6.

Considérons, pour commencer, l'histogramme du follicule n° 6. On remarque que 70 % des enclaves sont de taille inférieure à 100 u2 ; le reste comprend des enclaves de taille supérieure à 340 u2 (25 %) et des enclaves de taille intermédiaire (5 %). La quasi-absence d'enclaves moyennes (déjà regroupées pour former l'antrum) et l'opposition très nette entre petites et grosses enclaves apparaissent comme carac-téristiques d'une structure à antrum.

Le fait qu'il ait fallu plusieurs classes pour éliminer l'antrum tient à sa forme irrégulière et à l'arbitraire du découpage des classes : on voit sur la photo du

follicule n° 6 (Pl. 1) que le plus grand cercle inscriptible dans l'antrum (517 u2
d'après l'histogramme) est loin de le recouvrir entièrement. On retiendra cependant
la présence des deux modes dans l'histogramme d'un follicule à antrum développé.

A l'opposé, une granulosa dépourvue d'antrum au sein d'un petit follicule (le n° 1,
par exemple) se caractérise par un histogramme en cloche (Pl. 1) ne prenant en
compte que des enclaves de petite taille.

Entre ces deux formes de courbe se situe le processus de formation de l'antrum.
L'histogramme du follicule n° 4 (Pl. 1) présente déjà un embryon de deuxième mode
différencié alors qu'aucune enclave n'atteint la taille de 500 u2. Notons également
l'aspect de l'histogramme du follicule n° 2 (Pl. 1) : il présente un deuxième
maximum pour la dernière classe, mais le minimum précédent est trop important pour
que l'on puisse assimiler l'ensemble à une structure de granulosa à antrum.

On peut, à l'aide des courbes précédentes, tenter un classement des follicules par
ordre de maturité croissante de l'antrum :
- Follicule n° 1 - 2   : Pas d'antrum
- Follicule n° 3 - 4   : Développement de l'antrum
- Follicule n° 5 - 6   : Antrum développé.

On remarquera que le classement des follicules par ordre de taille absolue ne suit
pas rigoureusement celui des follicules en fonction du développement de l'antrum.

CONCLUSION

Nous avons essentiellement cherché, dans cette étude, à présenter une nouvelle
méthode d'approche du processus de la maturation du follicule. Il est évident qu'une
telle méthode peut, par son caractère quantitatif, dépasser la simple description
des vides intercellulaires. C'est ce que nous avons montré, encore trop qualitative-
ment, à la fin du dernier paragraphe sur un très petit nombre d'échantillons.
Par ailleurs, la coloration Feulgen ne colore pas les cytoplasmes cellulaires et,
dans l'analyse précédente, nous n'avons pas pu distinguer ce qui était propre aux
espaces intercellulaires et aux éventuels gonflements et vacuolisations du cytoplasme des cellules de la granulosa.
Le but que nous nous étions proposé au départ, à savoir, la recherche de l'influence
des hormones sur la formation de l'antrum, nécessite un travail de nature statistique pour la comparaison des différentes sections d'un même follicule, la comparaison de différents follicules du même animal, puis de plusieurs animaux d'un même
échantillon et, enfin, d'animaux normaux et traités aux hormones.
Les histogrammes des granulométries décrits ici constituent la base de ce travail
comme une série de données caractéristiques d'un follicule.
On notera également la rapidité, l'analyseur de texture donnant l'histogramme à
partir d'une photo en quelques secondes.

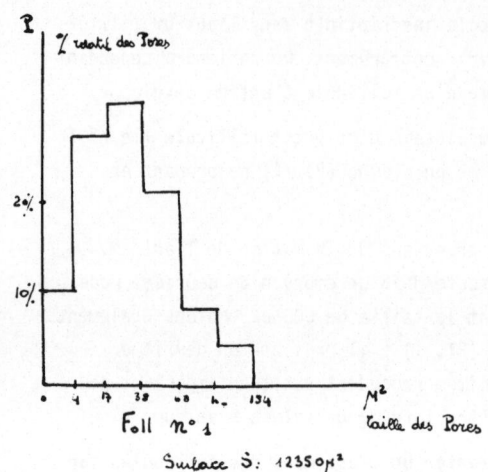

Foll n° 1

Surface S: 12350μ²

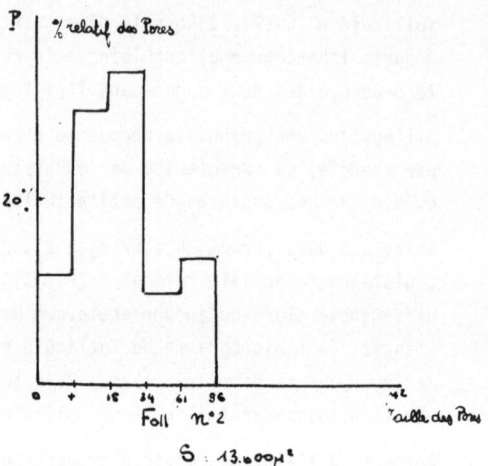

Foll n°2

S : 13.000μ²

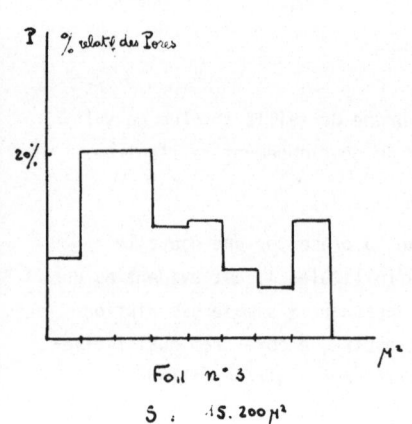

Foll n° 3

S : 15.200μ²

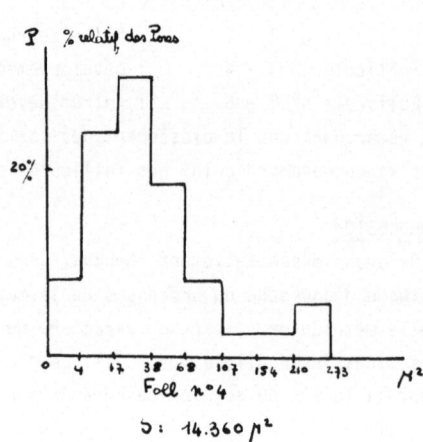

Foll n°4

S: 14.360 μ²

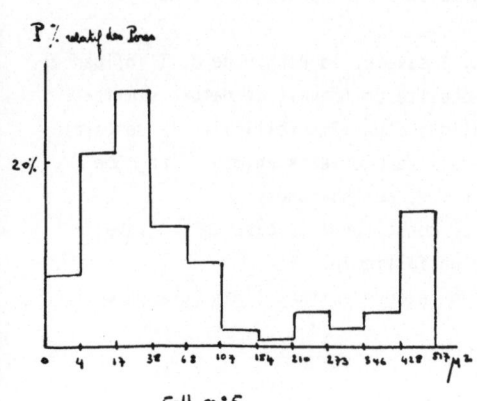

Foll n°5

S: 14170μ²

Planche 1

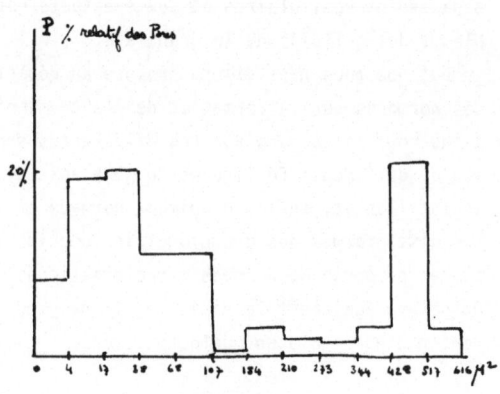

Foll n° 6

S: 16900μ²

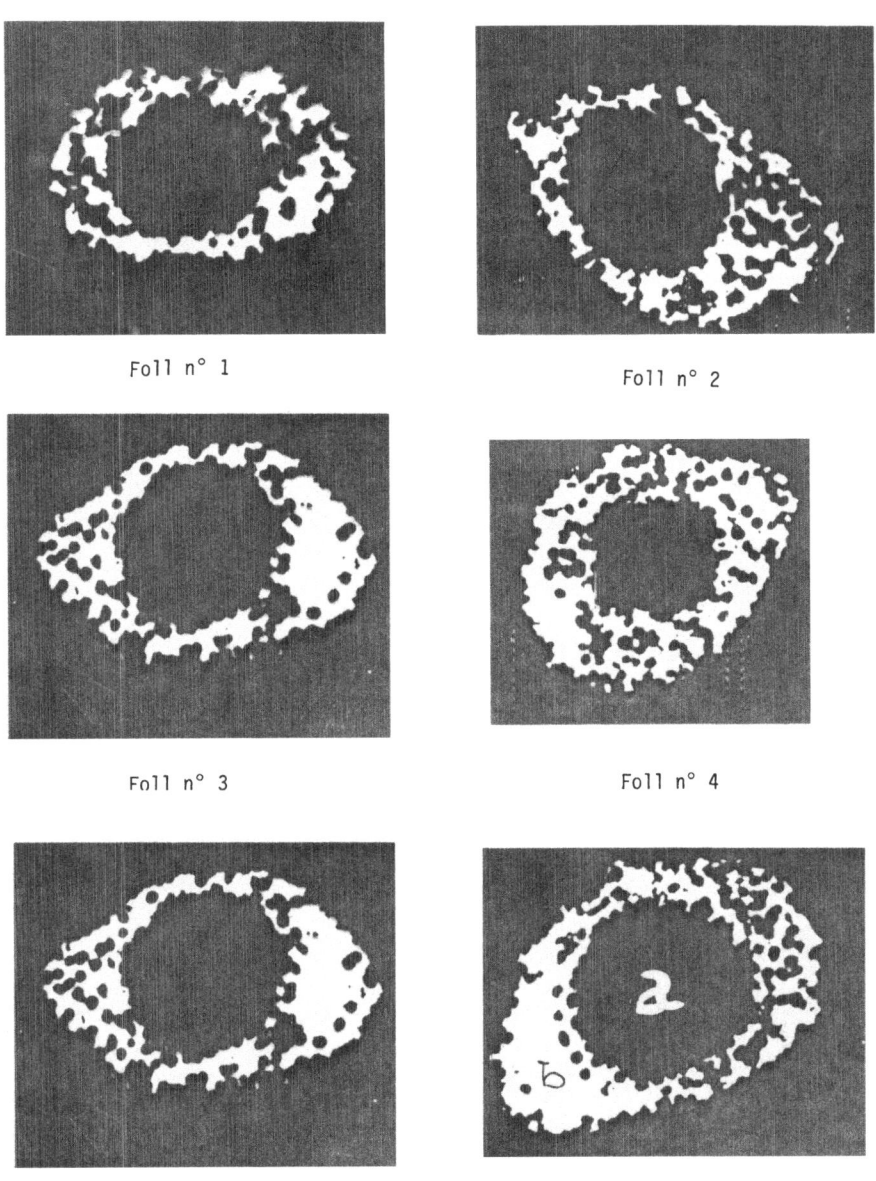

Foll n° 1

Foll n° 2

Foll n° 3

Foll n° 4

Foll n° 5

Foll n° 6

Photographie des reproductions des 6 follicules par l'écran de visualisation de l'analyseur de textures (1 cm reprêsente enciron 25   )

Exemple du follicule n° 6 - (a) ovocyte - (b) antrum et cellules de la granulosa.
Les enclaves correspondent aux taches blanches et les cellules aux taches noires.

Planche 2

STEREOLOGICAL USES OF THE AREA TANGENT COUNT

by R. T. DeHoff

The area tangent count, first suggested as a stereological descriptor by F. N. Rhines in the decade of the fifties, is performed on any two dimensional structure that contains lineal features. A test line is swept across some delineated field of area A_0, and a count is made of the number of times the line forms a tangent with some selected class of lineal features. The working stereological parameter, T_A, is obtained by dividing this count of tangent points by the area sampled, A_0.

For the purpose of the measurement, it is useful to distinguish two classes of lineal features in the plane: (1) those which are part of the boundary of some areal features, such as particles, nuclei, or some phase, and thus serve to divide the area into two classes of regions, i.e., those inside the boundary, and those outside it; and, (2) lineal features which are not part of such a boundary.

In the first case, it is possible to establish a convention which separates the boundary into two classes of arc: <u>convex</u> boundary segments, along which the tangent vector rotates toward the inside of the bounded area; and, <u>concave</u> segments, where it rotates toward the outside.* It is thus possible to distinguish, and separately tabulate, tangents with convex and concave segments; designate these counts as T_{A+} and T_{A-}, respectively. The net area tangent count,

$$T_{Anet} \equiv T_{A+} - T_{A-} \tag{1}$$

is stereologically useful, since it can be shown to be related to three dimensional geometry for this class of structures.[1,2] Figure 1 illustrates the net area tangent count.

Examples of lineal features on the plane which do not form part of a dividing boundary include the traces on a section of the surfaces that divide cells in a space-filling cell structure, such as a polycrystalline grain structure. Since both sides of the trace are equivalent phase forms or cell types, the concepts of the "inside" and "outside" of such a line do not apply, convex and concave segments cannot be defined, and thus cannot be distinguished in a tangent count. Similarly, lineal features that are projections of space curves onto the field of view and analysis cannot contain definable convex and concave segments, since they do not bound two dimensional areas in the projected image. Figure 2 illustrates the tangent count for the latter case. A direct stereological relationship holds for tangent counts made on projections of space curves;[3,4] a stereological meaning of the tangent count performed on a cell structure exists, but is much less direct.

---

*Note that which phase is defined to be inside, and which outside the boundary is purely a matter of convention. If the choice is reversed, the convex and concave segments will be interchanged. Usually, the phase present in the lesser amount is defined as inside the boundary.

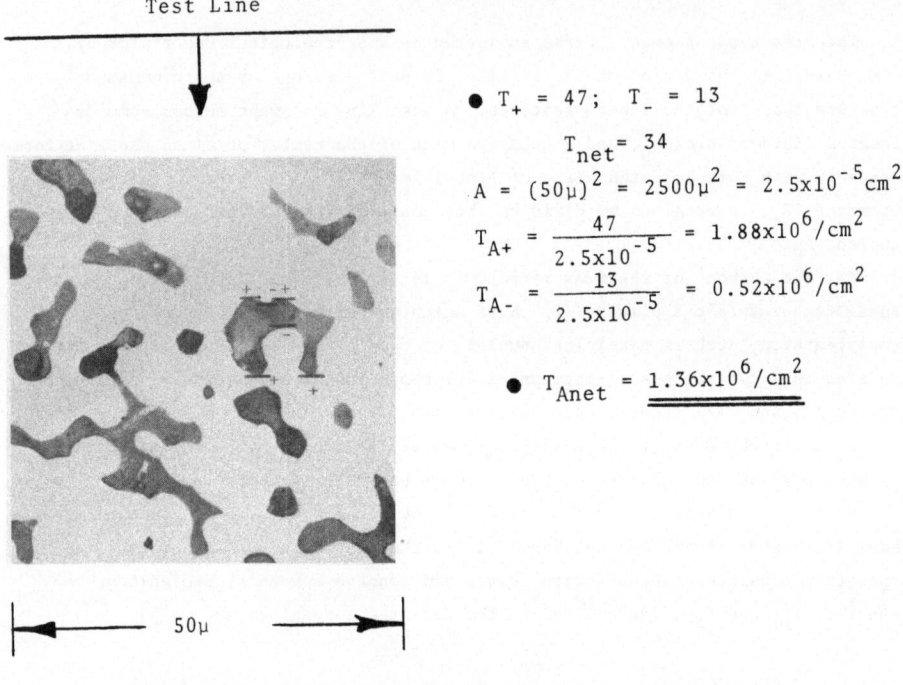

Test Line

- $T_+ = 47;\quad T_- = 13$

$$T_{net} = 34$$

$$A = (50\mu)^2 = 2500\mu^2 = 2.5 \times 10^{-5} cm^2$$

$$T_{A+} = \frac{47}{2.5 \times 10^{-5}} = 1.88 \times 10^6 / cm^2$$

$$T_{A-} = \frac{13}{2.5 \times 10^{-5}} = 0.52 \times 10^6 / cm^2$$

- $T_{Anet} = \underline{\underline{1.36 \times 10^6 / cm^2}}$

50μ

Figure 1. Procedure for measuring the net area tangent count.

If the surfaces in the three dimensional structure contain edges, sections will exhibit corners or vertices where the sectioning plane intersects an edge. Separate counts of tangents formed at such corners yield stereological information about the edges in the three dimensional structure. As in the case of smooth segments of boundaries, it is necessary to distinguish tangents with "convex" and "concave" corners;[1] the net tangent count has the most direct stereological relationship with the geometry of the edges in three dimensions.[3,4]

While the stereological meaning of the area tangent count has been known for more than a decade,[1] little use has been made of this measurement.[5-9] In the most general cases, the information supplied by the tangent count emerges in the form of curvature integrals. For tangents on sections through two phase structures,[1,2]

$$T_{Anet} = \frac{1}{\pi} M_V \qquad (2)$$

where

$$M_V \equiv \frac{1}{V_0} M = \frac{1}{V_0} \iint_S \frac{1}{2}(\kappa_1 + \kappa_2) dS \qquad (3)$$

κ_1 and κ_2 are the local principal normal curvatures,[10] $\frac{1}{2}(\kappa_1 + \kappa_2)$ is the local mean curvature, S is the surface area, M is the "integral mean curvature," V_0 is the

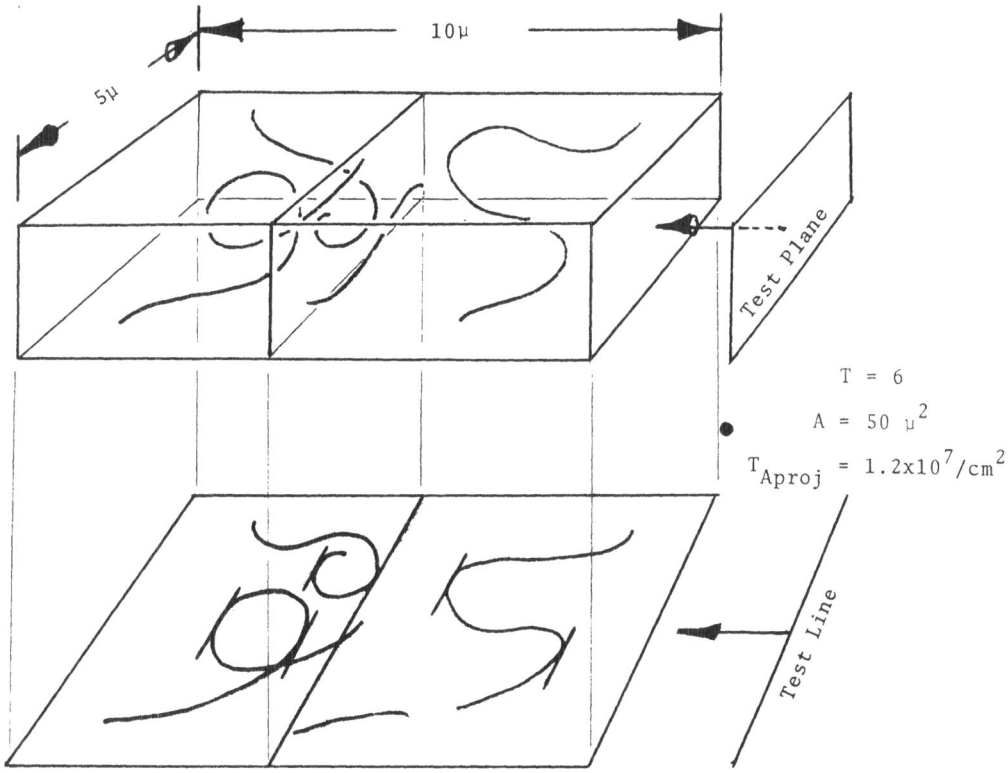

Figure 2.   Relation of the area tangent count on a projected image of lineal features to a volume tangent count.

sample volume, and M_V is thus the "integral mean curvature per unit volume" for the surfaces being analysed. If the surfaces have edges, the tangent count may be separately applied to sections through these edges, which appear unambiguously as corners on the section. Equation (2) also holds in this case, but the value of M_V must be interpreted as[3,4]

$$M_{Vedge} = \frac{1}{V_0} M_{edge} = \frac{1}{V_0} \cdot \frac{1}{2}\int_L \chi d\ell \tag{4}$$

where χ is the local dihedral angle at a point on the edge, and L is the length of edges in the sample volume. For tangents on projected images of lineal features (space curves), the relationship is[3,4]

$$T_A(proj) = \frac{t}{\pi}\mathcal{X}_V \tag{5}$$

$$\mathcal{X}_V \equiv \frac{1}{V_0}\mathcal{X} = \frac{1}{V_0} \int_L k d\ell \tag{6}$$

where k is the local curvature, defined to be positive at all points on the line, L is the line length, and $\mathcal{K}_V$ is the "total curvature of line per unit volume" of structure. The thickness of the slice or foil being projected, t, must be separately determined. These curvature integrals are unfamiliar concepts, and have no intuitive basis for visualizing their geometrical significance as do more familiar properties such as area of surface or length of line.

The purpose of the present paper is to provide some insight into the geometrical significance and potential utility of the three dimensional information provided by the area tangent count. Some illustrations of easily visualized geometrical meaning are supplied for special classes of structures. The role of integral mean curvature in shape description is explored. The significance of the tangent count for cell structures is described. Potential uses of the total curvature of lineal features in defining the scale of such structures are examined. Perhaps some of the insights provided will encourage application of the area tangent count in research where its potential utility has not yet been fully appreciated.

Two Dimensional Structures

Consider a collection of curves in a two dimensional field that has an area A_0. The tangent count reports the total rotation, Θ, of the tangent vector at a point as it traverses the collection of curves.[2] If convex and concave segments are distinguishable, separate tangent counts of each of these classes yield separate measures of the tangent rotation for convex and concave arc, Θ_+ and Θ_-. Thus,

$$T_A = \frac{1}{\pi A_0} \Theta \tag{7}$$

$$T_{A+} = \frac{1}{\pi} \Theta_{A+} \quad ; \quad T_{A-} = \frac{1}{\pi} \Theta_{A-} \tag{8}$$

$$T_{Anet} = \frac{1}{\pi} \Theta_{Anet} \tag{9}$$

where Θ_{Anet} is $(\Theta_{A+} - \Theta_{A-})$.

It can further be shown[2] that the total rotation of the tangent is equal to the total curvature of the lines in the area:

$$\Theta = \int_L k d\ell \tag{10}$$

Since the length of lines in the structure can be separately determined,[2,11,12]

$$L_A = \frac{\pi}{2} P_L \tag{11}$$

where P_L is the standard line intersection count in stereology, the average curvature can be estimated from

$$\bar{k} \equiv \frac{\int_L k d\ell}{\int_L d\ell} = \frac{\Theta}{L} = \frac{\Theta_A}{L_A} = 2 \frac{T_A}{P_L} \tag{12}$$

Equation (12) may be used to estimate the average curvature of all arc, of convex and concave arc separately, or the average curvature of all arc, with convex arc counted as positively curved, and concave as negatively curved.

The net tangent count for bounding curves may be shown to be simply related to the number of closed loops of particle outline bounding the features, and to the Euler characteristic of the areas they bound, which is simply related to their connectivity:[13]

$$T_{Anet} = N_{Anet} = C_A \qquad (13)$$

where N_{Anet} is the difference between the number of closed loops bounding the areas (N_{A+}) and those bounding holes in them (N_{A-}). In some applications, particularly when the phase of interest is present in minor amount, it may be easier to count N_A than T_A. More usually, the tangent count has an advantage because, while closed loops may cross the boundary of the field, or may cross several fields, a tangent point is either inside the area of analysis, or outside of it. Thus, the tangent count is immune to edge-of-field errors and corrections.

The Net Tangent Counts on Sections Through Three Dimensional Structures

The integral mean curvature, which is the general descriptor supplied by the net tangent count, has a simple geometric significance for certain classes of closed surfaces. Its meaning for convex bodies, tubules (one dimension large comparison to the other two), and muralia (two dimensions large in comparison to the other one) is explored in this section, together with the role of the integral mean curvature in defining shape factors for classes of structures.

Convex Bodies. For convex bodies, Minkowski's formula applies:[14]

$$M = 2\pi \overline{D} \qquad (14)$$

where $\overline{D}$ is the mean tangent diameter, obtained by averaging over the sphere of orientation the distance between parallel tangent planes for the body. For a collection of convex bodies,[2]

$$M_V = 2\pi \ N_V \ \overline{\overline{D}} \qquad (15)$$

where N_V is the number of particles per unit volume of structure, and $\overline{\overline{D}}$ is the average of the mean tangent diameter for all of the features in the structure. The potential utility of Equation (15) is undermined by the fact that N_V is generally not separately accessible to stereological estimation, so that $\overline{\overline{D}}$ is also inaccessible.

Relationships (14) and (15) are also valid for convex bodies with edges, such as polyhedra, cylinders, cones, etc., provided M is interpreted to include the contribution from the edges defined in Equation (4). This relationship greatly simplifies computation of $\overline{D}$ for such features.

Tubules. A crumpled string, or a bowl of spaghetti, are examples of tubule structures. A convex tubule would be called a rod. A feature may be classed as a tubule if the average radius of its cross section is negligible in comparison to the radius of curvature of its centerline. For features that satisfy this requirement, it can be shown that[15]

$$M_V(\text{tubules}) = \pi \; L_V(\text{tubules}) \tag{16}$$

where L_V is the total length of tubules in unit volume. Thus, the area tangent count reports the length of tubules.

Muralia. A crumpled sheet of newspaper or cloth is a muralium. Any feature that is thin in one dimension, but extensive in the other two is a muralium; a convex muralium would be called a plate. In order to be classed as a muralium, it is necessary that the thickness of the sheet be negligible in comparison to the magnitudes of the principal radii of curvature of the bounding surfaces. In this case, because the mean curvatures on the faces of the sheet are essentially equal, but of opposite sign, the integral mean curvature comes primarily from the edges of the sheet or muralium. It can be shown that, for such features, the integral mean curvature is simply related to the perimeter of the edge of the sheet:

$$M_V(\text{muralia}) = \frac{\pi}{2} \; L_V(\text{edges}) \tag{17}$$

For convex muralia, i.e., for plates, the mean lineal intercept,[2,11,12]

$$\overline{\lambda} = \frac{4V_V}{S_V} = 2\overline{t} \tag{18}$$

reports the average thickness of the sheet, $\overline{t}$, but gives no information about the dimensions in the plane of the sheet. It can be shown that the reciprocal of the average mean surface curvature, $\overline{H}$, is proportional to the mean lineal intercept in the plane of the plate:[15]

$$\overline{\lambda}_p = \frac{\pi^2}{4} \frac{1}{\overline{H}} = \frac{\pi^2}{4} \frac{S_V}{M_V} \tag{19}$$

For muralia that are not flat plates, this relation gives a mean curved intercept length in the surface of the muralia.

Thus, for muralia, the integral mean curvature reports the length of edges, and, together with the surface area, gives a measure of size in the plane of the sheet.

Network Structures. A connected network of tubules, i.e., a web, consists of nodes and branches. It may be argued that the contributions to M_V from the nodes of the network is negligible in comparison to that from the branches; under these conditions, Equation (16) applies to all but a negligible part of the elements of the network, and the integral mean curvature reports the total length of the elements of the network.

$$M_V(\text{tubule network}) = \pi \; L_V(\text{branches}) \tag{20}$$

A connected muralium network may be visualized by imagining a cell structure, and endowing the cell boundaries with a thin phase or membrane. Such a muralium has no edges, but does have internal lineal features, namely the triple lines from which the membrane is stretched. The surfaces at these lineal features are negatively curved with respect to the connected network; the integral mean curvature essentially reports the total length of these triple lines in the network:

$$M_V(\text{muralium network}) = -\frac{\pi}{2} L_V(\text{triple lines}) \tag{21}$$

An incomplete muralium network may be visualized as a membrane that covers the cell boundary network incompletely; the membrane may contain holes, or some faces may not be covered. Human and animal lungs are examples of incomplete muralium networks. In this case, both the triple lines of the network and the lineal boundaries of the holes contribute to and account for the integral mean curvature:

$$M_V(\text{incomplete muralium network}) = \frac{\pi}{2}(L_V(\text{edges}) - L_V(t.1)) \tag{22}$$

Shape Factors. Shape factors are unitless parameters that are formed from ratios of global properties, such as V_V, S_V and M_V. It is essential that such shape factors contain the integral mean curvature: V_V and S_V contain no shape information. Clearly, particles of any shape may have an arbitrary value of V_V; similarly, one can choose an arbitrary shape and V_V, and still assign S_V arbitrarily, just by changing the scale of the system. Thus, these two quantities can take on any values for any shape, accounting only for the amount of the second phase and the scale of the structure. A third independent parameter is required, at a minimum, in order to be able to discriminate shapes on the basis of global properties alone. The integral mean curvature is an obvious candidate for this third quantity.

In order that the numerical value of the shape factor be independent of the scale of the system, it is necessary that it be unitless. While other candidate factors can be defined in terms of the available global properties, perhaps the most commonly used is expressed variously as[9,11,16]

$$\overline{H\lambda} = \frac{4M_V V_V}{S_V{}^2} = \frac{\pi^3}{8}\frac{N_A A_A}{L_A{}^2} = \pi\frac{N_A P_A}{2P_L{}^2} \tag{23}$$

Values of this shape factor have been computed for a variety of models for particle shape; some of these results will now be explored.

Consider first a model which describes the structure as composed of ellipsoids of revolution that are all the same size and shape. Values for $\overline{H\lambda}$ computed for such a structure are shown as the top curve in Figure 3, plotted as a function of the axial ratio of the generating ellipse. The first difficulty encountered with $\overline{H\lambda}$ as a shape factor is that, even for this simple model, the function is double valued; i.e., for values in the range from 1.1 to 1.3, the corresponding shape may be either an oblate or a prolate ellipsoid. This problem is easily circumvented at large ellipticities, since whether the structure consists of elongated or flattened particles can be deduced by relatively cursory inspection. For aspect

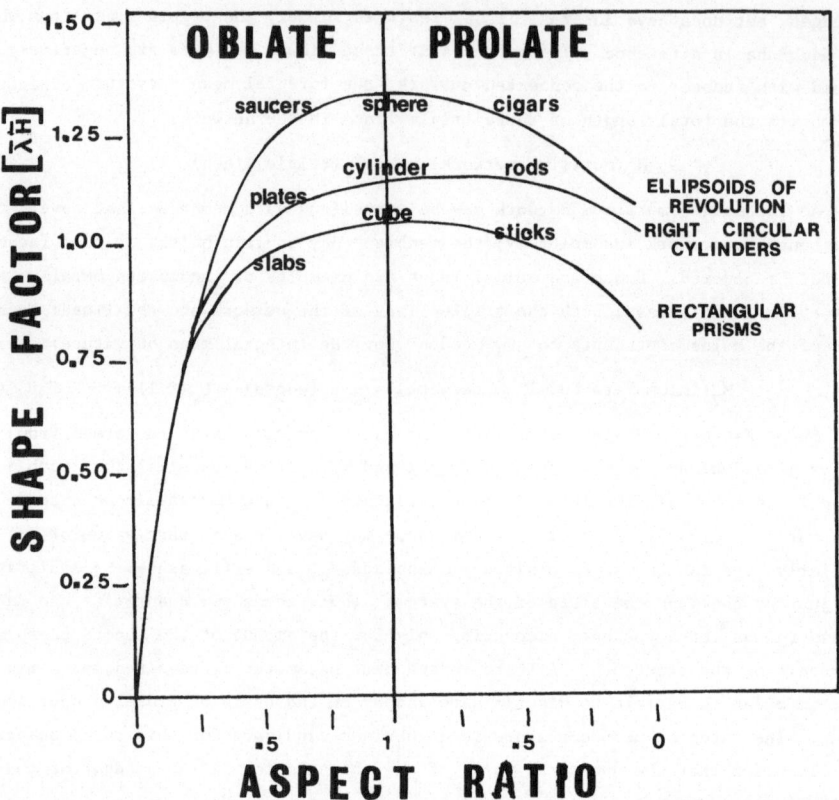

Figure 3. Dependence of $\overline{H\lambda}$ on aspect ratio for constant sizes ellipsoids, cylinders and prisms.

ratios between 0.5 and 1.0, the problem is not trivial, particularly when it is recognized that the model proposed (constant size, constant shape), will not be strictly obeyed for any real structure.

Figure 3 also illustrates that the shape parameter proposed is sensitive to the model chosen to describe the shape, with rectangular parallelapipeds, right circular cylinders and ellipsoids of revolution yielding significantly different $\overline{H\lambda}$ values for the same axial ratio. While this result provides some potential basis for using $\overline{H\lambda}$ to discriminate between different shape models, it also provides some insight into the effect that departures from a particular model shape might have on $\overline{H\lambda}$ for a real structure.

It is important to keep in mind that, although $\overline{H\lambda}$ is unitless, its value for a given system is not independent of the particle size distribution, even if all of the particles are the same shape. Specifically, if μ_i is the ith moment of the particle size distribution (μ_1, μ_2 and μ_3 are, respectively, the mean, mean square, and mean cube particle sizes), then $\overline{H\lambda}$ may be written

$$\overline{H\lambda} = (\overline{H\lambda})_0 \frac{\mu_1 \mu_3}{\mu_2^2} \tag{24}$$

where $(\overline{H\lambda})_0$ is the value for a nondispersed system of particles. The sensitivity of $\overline{H\lambda}$ to dispersion of the particle size distribution is shown in Figure 4, in which

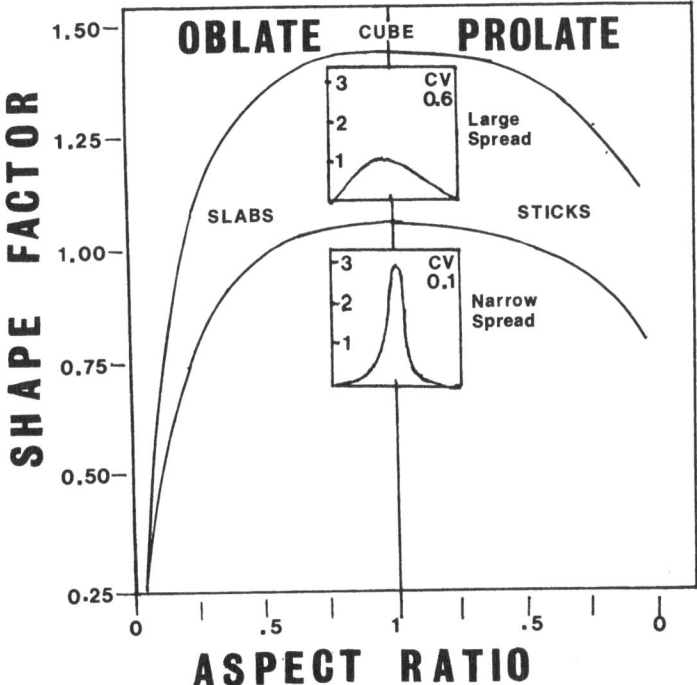

Figure 4. Sensitivity of the $\overline{H\lambda}$ product to particle size distribution.
the dependence of $\overline{H\lambda}$ upon axial ratio for particles assumed to be rectangular parallelapipeds is shown for two different particle size distributions. The lower curve corresponds to a narrow, almost constant, size distribution (coefficient of variance equal to 0.1), the upper curve to a broad size distribution (with C.V. equal to 0.6). It is clear that, without auxiliary information about the size distribution, a value of $\overline{H\lambda}$ in the range of, say, 1.2 may describe a broad range of shapes and size distributions.

Values of the $\bar{H}\bar{\lambda}$ product may also be computed for a variety of network models. Computations for two such models are shown in Figure 5. The network is constructed

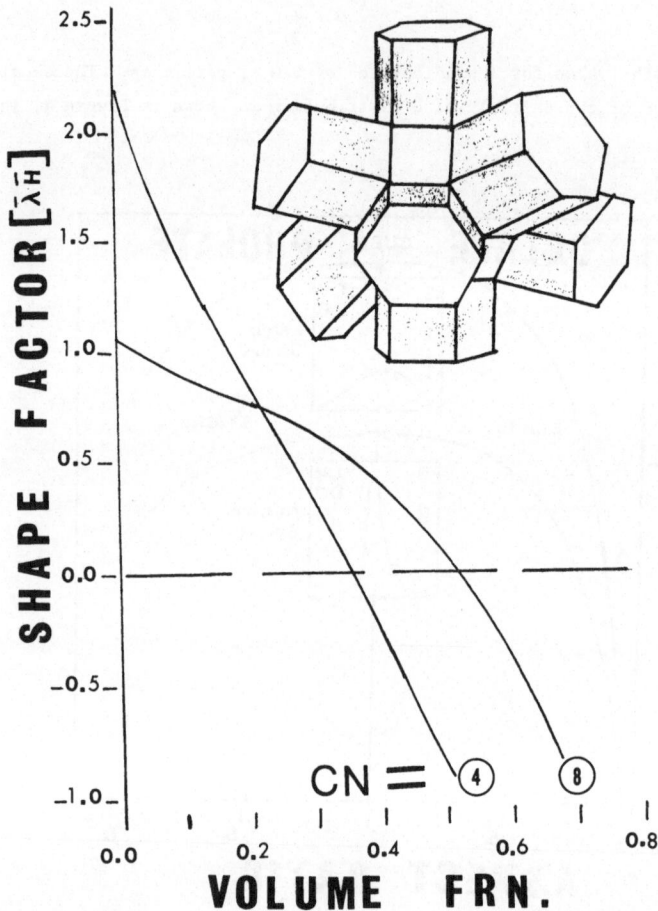

Figure 5. Variation of the $\bar{H}\bar{\lambda}$ shape factor with volume fraction and coordination number for network structures.

from a unit cell composed of solid and porosity (or, alternatively, any two phases). The cell shape is a Kelvin tetrakiadecahedron, chosen because it can be used to fill space. The node is a concentric tetrakiadecahedron; the branches are rectangular parallelapipeds, with square cross sections on square faces, and hexagonal cross sections on the hexagonal faces. The model is versatile, because a variety of networks may be constructed by varying the number of faces connected by prisms (the coordination number), the volume fraction of the solid phase, and the scale of the system. The $\overline{H}\lambda$ product, being unitless, is independent of the scale of the system. Figure 5 shows its dependence upon volume fraction for two different models; CN = 4 is constructed by occupying only four of the eight hexagonal faces with connecting prisms, arranged in a tetrahedral coordination. For the CN = 8 model, all eight faces are connected. Significant differences exist over most of the volume fraction range, so that $\overline{H}\lambda$ might be used to give insight into the coordination number of the network. However, the sensitivity of the model to departures from the shape suggested, and to variations in cell size distribution, has not been assessed for real systems.

Thus, for a variety of reasons, the $\overline{H}\overline{\lambda}$ product, and other shape factors computed from global properties, are inadequate to discriminate between shapes in the absence of additional information. Nonetheless, without the integral mean curvature, the global properties essentially yield no insight into particle shape. It is an essential first step toward understanding, and ultimately deducing, structural shapes.

The Total or Absolute Tangent Count of Sections

In those cases where the inside and outside of a surface cannot be defined, e.g., for the boundaries of space-filling cell structures, the net tangent count cannot be determined. Even in cases where the discrimination may be made, it may be useful for some purposes to ignore the difference between convex and concave tangent counts. The absolute tangent count for such a structure is illustrated in Figure 6.

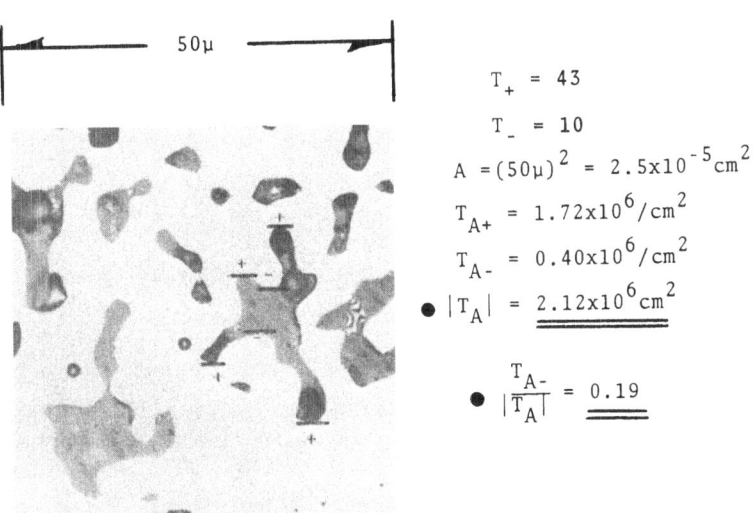

$$T_+ = 43$$
$$T_- = 10$$
$$A = (50\mu)^2 = 2.5 \times 10^{-5} \, cm^2$$
$$T_{A+} = 1.72 \times 10^6 / cm^2$$
$$T_{A-} = 0.40 \times 10^6 / cm^2$$
$$\bullet \; |T_A| = \underline{\underline{2.12 \times 10^6 \, cm^2}}$$

$$\bullet \; |\frac{T_{A-}}{T_A}| = \underline{\underline{0.19}}$$

Figure 6. Procedure for measuring the absolute tangent count.

In this case, the total number of tangents per unit area is determined:

$$|T_A| \equiv T_{A+} + T_{A-} \tag{25}$$

While this quantity is not simply related to the three dimensional geometry from which the section is obtained, it is a potentially useful parameter, containing information that is independent of that in T_{Anet}.

The absolute tangent count is not simply proportional to the absolute value of the integral mean curvature, nor to the integral of the absolute value of the local mean curvature. While the latter is true for convex and concave surface elements, complications appear for saddle surface elements (those elements for which the principal mean curvatures have opposite signs) in the system. For such an element, the direction of the asymptotic line, defined as the direction for which the normal curvature is zero,[10] determines the condition for which the curvature changes direction, Figure 7. If θ_0 is the angle between the asymptotic line for a saddle

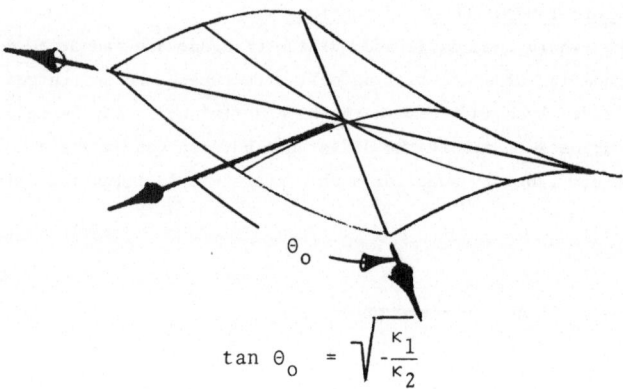

$$\tan \theta_0 = \sqrt{-\frac{\kappa_1}{\kappa_2}}$$

Figure 7. On saddle surface elements, the asymptotic direction is the direction of zero normal curvature.

element and one of its principal normal directions, then it can be shown that

$$|T_A| = \frac{1}{\pi V_0}\{M_{convex} - M_{concave} - M_{saddle} + \frac{4}{\pi}\iint_{saddle}\theta_0 dM + \frac{1}{\pi}\iint_{saddle}\sin 2\theta_0(\kappa_1 - \kappa_2)\} \tag{26}$$

Thus, all that can be said for the stereological meaning of the absolute tangent count is that it is not simple, at least in the general case.* Nonetheless, though

---

*If the saddle surfaces are minimal surfaces, for which $\kappa_1 = -\kappa_2$ at each point, the absolute tangent count indeed yields the absolute value of the integral mean curvature:

$$|T_A| = \frac{1}{\pi V_0}\{M_{convex} - M_{concave}\}$$

the stereological meaning is obscure, the absolute tangent count is a potentially useful parameter, because it contains information not reported by T_{Anet}.

The Tangent Count on Projected Images

Projected images in stereology usually arise in the examination of thin sections or foils viewed in a transmission microscope. Two classes of lineal features appear on such images: (1) boundaries of the projected areal images of particles of a second phase; or, (2) projections of lineal features (space curves) in the section or foil. Information supplied by the tangent count is different in these two cases.

Particle Boundaries. In this case image information obtained in transmission microscopy is less directly related to the global geometric properties of the three dimensional structure than information obtained on a planar section. Potential contributions to the image from truncation (projections from particles with centers outside the slice but cut by the top or bottom surface of the slice) and particle overlap.

If overlap is negligible, and the particles are convex, the tangent count yields[17,18]

$$T_A = \frac{1}{\pi}M_V + 2N_Vt \qquad (27)$$

where N_V is the number of particles per unit volume and t is the thickness of the foil or slice being analysed. In order to extract the three dimensional information, it is necessary to determine T_A on a series of images obtained from samples of different thickness. Equation (27) requires that a plot of T_A versus thickness should be linear with a slope equal to N_V and an intercept equal to M_V/π. Alternatively, for small thickness values (t << $\overline{D}$, the average particle size) the second term contributes negligibly to T_A, and Equation (27) becomes identical to Equation (2) for planar sections.

For a given thickness, overlap increases with the volume fraction of the particles being studied; for a fixed structure, it increases with the sample thickness. If overlap is a significant problem, the only available strategies for extracting three dimensional information require that the structure be modelled, and the model parameters be estimated from section information. Such models usually assume a specific particle shape, with the particle centers randomly distributed in space. A fairly general model for this purpose has been proposed by Miles.[17]

Space Curves. Space curves in a slice project as curved lines on the image, Figure 2. With the usual stereological assumptions about the representativeness of the image, a tangent count performed on such lineal features reports the total curvature of the collection of space curves, Equation (5). It can be shown that the total curvature of a space curve is equal to the total angle of rotation of the tangent vector to the curve; for a collection of space curves,

$$\mathcal{K}_V = \Theta_V = \frac{\Theta}{V_0} \qquad (28)$$

where, more precisely, Θ is the total length of the spherical image traced out by the tangent vector on the sphere of orientation by a point as it traverses the collection of curves contained in the sample volume, V_0.

It may be useful to define the parameter $\Theta/2\pi$, which is the number of times the spherical image traces out a length equal to one rotation on a great circle. A measure of the scale of the lineal structure may then be obtained by dividing the length of the lineal features by the number of rotations of the tangent in the set of curves:[4]

$$\overline{L_\Theta} = 2\pi \frac{L_V}{\Theta_V} \tag{29}$$

where L_V is obtained by applying the standard stereological relation,[11]

$$L_V = \frac{2}{t} P_L(proj) \tag{30}$$

where $P_L(proj)$ is a line intercept count performed on the projected image. Thus, in terms of the measured parameters,

$$\overline{L_\Theta} = 4 \frac{P_L(proj)}{T_A(proj)} \tag{31}$$

Summary

The net area tangent count on a section gives the most direct information about the three dimensional structure being sampled, yielding the integral mean curvature per unit volume. The integral mean curvature is related to mean tangent diameter for convex particles, total length for a collection of tubules, and total perimeter for a collection of muralia. Similar length information is obtained for tubule and muralia networks.

While the integral mean curvature is indispensable in the description of shape with global properties, exploration of the variables affecting shape factors defined from global properties reveals that combinations of V_V, S_V and M_V are inadequate, even for relatively simple model systems. Additional information, in the form of other global measurements or more detailed descriptors, is required in the quest for the adequate description of shape.

The absolute tangent count is a candidate descriptor that may provide further insight into particle shape, but its stereological relation to the three dimensional geometry is complex.

The area tangent count on a projected image of a two phase structure may provide information about both the integral mean curvature and the number of particles per unit volume, provided the structure is simple enough and measurements are made on slices of different thickness. For more complex structures, particle overlap becomes a significant problem, and stereological information can only be extracted through geometric models of the structure.

The area tangent count applied to projected images of lineal features yields the total curvature of the collection of space curves. This information can be combined with estimates of the total length of these lines to provide a measure of the scale of the lineal structure.

References

1. DeHoff, R. T. (1967): _Trans. Met. Soc. AIME_, 239, 610.

2. DeHoff, R. T. and Rhines, F. N. (1968): _Quantitative Microscopy_, McGraw-Hill Co., Inc., New York, N.Y.

3. DeHoff, R. T. (1976): _Fourth Intl. Congress for Stereology_, Natl. Bur. Stds. Spec. Pub. 431, 29.

4. DeHoff, R. T. (1977): submitted to _Jnl. of Microscopy_.

5. Gil, J. and Weibel, E. R. (1972): _Respiration Physiology_, 15, 190.

6. DeHoff, R. T. and Whitney, E. Dow (1976): _Proceedings Intl. Mats. Symp._, Berkeley, Ca.

7. Walther, G. C. (1976): _Kinetics of TiB Formation_, Doctoral Dissertation, University of Florida.

8. Gokhale, A. M. (1977): _A Study of Global Evolution of Microstructures_, Doctoral Dissertation, University of Florida.

9. Rhines, F. N., DeHoff, R. T. and Kronsbein, J. (1969): _A Topological Study of the Sintering Process_, Final Report, AEC Contract No. AT-(40-1)-2581.

10. Struik, D. J. (1950): _Lectures on Classical Differential Geometry_, Addison-Wesley Press, Cambridge, Mass.

11. Underwood, E. E. (1970): _Quantitative Stereology_, Addison-Wesley Press, Cambridge, Mass.

12. Weibel, E. R. and Elias, H. (1967): _Quantitative Methods in Morphometry_, Springer-Verlag, Berlin.

13. Santalo, L. A. (1967): _Studies in Mathematics, Volume 4: Studies in Global Geometry and Analysis_, Prentice-Hall, Inc., 147.

14. Minkowski, H. (1903): _Math. Ann._, 57, 447.

15. DeHoff, R. T. (1975): _Metallography_, 8, 71.

16. Underwood, E. E. (1976): _Fourth Intl. Congress for Stereology_, Natl. Bur. Stds. Spec. Pub. 431, 91.

17. Miles, R. E. (1976): _Fourth Intl. Congress for Stereology_, Natl. Bur. Stds. Spec. Pub. 431, 3.

18. Cahn, J. W. and Nutting, J. (1959): _Trans. Met. Soc. AIME_, 215, 526.

The importance of proper model specification in stereology

R.E. Miles

Dept of Statistics (I.A.S.), Australian National University,
P.O. Box 4, Canberra, A.C.T. 2600, Australia   and
†Anatomisches Institut der Universität Bern, Bühlstrasse 26,
Bern 9, Switzerland

ABSTRACT

It is desired to estimate, for a given phase Y of a 2-phase 3-dimensional

opaque specimen X, its volume fraction V_V, surface fraction S_V and integral

of mean curvature fraction K_V. Various estimates of these resulting from a

random plane section (or sections) of X itself (the RESTRICTED case) are

compared on the basis of bias, mean square error and likely feasibility in

practice.

The same problem is considered in the case where the examinable specimen

is but a random part of a much larger body Z, for which the corresponding

stereological ratios are to be estimated (the EXTENDED case). It is seen that,

although the same stereological formulae appertain, the sampling strategy is

entirely different.

So far, the material of interest has been DETERMINISTIC. However, often in

practice it is clear that it may be supposed RANDOM – the realization of a

homogeneous spatial stochastic process. Then the deterministic extended case

theory extends, with the great practical simplification that the specimen X

need only be arbitrary relative to the material (not uniform random as before).

This theory seems relevant to the current controversy among stereologists

as to the desirability of a completely standardized notation and nomenclature.

It shows that V_V, S_V and K_V may have a number of different interpretations,

so that further specification is essential. Thus, while admitting the advantages

---

† Financial support from the Roche Research Foundation, Basle, during the
  preparation of this paper is gratefully acknowledged.

to be gained from such standardization, it does seem more important that stereo-logical authors, early in their papers, clearly describe the type of model assumed, including a precise definition of the notation and a proper (proba-bilistic) specification of the sections (or projections) taken.

<u>Introduction</u>   Stereology may rather generally be defined as the inference of spatial geometrical structure upon the basis of partial information, such as plane sections or projections onto planes. Basic stereology is concerned with the estimation of simple geometrical properties, such as embedded volume and surface area, and this is the concern of this paper.

The main object is to present a brief, elementary and hopefully lucid account of three different theoretical bases for basic stereology. Any proper such theory must describe the material of interest, especially its precise extent, in sufficient detail and, having done that, properly and completely specify the random test set of the material being sampled. The only type of test set we consider is a planar section. Although much of the theory presented here has been developed in greater detail in a series of papers by myself and P.J. Davy [1, 10, 11], it seemed worthwhile preparing for this interdisciplinary symposium a more elementary version, highlighting the contrasting alternative models.

Part I. RESTRICTED DETERMINISTIC MODELS

Here the specimen X, the sole object of interest, is supposed fixed (i.e., 'deterministic' = non-random), bounded (i.e., may be contained in a sphere of finite radius), and comprises two complementary components or phases, denoted by Y and Y'. Thus in mathematical notation, $X = Y \cup Y' = X$ is the <u>union</u> of Y and Y'. The shape and size of X being assumed directly measurable, our main interest is to investigate, by means of plane sections of X, the (Y, Y') structure of X.

Writing S for the entire three-dimensional space, and X^c for the <u>complement</u> of X ('not X'), we have $S = X^c \cup Y \cup Y'$. We suppose X

contains no 'holes'; if it does, and we denote their union by Y", then

$X = Y \cup Y' \cup Y'' = Y \cup Y'''$, where $Y''' = Y' \cup Y''$; in this way the theory

below extends to the case in which holes are allowed.

An obvious characteristic of interest is $V(Y) = V(X) - V(Y')$, where V

denotes 'volume of'. Other basic characteristics relate to the interface I

between Y and Y', which we assume to be smooth, with possibly a finite length

of sharp, yet smooth, edge. Writing ∂Y for the boundary or surface of Y,

clearly

$$I = \partial Y \cap \partial Y'$$

- I is the <u>intersection</u> of ∂Y and $\partial Y'$. The most obvious characteristic of I

is S(Y), the total area of I. We adopt what at first sight appears a poor

notation to achieve compact uniformity of notation below ('S(I)' would

evidently be more natural than S(Y); moreover, S(Y) as we use it depends also

upon Y'). Reversing roles, $S(Y') = S(Y)$.

To complete the specimen specification, we assume X is 'opaque', that is,

the interior of X is invisible to the investigating stereologist, and that

∂X carries no useful component information or, if it does, this information gives

no clue as to the internal (Y,Y') structure of X. It might be said that X

has been hidden under a coat of un-strip-able paint. This may seem somewhat

unrealistic, but it is necessary to obtain a nice mathematical model. In parts

II and III such information, which only has nuisance value in this part, may have

great practical value. As mentioned, it is assumed that all measurements of the

external shape and size of X can be freely made.

X may be sectioned by any plane T, yielding the section $X \cap T$, in

which it is supposed that all detail of the planar sets $Y \cap T$ and $Y' \cap T$

is discernable. In fact, it is supposed any required measurement of these two

planar sets may be made, for example by means of an automatic image analyser.

Obvious section characteristics are the area $A(Y \cap T)$, $= A(X \cap T) - A(Y' \cap T)$,

and $B(Y \cap T)$, $= B(Y' \cap T)$, the length of the intersection curve $I \cap T$.

Note again the 'poor' notation - $B(Y \cap T)$ depends also on $Y' \cap T$ - adopted

to unify the main formulae below.

<u>Curvatures</u>   We now add another fundamental sectional characteristic, which depends for its definition on the fundamental notion of curvature of a smooth plane curve. Such a curve may be specified by the tangent angle ψ as a function of s, the distance along the curve from some origin. Needed also to complete the specification is, of course, the location of the origin (Fig. 1).

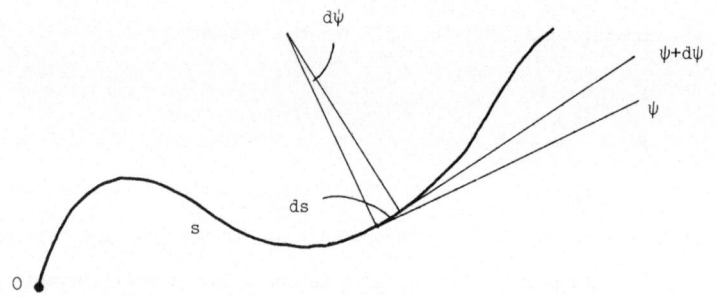

<u>Fig. 1</u>.   The $\psi = \psi(s)$ specification of a planar curve.

It is evident from Fig. 1 that the <u>curvature</u> at the point s,

$\kappa(s) = d\psi(s)/ds,$

is the reciprocal of the (signed) radius of curvature at the point, and hence measures the rate at which the curve is turning at the point; note it may be positive, zero (usually corresponding to a point of inflexion) or negative.

A few words must be included about 'oriented curves' and 'curve orientation'. Essentially these terms indicate that means exist to distinguish the two sides of the curve. For example, in a plane section of (Y,Y'), the interface curve is clearly oriented, for the observer may identify one component from the other. On the other hand, for a simple curve drawn on a sheet of paper, there is no way of distinguishing the two sides. Thus each unoriented curve corresponds to two oriented curves. All the curves considered in this paper are oriented, being plane sections of (Y,Y') interfaces. It only remains to specify the s, ψ coordinates in terms of (Y,Y'). We suppose that, as we move along the curve with s increasing, Y is on the left and Y' on the right; also, as in Fig. 1, ψ increases anticlockwise. With this convention, κ is positive when $Y(Y')$ is convex (concave) at the point in question. (Note that the same holds when the directions of both s and ψ are reversed.)

Now, integrating with respect to s along the curve, we get the total curvature (or 'integral of curvature') between points s_0 and s_1

$$C(s_0,s_1) = \int_{s_0}^{s_1} \kappa(s)ds = \int_{s_0}^{s_1} \{d\psi(s)/ds\} \, ds$$
$$= \int_{s_0}^{s_1} d\psi(s) = \psi(s_1) - \psi(s_0) \ ,$$

i.e., the net angle the tangent vector turns through between the initial and final points s_0 and s_1. C is not to be confused with the 'integral of absolute curvature' $\int |\kappa(s)| ds$ (see Chapter 10 by DeHoff in [2]), which is not considered in this paper. The definition of C extends easily to the case where the curve may have corners, the contribution of each corner being simply $\pm$ (π minus the angle at the corner), depending upon the orientation adopted. In the important special case of a simple closed curve, a curve which returns to its initial point and tangent without crossing itself on the way, C is simply $\pm 2\pi$, again depending upon the orientation; with the (Y,Y') orientation convention defined above, C is 2π or -2π according as the curve encloses Y or Y'. Now we are in a position to define the third fundamental characteristic of the oriented interface section I ∩ T. It is the sum C(Y ∩ T) of the total C values over all curves (and corners) in I ∩ T. It is important here that all such distinct curves are given the correct orientations, so that the corresponding contributions to C are given the correct signs. For example, when Y ∩ T comprises N separate Y- particles bounded by simple closed curves, C(Y ∩ T) = 2πN; if the particles contain altogether N' similar Y' particle holes, C(Y ∩ T) = 2π(N-N').

A, B and C as above defined constitute a fundamental set of planar characteristics, with length dimensions 2, 1 and 0, respectively. There is a corresponding set of four characteristics in space, the first two of which are V, S (dimensions 3, 2) and the remaining two are defined in terms of curvatures as follows. At a smooth point x of I, construct the normal n(x) to I outwards from Y, and a plane P through x and n(x). Now I ∩ P is a curve having a well-defined curvature at x, thanks to our convention regarding (Y,Y') curvature sign. Thus, as P is rotated about n(x), we obtain a curvature function κ(θ) over a range π. It turns out that, with appropriate choice of θ-origin,

$$\kappa(\theta) = \kappa_1 \cos^2\theta + \kappa_2 \sin^2\theta \qquad (0 \leq \theta < \pi)$$

where we may suppose $\kappa_1 \geq \kappa_2$. This is the classical Euler's theorem of differential geometry [13; p.81]. It implies in particular that, when $\kappa_1 \neq \kappa_2$, κ assumes its minimum and maximum values at orientations perpendicular to each other. κ_1, κ_2 are called the principal curvatures of the surface at x. In turn, $\frac{1}{2}(\kappa_1 + \kappa_2)$ and $\kappa_1 \kappa_2$ are called the mean curvature and gaussian curvature at x, respectively. The other two spatial characteristics are their surface integrals over I : the integral of mean curvature

$$K(Y) = \int_I \tfrac{1}{2}(\kappa_1 + \kappa_2)\,dS$$

and the integral of gaussian curvature

$$G(Y) = \int_I \kappa_1 \kappa_2 \, dS.$$

Note that, by exchanging the roles of Y and Y', i.e. the orientation, κ is reversed in sign, so that K(Y') = -K(Y) and G(Y') = G(Y).

In special cases, these characteristics reduce to better-known quantities:

(i)   If Y is convex, K(Y) = 2π H(Y), where H(X) denotes the mean caliper diameter of X, i.e. the average over all orientations of the total length of the orthogonal projection of X onto a line.

(ii)   If the interface I is in equilibrium under surface tension τ and a pressure difference Δp, then the mean curvature is constant over I, so that K(Y) ∝ S(Y). In fact, K(Y) = (Δp/2τ)S(Y).

(iii)   If Y comprises N 'particles' bounded by closed smooth surfaces, then G(Y) = 4πN. If these particles contain N' similar Y' holes, then G(Y) = 4π(N+N'); and so on. The sign difference between two (N−N') and three (N+N') dimensions should be noted.

The fundamental characteristics in two and three dimensions are summarized in the following table.

Length dimension	L^3	L^2	L	1
Plane		$A(Y \cap T)$	$B(Y \cap T)$	$C(Y \cap T)$
Space	$V(Y)$	$S(Y)$	$K(Y)$	$G(Y)$

'Fundamental' here is in the sense of integral geometry - in n dimensions there are $n+1$ corresponding characteristics [1]. The reader might investigate the simplest case $n = 1$.

It should be mentioned that there is a somewhat different set of characteristics, the respective dimensional contents, which are fundamental in the sense of solid geometry. In 3 dimensions they are: volume V of region, area S of surface, length L of curve, and number N of points. In n dimensions there are also $n+1$ of them, the first two of which coincide with the first two of the $n+1$ integral geometric characteristics. Fundamental formulae of stereology (see below) exist for them too [7; equation (2.16)]. Actually the fundamental formulae corresponding to both sets can be combined into a single formula, given in the Appendix of [1].

Sectioning planes. We need to be able to specify the (unoriented) sectioning planes mathematically. Three parameters are required to specify a plane T, e.g. (α, β, γ) if the plane has equation $(x/\alpha) + (y/\beta) + (z/\gamma) = 1$. The most useful parametrization is (p, θ, ϕ), where (θ, ϕ) are the spherical polar coordinates of the normal to the plane, and p is its perpendicular distance from the origin. This parametrization will be utilized in the Appendix, but in this main body of the paper we only need an arbitrary parametrization

$$T = T(t), \qquad \text{where} \quad t = (t_1, t_2, t_3).$$

Thus the set of planes in space corresponds to the set of points t in some three-dimensional set T. We are only interested in planes which hit X, thus in the subset $T(X)$ of T corresponding to this condition.

Assuming we may section X by any of the planes through it, which should we choose? This question is akin to that of sampling n members from a

population of size N, as in an opinion poll. The answer there is to choose a
random n-sample, usually in a uniform (i.e., with equal probabilities) way. By
so uniformly randomizing, all population members have an equal chance of being
chosen, and so contribute 'equally' to the final estimates. This suggests that
we should take a random plane section $\dot{T}$ of X, corresponding to a random point
$\dot{t}$ of $T(X)$, in such a way that different localities of X contribute 'equally'.
[In this paper, we 'dot' anything random, i.e. something that will assume a
sequence of (independent, identically distributed) random values, conforming to
a probability distribution, when repeated independently; e.g. the number on a die,
or the location of Buffon's needle.] In this way, the geometrical measurements
we make on sections become random, or random variables: $\dot{A}(Y \cap \dot{T})$, $\dot{B}(Y \cap \dot{T})$,
$\dot{C}(Y \cap \dot{T})$; each of which has a probability distribution. Each such probability
distribution depends on two things:

(i)   the geometry of X and Y,

(ii)   the distribution of $\dot{t}$ in $T(X)$.

(i) is given, but we assume we have complete control in (ii) and can choose any
one of the infinity of different distributions of $\dot{t}$ in $T(X)$. Thus which should
we choose? To be of any real utility, we need a $\dot{t}$-distribution defined not just
on specific $T(X)$'s, but on all $T(X)$'s for varying X.

It so happens that there exists a basic type of random plane section of X,
for all X, called an IUR (isotropic uniform random) plane section, which has
the important property that $Pr(\dot{T}$ hits $Y)$ does not depend upon the position of
Y within X, a property which uniquely determines the distribution of $\dot{t}$. In
fact [8, 10]

$$Pr(\dot{T} \text{ hits } Y) = H(Y)/H(X), \tag{1}$$

which clearly satisfies the above property. Another basic property of IUR
sections is that, if it is given that $\dot{T}$ hits Y then, conditional upon this
event, $\dot{T}$ is IUR in Y. We include in the Appendix a practical method of
generating IUR plane sections.

It is clear from these properties that any locality of X is hit 'with the
same probability' and 'in the same way' by an IUR $\dot{T}$. Hence since we are

interested in estimating characteristics which do not depend upon the position of Y within X, it is natural to suppose that an IUR section is the appropriate or best one to use. Surprisingly, this is usually not so, as we shall see.

Basic formulae. First we determine the mean value of $\dot{\alpha}(Y \cap \dot{T})$ for $\dot{T}$ IUR in X, where α denotes anyone of A, B or C. The corresponding values of $\dot{\alpha}(X \cap \dot{T})$ will then follow by simply taking $Y = X$ less its boundary $\partial X, Y' = \partial X$. Let dV be a small volume element of X, and $d\dot{\alpha}$ be the contribution of dV to $\dot{\alpha}$; this contribution is zero if $\dot{T}$ does not hit dV. Now we have the mean value

$$
\begin{aligned}
E(\dot{\alpha}) &= E\left(\int_X d\dot{\alpha}\right) && \text{(an expectation of a sum is always the} \\
& && \text{sum of the expectations)} \\
&= \int_X E(d\dot{\alpha}) && \text{(changing the order of integration)} \\
&= \int_X E(d\dot{\alpha}|\dot{T} \text{ hits } dV)\; \Pr(\dot{T} \text{ hits } dV), && (2)
\end{aligned}
$$

where $|$ denotes 'conditional upon', the last equality being a basic probability decomposition rule [4; Chapter 5]. Now by (1)

$$
\Pr(\dot{T} \text{ hits } dV) = H(dV)/H(X) \tag{3}
$$

while, by basic geometrical probability (or integral geometry) [1, 10],

$$
E(d\dot{\alpha}|\dot{T} \text{ hits } dV) = \gamma(dV)/H(dV) \tag{4}
$$

where the three (α, γ) pairings are

α	A	B	C
γ	V	$\frac{\pi}{4}S$	K

Substituting (3) and (4) into (2), we have

$$
\begin{aligned}
E(\dot{\alpha}) &= H(X)^{-1} \int \gamma(dV) \\
&= \gamma(Y)/H(X).
\end{aligned} \tag{5}
$$

Estimators and estimation. We shall now introduce the notion of 'estimators' and their desirable properties, and then show that (5) is our first stereological result. Let θ denote an unknown parameter, for example $V(Y)$, $S(Y)$, $K(Y)$ or $G(Y)$ in our model. We are at liberty to adopt any random variable $\dot{x}$ resulting from some random experiment, e.g. measurements on our random plane section $\dot{T}$, as an estimator of θ. Thus the term 'estimator' by itself is almost without substance. Substance is provided by properties of estimators or, more precisely, properties of their distributions in relation to the unknown θ (see Fig. 2).

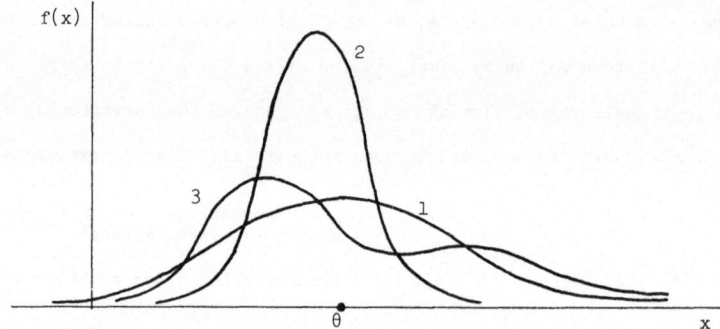

Fig. 2. The probability densities of three estimators of an unknown parameter θ.

These properties are generally expressed in terms of the random variable

$$\dot{e} = \dot{x} - \theta,$$

the error committed in using $\dot{x}$ as an estimator of θ. The associated bias is the average error

$$b = E(\dot{e}) = E(\dot{x}-\theta) = E(\dot{x})-\theta$$

which may be positive, zero or negative; $\dot{x}$ is an unbiased estimator of θ if $b = 0$, a highly desirable estimator property. If $\dot{x}$ is biased, $\dot{x}-b$ (whose distribution is that of $\dot{x}$ shifted b to the left) is unbiased, but of course b is usually unknown.

Thus (5) furnishes our first stereological result: an unbiased estimator or $\gamma(Y)$ is $H(X)\dot{\alpha}(Y \cap \dot{T})$ with respect to an IUR plane section of X. In practice first $H(X)$ is measured externally on X, then after sectioning $\dot{\alpha}(Y \cap \dot{T})$ is measured on the section. Here, and below, we find we can estimate $V(Y)$, $S(Y)$, $K(Y)$ but not $G(Y)$. It seems likely that $G(Y)$ cannot be estimated by means of a single random plane section, but I am not aware of a proof. $G(Y)$ can be estimated from the information on a random wedge section close to the edge [12].

Bias by itself is insufficient as a criterion of a good estimator, since it says nothing about the spread of the distribution of $\dot{x}$ about θ, which is clearly to be reduced as far as possible. The most useful measure of spread is the <u>mean</u> <u>square</u> <u>error</u>

$$MSE = E(\dot{e}^2) = E\{(\dot{x}-\theta)^2\},$$

which equals the variance $Var(\dot{x})$ of $\dot{x}$ if $\dot{x}$ is unbiased. The MSE is always non-negative, and a zero MSE implies that $\dot{x} = \theta$ with probability 1 - the perfect estimator! By Schwartz's Inequality, $MSE \geq b^2$, so each estimator corresponds to a point above the parabola in Fig. 3. By our criteria, $\dot{x}_1$ is

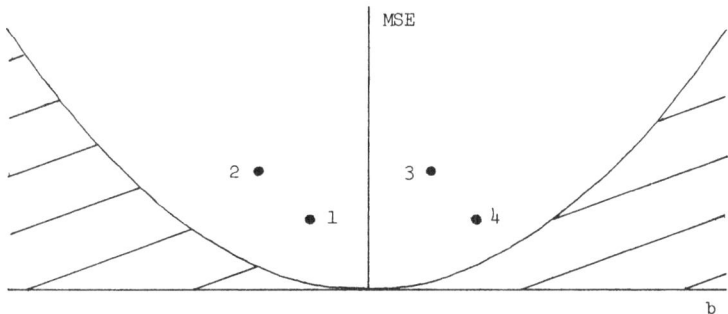

Fig. 3. The possible joint values of (b, MSE)

preferred to $\dot{x}_2$ if $|b_1| < |b_2|$ and $MSE_1 < MSE_2$ (left side of Fig. 3), but $\dot{x}_3$ may or may not be preferred to $\dot{x}_4$ if $|b_3| < |b_4|$ but $MSE_3 > MSE_4$ (right side of Fig. 3). For example (Fig. 2), a biased estimator (2) may be preferred to an unbiased estimator (1).

Let us briefly consider replication of the experiment n times (independently), and consider properties of the natural resulting arithmetic mean estimator

$$\bar{\dot{x}} = (\dot{x}_1 + \ldots + \dot{x}_n)/n$$

of θ. Its bias

$$b_n = E(\bar{\dot{x}} - \theta) = b_1 \tag{6}$$

but its MSE

$$MSE_n = E\{(\overset{\cdot}{\overline{x}} - \theta)^2\} = n^{-2}E[\{\Sigma(\dot{x}_i - \theta)\}^2]$$
$$= n^{-1} MSE_1 + (1 - n^{-1}) b_1^2 \qquad (7)$$

which tends to b_1^2 as $n \to \infty$. Thus, when independent replication is feasible, unbiasedness may well assume a much greater importance than low MSE.

With these criteria in mind, let us return to our geometrical model. We must clearly ask how sizeable is the MSE of $H(X)\dot{\alpha}(Y \cap \dot{T})$ as an unbiased estimator of $\gamma(Y)$. Intuitively it will be quite large, since much of the variation of $H(X)\dot{\alpha}(Y \cap \dot{T})$ stems from the considerable variation of $\dot{A}(X \cap \dot{T})$. The obvious way of reducing this source of variation is to consider a ratio estimator like $\dot{\alpha}(Y \cap \dot{T})/\dot{A}(X \cap \dot{T})$. 'Dividing' (5) by the special use $E\{\dot{A}(X \cap \dot{T})\} = V(X)/H(X)$ of (5), we are led to suspect that the random variable estimator $\dot{\alpha}(Y \cap \dot{T})/\dot{A}(X \cap \dot{T})$ is 'of the same order' as the unknown constant $\gamma(Y)/V(X)$. This corresponds to the fundamental formulae of stereology (FFS) '$\alpha_A = \gamma_V$', which is written in the conventional stereological shorthand $(\alpha_A = \alpha/A, \gamma_V = \gamma/V)$. However, such division is impermissible. In fact, the bias of $\dot{\alpha}_A$, with respect to an IUR plane section, as an estimator of γ_V is

$$b = E(\dot{\alpha}_A) - \gamma_V$$
$$= \{E(\dot{\alpha}_A)E(\dot{A}) - E(\dot{\alpha})\}/E(\dot{A})$$
$$= -\{H(X)/V(X)\} Cov(\dot{A}, \dot{\alpha}_A),$$

where the covariance $Cov(\dot{x}, \dot{y}) = E[\{\dot{x} - E(\dot{x})\}\{\dot{y} - E(\dot{x})\}]$ is a measure of the correlation of $\dot{x}$ and $\dot{y}$. Thus b may be positive or negative, depending upon the geometry of X and Y. For example, if Y is concentrated in the periphery of X, $\dot{\alpha}_A$ will tend to be large when $\dot{A}$ is small, yielding a negative covariance and positive bias. Conversely, if Y is central within X, the bias may be expected to be negative.

Note the two ways in which the estimator $\dot{\alpha}_A$ of γ_V may be viewed:

(i) as a simple estimator of the 'density' γ_V,

(ii) as an estimator of $\gamma(Y)$, where we assume there is little difficulty in directly measuring $V(X)$.

Other types of random plane section. An IUR plane section of X corresponds to a particular probability density $f_X(t)$ on $T(X)$, whose explicit form is easily derived [8, 10], but not required here. This may be used as a 'base' density, since the density of any other type of random plane section of X is $f_X(t)$ weighted by a weight function $W(t) \geq 0$ on $T(X)$. Thus we define a W-weighted plane section of X to be one having density

$$f_X^{(W)}(t) = c \ W(t) \ f_X(t)$$

on $T(X)$. Since probability densities must integrate to 1 (unit probability),

$$c = 1/\iiint W(t) \ f_X(t)dt = 1/E(\dot{W})$$

where here and elsewhere undecorated 'E' denotes an expectation with respect to an IUR section. Denoting an expectation with respect to a W-weighted section by $E^{(W)}$ (thus $E^{(1)} = E$), we have for any random variable sectional measurement $\dot{x}(t)$

$$E^{(W)}(\dot{x}) = \iiint x(t) \ W(t) \ f_X(t)dt/E(\dot{W})$$

$$= E(\dot{x}\dot{W})/E(\dot{W}).$$

Hence the bias in using $\dot{\alpha}_A$ with respect to a W-weighted section as an estimator of γ_V is

$$b^{(W)} = E^{(W)}(\dot{\alpha}_A) - \gamma_V$$

$$= \frac{E(\dot{\alpha}\dot{W}/\dot{A})}{E(\dot{W})} - \frac{E(\dot{\alpha})}{E(\dot{A})}$$

$$= \text{Cov}\left\{\frac{\dot{W}}{E(\dot{W})} - \frac{\dot{A}}{E(\dot{A})}, \ \dot{\alpha}_A\right\} . \tag{8}$$

Thus the bias is zero for A-weighted random plane sections, for all X and all Y within X. That is,

$$E^{(A)}(\dot{\alpha}_A) = \gamma_V$$

for all X and all Y within X, a rigorously correct form of the FFS. The practical details of how to generate an A-weighted section are given in the Appendix. Suffice it to say here, the difficulty is of the same order as that of generating an IUR section. An advantage of A-weighted over IUR sections is that grazing sections (small A) are much less common.

$\dot{\alpha}_A$ for an IUR section could yet be preferred to $\dot{\alpha}_A$ for an A-weighted section if its MSE were sufficiently smaller. We have

$$\text{MSE } \dot{\alpha}_A - \text{MSE}^{(A)}\dot{\alpha}_A = E\{(\dot{\alpha}_A - \gamma_V)^2\} - \frac{E\{\dot{A}(\dot{\alpha}_A - \gamma_V)^2\}}{E(\dot{A})}$$

$$= - \text{Cov}\{\frac{\dot{A}}{E(\dot{A})}, (\dot{\alpha}_A - \gamma_V)^2\}. \tag{9}$$

For 'homogeneously and isotropically distributed' Y within X, we expect that the larger the section area $\dot{A}$, the more accurately will the FFS hold. Hence the expectation for such homogeneous and isotropic media is that the right side of (9) will tend to be positive, so that $\text{MSE}^{(A)}\dot{\alpha}_A < \text{MSE}\dot{\alpha}_A$.

Now a brief comparison of the three estimators of γ_V derived above:

(i) $\{H(X)/V(X)\}\dot{\alpha}(Y \cap \dot{T})$ (IUR section)

(ii) $\dot{\alpha}_A$ (IUR section)

(iii) $\dot{\alpha}_A$ (A-weighted section)

(i) is unbiased but usually has an unacceptably large MSE. This MSE should usually be reduced by using the ratio estimator (ii), at the cost of introducing bias. Unbiasedness is recovered in (iii), with the expectation of even further reduced MSE. Hence, since A-weighted sections are reasonably easy to generate, (iii) should usually be the preferred estimator.

There are in fact positions of T for which particular cases of $\alpha_A = \gamma_V$ hold exactly. If we ascribe probability 1 to such a position, then we have the perfect estimator (b = MSE = 0)! However, this might be described as 'begging the question', for we have no way of knowing such positions from the purely external information about X available. Moreover, we seek universality: theory valid for all X, and all Y within X.

Replication. Only in the case of a sufficiently homogeneous and isotropic specimen (X, Y) can an estimator based on a single random plane section of X be expected to be particularly accurate. Hence often we should like to be able to glue X back together, take another random section, glue it back together, and so on. For theoretical simplicity, just as we assumed ∂X bore no (Y,Y') information, suppose glueing is feasible. How then should we take the sections and combine the measurements into a final overall estimator?

The obvious way is to take n independent estimators $\dot{\alpha}_A(i)$ $(i = 1,\ldots,n)$ of the type (iii) above, and take their arithmetic mean

$$\dot{\alpha}'_A = n^{-1} \sum_i \dot{\alpha}_A(i).$$

By (6) and (7), this estimator is also unbiased, with MSE (= variance) $\propto n^{-1} \to 0$ as n increases. However, this estimator does not seem quite satisfactory, because it ascribes equal weight to all the $\dot{\alpha}_A(i)$, regardless of whether the corresponding $\dot{A}(i)$ is large or small. We should prefer to ascribe greater weight to those sections with larger $\dot{A}(i)$. A natural way of so doing is to ascribe $\dot{A}(i)$ itself as weight to each section - then each measurement $\dot{\alpha}(i)$ contributes to the final estimator independently of the actual section size it happens to be on. This estimator is

$$\dot{\alpha}''_A = \sum \frac{\dot{A}(i)}{\Sigma \dot{A}(i)} \dot{\alpha}_A(i) = \frac{\Sigma \dot{\alpha}(i)}{\Sigma \dot{A}(i)} \quad .$$

We must specify the joint distribution of $\dot{T}(1),\ldots,\dot{T}(n)$. The most general one is independent IUR weighted by some function $W = W\{t(1),\ldots,t(n)\}$. With this joint plane distribution, the bias of $\dot{\alpha}''_A$ as an estimator of γ_V is

$$\mathrm{Cov}\left\{ \frac{\dot{W}}{E(\dot{W})} - \frac{\Sigma \dot{A}(i)}{nE(\dot{A})} , \frac{\Sigma \dot{\alpha}(i)}{\Sigma \dot{A}(i)} \right\} ,$$

generalizing (8). Thus unbiasedness holds when the n planes are taken $\{\Sigma A(i)\}$-weighted. This means, in particular, that the n section planes are dependent; hence that the generation of such planes is extremely difficult, if not impossible, in practice.

However, there turns out to be a whole class of joint plane distributions for which $\dot{\alpha}''_A$ is unbiased. For when

$$W = \Sigma \lambda_i A(i)$$

with any non-negative constant weights λ_i having non-zero sum, the bias is

$$\mathrm{Cov}\left\{ \frac{\Sigma \lambda_i \dot{A}(i)}{(\Sigma \lambda_i)E(\dot{A})} - \frac{\Sigma \dot{A}(i)}{nE(\dot{A})} , \frac{\Sigma \dot{\alpha}(i)}{\Sigma \dot{A}(i)} \right\}$$

which, by symmetry,

$$= \frac{1}{n} \, \text{Cov} \left\{ \frac{(\Sigma \lambda_i) \Sigma \dot{A}(i)}{(\Sigma \lambda_i) E(\dot{A})} - \frac{\Sigma \dot{A}(i)}{E(\dot{A})} \, , \, \frac{\Sigma \dot{\alpha}(i)}{\Sigma \dot{A}(i)} \right\} \, = \, 0.$$

When only one of the λ_i is non-zero, the planes become independent and generateable, as when $\dot{T}(1)$ is $A(1)$-weighted and $\dot{T}(2),\ldots,\dot{T}(n)$ are each IUR. This beautiful result, first observed by Midzuno [6], is of great generality, applying whenever one has a sequence of independent identically distributed ratio estimators. For example, it may be used for estimating, by 'point-counting' with an overlay grid of points, the value of A_A for the actual planar sections we have so far implicitly assumed could be mensurated exactly. In theory, if not in practice, this procedure should be used in ratio measurements made by automatic image analysers.

For small n, the Midzuno estimator would seem to be the best usually available. However, for large n, there is little to choose between any of these estimators, for both the bias and the MSE are of order n^{-1}; then the bias is negligible compared to the standard error $\sqrt{MSE}$, and so may be ignored.

In the last two parts, essentially the model of Part I is reversed: the specimen becomes random, and the section deterministic.

Part II. EXTENDED DETERMINISTIC MODELS

Here the body Z of material of interest is still deterministic (non-random), and comprises two components Y and Y' $(Z = Y \cup Y')$. We wish to estimate the ratios $\gamma_V = \gamma(Y)/V(Z)$.

The difference with Part I is that the specimen X available for examination, and upon which estimates are to be based, is but a portion sampled from Z. With no assumptions made about the spatial distribution of Y within Z, we are left with few options: with the shape and size of X fixed but quite arbitrary, the location and orientation of X within Z must be IUR, i.e. the centroid of X must be uniformly random over Z and (independently) the orientation of X must be isotropic random. It should be noted that IUR has somewhat different meanings here and above: isotropy of orientation is in terms of 3 parameters (2 above), uniformity of location 3 parameters (1 above). Thus the specimen X is in fact a random set $\dot{X}$ with random (Y, Y') constitution. Finally, to render edge effects due to $\dot{X}$ intersecting ∂Z negligible, we must suppose that the size of $\dot{X}$ is very much smaller than that of Z (actually this requirement can be avoided: see [11]).

In practice, to 'break up' Z and choose one of the resulting fragments at random is usually invalid as a method of sampling an IUR $\dot{X}$, for the breakage surfaces will usually depend in some way upon the component interfaces, so that the fragment will not be IUR.

Although so far we have had to be rather restrictive, the surprising thing is that, having satisfied the above conditions, the sampling of the specimen $\dot{X}$ is very much easier than in Part I, where we had to generate a specific type of random plane section. Here we have, so to speak, already accomplished the difficult sampling by arranging that $\dot{X}$ is IUR in Z. The crux is that IUR implies that, if T is an arbitrary planar section of $\dot{X}$ (i.e., chosen in any way which takes no account of the random (Y, Y') structure of $\dot{X}$), then $\dot{X} \cap T$ is equivalently an IUR plate $\dot{U}$ of fixed shape and size in Z. Analogous to (2),

$$E\{\dot{\alpha}(Y \cap \dot{U})\} = \int_{Z} E(d\dot{\alpha} | \dot{U} \text{ hits } dV) \, Pr(\dot{U} \text{ hits } dV).$$

The value of the conditional expectation is again given by (4), and it is fairly easily shown that

$$Pr(\dot{U} \text{ hits } dV) = A(\dot{U})H(dV)/V(Z).$$

Hence

$$E\{\dot{\alpha}(Y \cap \dot{U})\} = A(\dot{U})\gamma(Y)/V(Z).$$

That is, for arbitrary planar sections of X,

$$E(\dot{\alpha}_A) = \gamma_V.$$

Thus the FFS again hold, but for a different model!

The arbitrariness of T gives us great freedom, meaning in particular that we may polish the surface of $\dot{X}$ in some convenient place down to a plane face and take this as section. It is emphasized that the arbitrariness of T means that we must not be influenced in any way by what (Y, Y') structure we 'see', in deciding when to stop polishing.

Since all arbitrary sections yield unbiased forms of the FFS, we naturally turn our attention to the MSE (= variance here) of $\dot\alpha_A$. In the simplest (but typical) case, $\alpha = A$, the corresponding FFS being the 1847 Delesse [3] relation $E(\dot A_A) = V_V$, we have the expression, for a particular sized and shaped plate U of section

$$\text{MSE}^U(\dot A_A) = \int_0^\infty f^U(r)\rho(r)\,dr \qquad (10)$$

where

$\rho(r) = $ Pr(the 2 end points of a 'Buffon needle' of length r positioned IUR in Z both lie in component Y) $- V_V^2$

and

$f^U(r)$ is the probability density of the distance between 2 independent uniform random points in U.

We seek to choose U from the possible sets $X \cap T$ (for varying T) so as to minimize the right side of (10). Since $\rho(0) = V_V(1-V_V)$, and usually $\rho(r) \downarrow \rho(\infty) = 0$ as r increases, we should choose a U whose associated density $f^U(r)$ is concentrated on as large r-values as possible. Thus U should be chosen not only large in area, but also 'extensive'. We conclude that, even though the FFS hold unchanged, the optimum method of sampling the specimen is entirely different from that of Part I.

Of course, the great practical problem involved in this theory is that of actually securing an IUR $\dot X$ from Z. Presumably if it can be done, $V(Z)$ can also be measured, so that this method allows the estimation of both γ_V and of $\gamma(Y)$. (The estimation of $\dot\gamma(\dot X)$ is of no interest, because X is a random set.) In the final part, we discuss a natural way out of this problem.

Part III. RANDOM MODELS

In Part II the random location of the specimen $\dot X$ within the body Z of interest endows the specimen with <u>random structure</u>. The same is achieved by supposing the specimen is a fixed subset in space, but that the material itself is random - in technical language a 2-state (i.e., Y and Y') stochastic process (SP) in S. In order to be able to define the corresponding characteristics γ_V for the SP, it is necessary to suppose that the SP is homogeneous and isotropic, i.e. invariant under arbitrary translations and rotations of the coordinate frame (this implies that the SP must extend throughout S, a fact that considerably raises the mathematical level of difficulty, but which also serves to eliminate edge effects!). With the additional condition of ergodicity (which effectively prevents our SP being a random mixture of other homogeneous SP's), we have that $\dot{\gamma}_V(r)$ for the SP, defined as in Part I over a sphere of radius r and arbitrary fixed centre, tends almost surely as $r \to \infty$ to a constant, which we naturally define to be γ_V, an 'ergodic limit'. For brevity, write HIESP for a homogeneous and isotropic ergodic SP. The 'natural stereology' for a HIESP is an arbitrary plane section, which itself is a HIESP over the entire two-dimensional space. It may be proved that, with α_A defined in this plane analogously to γ_V in S,

$$\alpha_A = \gamma_V,$$

yet another version of the FFS. No expectation E is required here because both sides are constant ergodic limits.

However, we are concerned with the case in which the specimen X is the realization of the HIESP within an arbitrary fixed subset X (not random, as in Part II). Not only is the (Y,Y') structure of the specimen X random but, due to the homogeneity and isotropy, the 'probability distribution' of its structure relative to the geometry of X is independent of the location or orientation of X in S. As usual, we section X by an arbitrary plane T and, just as in Part II, we may prove for X $\cap$ T

$$E(\dot{\alpha}_A) = \gamma_V \qquad (11)$$

yet another instance of the FFS. The MSE (= variance) has a similar expression to (10) and a similar minimization theory. Since the Z of Part II has become in effect an infinite region of a SP here, (11) yields only an estimator of γ_V for the SP - nothing else. For example, to estimate $\gamma(X)$ we must clearly revert to a theory like that of Part I.

The basic difference to Part II is that X is arbitrary (easy to effect) and
not IUR (often extremely difficult to effect). The cost of this advantage is the
change from no assumption regarding (Y,Y') structure to the assumption that it
constitutes a HIESP. Often the validity of this assumption is 'self-evident' from
the 'appearance' of the material, e.g. in 3 dimensions the particulate structure
of a metal, or in 2 dimensions the surface of flat sand on a sea-shore. Other
materials (e.g. liver or lung at the electron microscopic level) may appear to be
'homogeneous' but, with any doubt at all, it is probably best to proceed as in
Part II, preferably with independent replication of sections, which enables an
estimate of the error variance [9]. Even if the assumption is satisfied,
sufficient of the material must be incorporated into the specimen, so that the
ergodic limits are sufficiently closely approached. Again, even if the material
appears to be homogeneous, there is a danger that there may be a very gradual
trend; accounting for it will necessitate departing from a totally arbitrary
sampling, e.g. by sampling from a good selection of widely dispersed localities,
arbitrarily in each. For further details of the general use of HIESP models, the
reader is referred to Matheron [5].

Conclusion. We have seen that rather different models all give, at least super-
ficially, the same result (the 'fundamental formulae of stereology'). However,
the sampling procedures are radically different in all cases. The moral is that
the stereologist should think very carefully which of these models is the
appropriate one for him (or if none is, he should similarly formulate and develop
yet another model), and sample accordingly. The latter may well mean that tables
of random numbers must be used to generate uniform or isotropic random quantities.
It is especially important that he should properly define all variables, e.g. is
γ_V based on the specimen, a larger body, or an HIESP?

Finally, a comment regarding the current controversy among stereologists
about the desirability of a standardized notation in stereology. Of course this
would be nice, especially as the discipline is of interest to workers in widely
diverse fields, but it seems to me much more important that all notation used
be defined precisely. Again, anyone using a stereological formula should fully

understand the definitions of the quantities involved - if he can do that, then he can surely have no problems over what particular notation is used.

APPENDIX

How to generate IUR and A-weighted random plane sections in practice.

In these stochastic constructions, we need a specific parametrization of planes rather than the general $t = (t_1, t_2, t_3)$ above. Let n be the perpendicular from the origin O to the plane, regarded as a vector, and let $p(\geq 0)$ be its length. Suppose n makes an angle θ with the z-axis $(0 \leq \theta \leq \pi)$, and that the perpendicular projection n' of n onto the (x,y)-plane makes a clockwise angle ϕ with Ox $(0 \leq \phi < 2\pi)$. Then (θ, ϕ) are spherical polar coordinates and the parametrization (p, θ, ϕ) of planes permits the following stochastic constructions.

IUR plane through X:

1. 'Embed' X (tightly) in a (fictitious) sphere $Q(r)$ of radius r, the centre of which is taken as origin.

2. Select an isotropic random direction $(\dot{\theta}, \dot{\phi})$ as follows: (a) $\dot{\phi}$ is uniform on $(0, 2\pi)$ and hence may be chosen directly by random numbers, (b) $\dot{\theta}$ has density $\frac{1}{2}\sin\theta$ on $(0, \pi)$, and hence may be chosen by random numbers (specifically $\cos^{-1}(1 - 2\dot{x})$, where $\dot{x}$ is uniform on $(0,1)$).

3. $\dot{p}$ is chosen uniform on $(0, r)$.

4. Now the plane $(\dot{p}, \dot{\theta}, \dot{\phi})$ hits $Q(r)$ - accept it if it also hits X.

5. If it does not hit X, repeat the steps 1-4 until the plane does hit X - this plane is then IUR in X.

A-weighted plane through X:

1. Embed X (tightly) in a rectangular box R of sides a, b, c, and take axes $Oxyz$ parallel to the box edges.

2. Generate a uniform random point $(\dot{x}, \dot{y}, \dot{z})$ in R by random numbers.

3. Accept this point if it falls in X. Otherwise keep generating uniform random points in R until one falls in X, which accept.

4.   Through this point construct an isotropic random plane, by taking the
     spherical polar coordinates $(\dot\theta,\dot\phi)$ of its normal as in the IUR step 2 above.
     This plane is A-weighted through X.

REFERENCES

[1] DAVY, P.J. & MILES, R.E.   Sampling theory for opaque spatial specimens.
        J.R. Statist. Soc. B 39 (1977) 56-65.

[2] DEHOFF, R.T. & RHINES, F.N. (eds.) Quantitative Microscopy. McGraw-Hill,
        New York, 1968.

[3] DELESSE, M.A. Procédé mécanique pour déterminer la composition des roches.
        C.r.hebd.Séanc.Acad.Sci., Paris 25 (1847) 544.

[4] FELLER, W. An Introduction to Probability Theory and its Applications, Vol.I
        Wiley, New York, 1957, 2nd Edition.

[5] MATHERON, G. The Theory of Regionalized Variables and its Applications.
        Les Cahiers du Centre de Morphologie Mathématique de Fontainebleau,
        École Nationale Supérieure des Mines de Paris, no. 5, 1971.

[6] MIDZUNO, H. On the sampling system with probability proportionate to sum of
        sizes. Ann. Inst. Stat. Math. 2 (1951) 99-108.

[7] MILES, R.E. Multidimensional perspectives on stereology, J.Microscopy 95
        (1972) 181-196.

[8] MILES, R.E. On the information derivable from random plane and line sections
        of an aggregate of convex particles embedded in an opaque medium.
        Proc. 4th Conference on Probability Theory (Braşov, September, 1971)
        (1973) 305-317, Editura Academiei Republicii Socialiste Romania.

[9] MILES, R.E. The fundamental role of independent replication, with stratified
        sampling if necessary, in stereology (in preparation).

[10] MILES, R.E. & DAVY, P.J. Precise and general conditions for the validity of
        a comprehensive set of stereological fundamental formulae. J. Microsc.
        107 (1976) 211-226.

[11] MILES, R.E. & DAVY, P.J. On the choice of quadrats in stereology. J.
        Microsc. 110 (1977) 27-44.

[12] MILES, R.E. & DAVY, P.J. The integral of gaussian curvature can be
        stereologically estimated by wedge sections. J.Microsc. (to appear).

[13] STRUIK, D.J. Lectures on Classical Differential Geometry. Addison-Wesley,
        Mass. 1961.

"ONE, TWO, THREE,... INFINITY"

J. SERRA

Centre de Morphologie Mathématique
Ecole Nationale Supérieure des Mines de Paris
35 rue St. Honoré, 77305, Fontainebleau - France.

ABSTRACT

This famous title remarkably summarizes the recent evolution of the
ideas in quantitative image analysis, and the impact of mathematical
morphology on it. "One", for the case of basic stereological measurements. In fact, they all involve an implicit image transformation.
"Two" appears when this transformation is preceded by another one, as
a linear erosion (laws of intercepts) or a covariance. With "Three",
we iterate two transformations before the measurements : bi-dimensional
openings, for instance. And immediately after, one jumps to infinity
with the pattern recognition algorithms, as the Skeletton.

In this evolution, the train of thoughts is given by the conditions
that the transforms have to satisfy. The quantization is associated
with four basic principles : invariance under the translations, compa-
tibility with homothetics, local knowledge, semi-continuity. Moreover,
each criterion corresponds to new logical constraints, like the granulometric axioms for example.

SET TRANSFORMATIONS AND THEIR ITERATIONS

Obviously, Gamow had not foreseen the following paper when he found
such an evocative title for his book, although it fits our purpose so
perfectly that we have "adopted" it. It fantastically describes - in
four words - the whole recent evolution of the methods in Texture Analysis, such as the quantitative description of histological slides for
example.

One may start the contemporary history of these methods and techniques
at the beginning of the 1960's. A cristallisation of various events
occurred. At approximately the same period, an interest for quantization came into the scope of micro-biology, petrography and metallography. The Society for Stereology and the first Q.T.M. date from this
time. For the first as well as for the second, to characterize quanti-

tatively a structure meant to assign a few well representative parameters to it. One knows that three of them are considered as the basic
ones, namely the area, the number of intercepts, and the number of particles, since they are significant in 3 D, and are respectively proportional to the volume, the surface area and the sum of the mean curvature. (12), (13), (14).

a - Step "one".

We shall call "one" this period. Here, "one" is not a more or less arbitrary label, but simply means that beyond each basic parameter, one
elementary image transformation is always realized.
As two examples of step one, consider the transformations respectively
associated with the number of intercepts, and with the connectivity
number (number of particles, minus number of their holes). In the first
case, shown on Fig. 1, the transformation consists in translating by

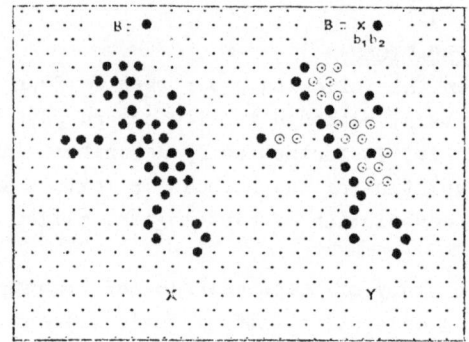

For all Figures :

- B : Structuring Element

- Transform Point Symbols :

  O   Point remaining from
      the initial set

  *   New point

  ●   Point removed

Fig. 1 : Initial set X and transform Y associated with the
intercepts.

a vector x a couple of adjacent points (i.e. structuring element : b_1 b_2)
on the digitalized image X. If $x + b_1$ belongs to the background X^c and
$x + b_2$ belongs to X, then the transform values "one" in $x + b_2$. When x
is swept everywhere, a new image Y is generated. The number of intercepts of X is the number of picture points of Y (i.e. its area).
A similar approach may be used for the connectivity number determination, by changing the structuring element and by performing two parallel elementary transformations. As one can see on Fig. 2 :

Y is the set of x such that $b_1 + x$, $b_2 + x \in X$ and $b_3 + x \in X^c$
Y' "   "   "   "   " $b_1' + x$, $b_2' + x \in X^c$ and $b_3' + x \in X$
The connectivity number is the difference between the areas of Y and Y'.

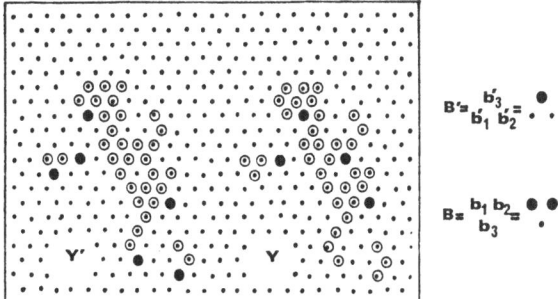

Fig. 2 : transforms involved in the connectivity number computation.

The results illustrated on Fig. 1 and 2 are classical. For simplicity's sake, we recall them in the digital case. Mutatis Mutandis, they can be transposed to the continuous space $\mathbb{R}^2$ (doublets and triplets of points lead to measures, and the sets X, Y have to belong to a rather regular class, such as the convex ring).

These figures show clearly how the set transformations were performed. In particular, it is instructive to notice that the smallest structuring element (in brief : str. el.) of zero dimension (i.e. the point), one (the doublet), and two (the triplet) respectively open the way to the areas, lengths, and number measurements on set X. This result turns out to be a derivation of the kinematic formula due to Van Blaschke and Santaló (2) which, of course, may be generalized to the n-dimensional space. From the stereological point of view, its main interest lies in the fact that it limits the knowledge of X accessible by sections of subspaces of a given dimension (Ex. : by using plane sections, the number of particles in $\mathbb{R}^3$ cannot be estimated).

b - Step "two"

The str. el. B that we just saw were infinitesimal ones. Let us now follow the same geometrical procedure, and use it for a larger B. Then we are able to interpret the classical "measurements" and to find new ones.

For example, imagine that we want to count the number $N(\ell)$ of chords with length ℓ , in the horizontal direction (linear granulometry). Here, the transformation will consist in marking the centers of all these chords, by transforming the initial set X into Y according to the structuring element B :

$$B = 0\ 1\ 1\ 1\ 0 \qquad\qquad \text{(here } \ell = 4)$$

with ℓ marked over the middle $1\ 1\ 1$.

Set Y is an intermediary step. For the final result, one needs to measure the area of Y, which equals $N(\ell)$, up to a calibration factor. This last measurement belongs to the category of elementary transformations defined in step "one".

Consider another example, the one of the covariance (Fig. 4)

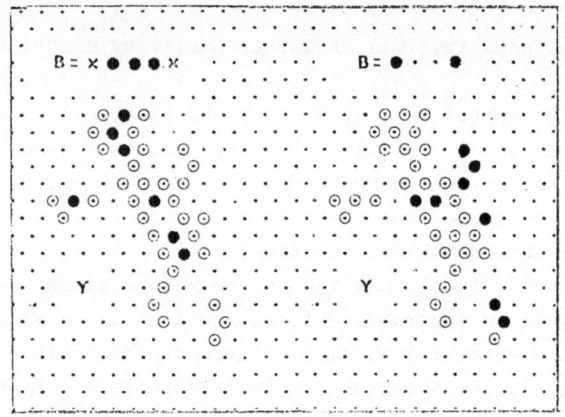

Fig. 3 : Chord distribution   Fig. 4 : Covariance
   transformation (size 4)   transformation (size 3)

The covariance analysis looks like linear granulometry. The procedure is the same, but with another structuring element B, which is now the group of the two points in x and x + h distant of h in the horizontal direction. They must both hit the set X, and then point x + h values one in the transform set Y.

As we see, two analyses as different as the covariance and the linear size distribution are now regrouped in a single approach. We shall denote by number "two" this new step. "Two", because a first image transformation precedes the one leading to the basic measurements. Notice that the two steps use the same kind of mapping, namely the "Hit or Miss Transformation", or H.M.T. (J. Serra, 1965). In brief, in the H.M.T., the space in which X lies is transformed point by point as follows. The structuring element B (B^1 and B^2) is centered at point x, which belongs to the transform Y of X if and only if the part B^1 is included in X, and the part B^2 included in X^c :

$$Y = \{x \; ; \; B_x^1 \subset X \quad \text{and} \quad B_x^2 \subset X^c\}$$

This transformation is reducible to set intersections of a simpler operation, the erosion, for which B^2 is the empty set (written $Y = X \ominus B$).

Why did we restrict ourselves to the H.M.T. ? We shall give below some deep logical reasons for this behaviour, but two other causes oriented this choice. First, classical stereology, as it was in 1965 (basic parameters, chord distribution,...) plus the covariance analysis, appear as particular cases of the synthetic H.M.T. Second, this transformation lends itself perfectly to digital experiments on scanned images (the first patent of the Texture Analyser, 1965, consists precisely of that)(9).

Such a similarity between the various morphological operations suggested the study of the H.M.T. characteristics in general. What are its algebraic properties? Its topological ones ? How do the edges of a frame affect its action ? How can we formalize the concept that a family of transforms tends towards a limit set ? According to which laws can we iterate successive H.M.T.s ? Can a transformation using a multidimensional structuring element be simplified as the product of unidimensional ones ? etc... etc... All these questions needed a new formalism (symbols, notations, theorems,...) which was carried out mainly by G. Matheron (1967) and for a lesser part by J. Serra (1969)(3),(4).

c - Step "three"

The last two questions above announce the next step, number "three". With it, two H.M.T. now precede the basic counts. If this new step had been reduced to one more iteration only, it would not be worth speaking about. In fact, it opens the door to the bi-dimensional transformations, and allows the birth of new criteria (this is normal : after unifying the former stereometry, the new approach had to prove its fecundity). As an example, let us mention (Fig. 5) the now well-known operation of "opening", or "closing", which dates from this period (3), (6). On Fig. 5, we have opened the background X^c by the regular hexagon B of size one (Fig. 5b, set X_B^c). Equivalently, we have closed the set X itself. The opening X_B^c is the set of all points of X^c covered by the translates of B, assumed to be inside X^c. This opening X_B^c is the product of a first erosion of the background X^c by B (Fig. 5 a), followed by a dilation of the new background, which gives X_B^c .

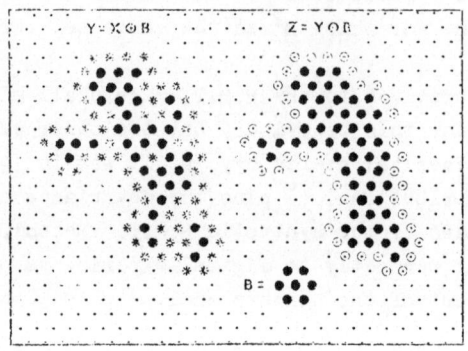

Fig. 5 : Erosion and opening of the background X^c

The concept of opening has been constructed in order to satisfy a group of strong morphological properties (which explains its success). First, it generalizes to the plane and to the space the chord distribution, weighted in measure : for B reduced to a segment, we come back to the law of linear intercepts. In fact, there is a stronger theorem : every granulometric algorithm defined on a set is necessarily a union of openings (6). In this sense, the opening is the basic descriptor of the size. Now, if we restrict ourselves to isotropic openings (i.e. B circle in R^2, or hexagonal (eventually dodecagonal) in the hexagonal raster), and if we notice that the sizing with respect to the openings is independent of the notion of connected particles, then the opening turns out to be an actual shape descriptor. It filters out the finest details independently of their directions, breaks the narrow bridges, clusters the particles according to their distances, etc... In brief, it reduces the shapes to their main features in a progressive and statistically robust way (the study of step 3 (openings, bidimensional analyses) and its technical realisation on the Texture Analysers (the Fontainebleau and Leitz ones) roughly date from 1970).

d - Step ..."Infinity"

From this point on, the next step, "infinity", was unavoidable. The keyboard of all the potential transformations is unbounded, whereas the number of possible parameters is extremely reduced, and so is the

train of thoughts given in the direction of image transforming. By
"infinity", we mean here, either an infinite iteration of the same
algorithm, or systematic modifications of a threshold on an image,
leading to series of sets included in each other ; or more, generally,
complex flow sheets of transformations, with logical loops, etc...
The reader is invited to refer to Mmrs Lantuejoul, Digabel, Beucher
and Meyer's papers in this book, which are all based on infinite transformations programs.

However, let us show with two pedagogical examples, how such infinite
distortions work.

First problem : a connected component extraction. Let X be the initial
set, and x_0 a given point belonging to X. One wants to extract the
connected component Y which contains x_0. If H_1 is the smallest regular
hexagon of the raster (see Fig. 6) Y obviously appears as the limit
of the induction system :

$$X_1 = x_0$$
$$- - - - -$$
$$X_{i+1} = (X_i \oplus H_1) \cap X$$

$$\boxed{Y = X_\infty}$$

since by construction, each point of Y belongs to the set X and is connected by a path to x_0.

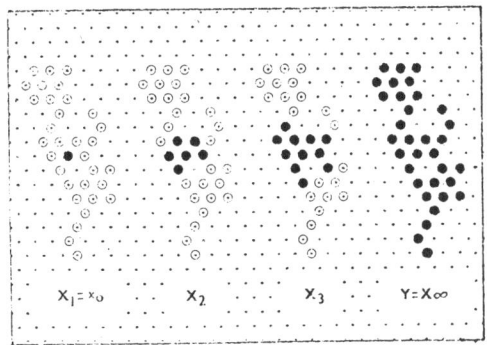

Fig. 6 : connected component extraction. The successive elementary dilations invade the particle Y which contains x_0, but cannot go out of Y.

The second example is relative to the convex hull (Y) extraction of a
given set (X). Let B^j be the following elementary triangles, deduced
from each other by turning clockwise :

$$
\begin{array}{l}
\text{x} \\
\text{x} \quad \text{o} = B^1 \\
\text{x}
\end{array}
\qquad
\begin{array}{l}
\text{x} \quad \text{x} \\
\text{x} \quad \text{o}
\end{array} = B^2 \ldots \ldots
\qquad
\begin{array}{l}
\text{x} \quad \text{o} \\
\text{x} \quad \text{x}
\end{array} = B^6
$$

The dilations are performed by three points and the results affected
to the fourth one (the small circle). The method consists in filling
up X, direction by direction, as follows :

$$X_1^1 = X \quad ; \quad X_{i+1}^1 = (X_i^1 \ominus B^1) \cup X \ldots \ldots X_\infty^1 = X^1$$

$$Y = \bigcup_{j=1}^{6} X^j$$

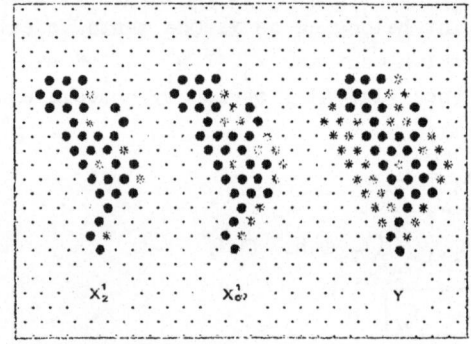

<u>Fig 7</u> : Hexagonal convex hull of X (X is the same set as on Fig. 6)

A similar, but a bit more complicated approach, would yield the dodecagonal convex hull. Some other iteration procedures can be found in
(5), proposed by Golay and K. Preston. Their papers are instructive
and their approach to the "infinity" step prior to ours, but they limit themselves to iterations of infinitesimal H.M.T. In fact, most of
the macroscopic transforms cannot be considered as iterations of elementary ones (Ex. : openings, covariance...). By the way, one will notice
that "infinity", in terms of digitalization, really means fifty, or
sometimes five or six...

<u>QUANTIZATION IN IMAGE ANALYSIS</u>

How could such a synthesis occur? For which reasons can notions coming
from such disparate fields as integral geometry (stereological parameters), geometrical probabilities (laws of intercepts), random functions
of order two (covariance), topology or graph theory (connected component algorithms) and some pattern recognition procedures (Golay-Preston

transformations), be so easily re-organized into a common approach?
This question suggests another one : what does "quantization" mean in
Image Analysis? Is it the fact that some numerical values are assigned
to an image? Surely not. All the qualitative classifications that use
standard pattern systems (in cytology for example) provide numbers :
there are classes n°1, n°2, etc.. although they are qualitative.
Basically, in physics, quantization means reproducibility. Imagine that
two biologists investigate, separately, the same tissue. Their approach
will be quantitative if it statistically leads them to the same histological description. Of course, this final statistical identification
implies dealing with some numerical values, but it also implies a few
logical conditions. In our opinion, four of these always have to be
satisfied :

a) Unvariance under translations

Since our two biologists must sample the tissue they are studying, they
have to work with locally floating coordinates. When they are looking
at the brain cortex, for example, they can roughly define layers parallel to the pial surface, with a 200-400μ thickness, but it is meaningless to take the exact location of each neuron into account, with an
accuracy of 1μ. Practically, one can consider the image transformation
for each neuron as being independent of the corresponding neuron location, up to the scale of the measuring field. In other words, if we
denote by X_h the translate of the set X (the neuron) by the vector h,
we can indifferently :
i) translate X and then transform the translate into the set $Y = \Psi(X)$
ii) transform X first, and translate the result only afterwards, which
   leads us to the first quantization principle :

(1) $$\Psi(X_h) = [\Psi(X)]_h$$

We shall illustrate the meaning of relation (1) with a brief counter-
example. $\Psi(X)$ is defined as follows : the subset of X intersected by
the first line of the mask Z is dilated by the segment of length one,..
line n° i by the segment of length i, etc.. (Fig. 8). Obviously, Ψ does
not satisfy the first quantization principle.

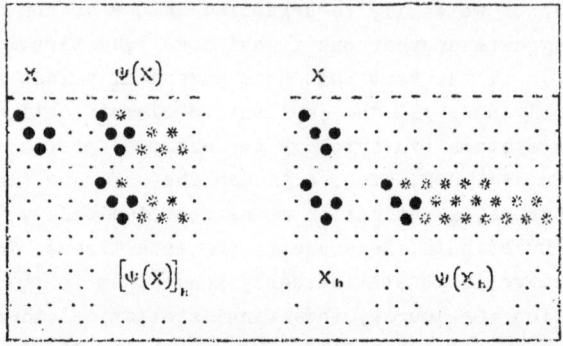

Fig. 8 : non-translation unvariant transformation. The closer the set
X is to the top of the mask, the less it is dilated.

b) Compatibility with the homothetics

Our two biologists never look at their objects at scale one, but must
necessarily work on magnified specimens. So, the transformations have
to be compatible with the homothetical changes.

As a counter example, consider isolated particles X_i and the transformation which eliminates the ones for which the quantity "Area + Perimeter" is larger than 10. If we magnify by 4, the areas are multiplied
by 16 and the perimeters by 4, which completely destroys any classification with the above algorithm.

This principle brings into play a family Ψ_λ of transformations depending upon the positive parameter λ (number 10 in the counter-example),
and may be written as follows :

$$(2) \qquad\qquad \Psi_\lambda(\lambda X) = \lambda \, \Psi_1(X)$$

where λX is the homothetic of X with a ratio λ. (As a consequence of
this principle, the morphological results are often plotted as curves
depending upon a size parameter ; the unvariant is the whole curve, up
to a scale calibration of the x-axis, and not a point on the curve : the
chords larger than 5μ are not the same as the one larger than 50μ).

c) Local knowledge

As a general rule, we do not see entirely the set X, but only the part
of it which is within a measuring mask, let us say Z. If X is made up
of bounded isolated particles, each of them can be completely surrounded

by a mask (a bias problem remains, since the larger particles have more chances to be caught),but when we look at a connected tissue, the edges of the mask limit the object itself. We do not know X, but only X ∩ Z, so we have to define a new zone Z' (depending upon X ∩ Z and on Z) in which Ψ(X) is actually known. In logical terms, we can say :

(3) $\forall Z$, $\exists Z'$: $[\Psi(X \cap Z)] \cap Z' = \Psi(X) \cap Z'$ (Z, Z' bounded sets)

One can easily construct a counter-example as follows : set X is a connected background (Fig. 9). For example, X^c is the union of separate small particles. Ψ(X) is defined as the union of all the points x ∈ X such that the largest chord goind through x has its direction between 0 and $\frac{\pi}{2}$. How can such a condition be checked, when the edges cut the background X ?

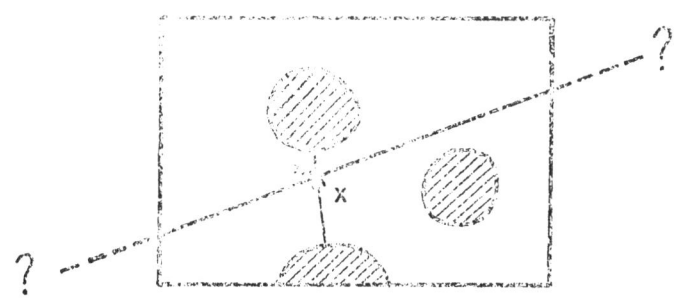

<u>Fig. 9</u> : example of an impossible local knowledge

d) <u>The upper semi-continuity</u>

The last principle is due to the power of resolution of the instruments. We cannot avoid some slight modifications at the border of the objects we threshold to construct the sets X. But we can try to make our transformations Ψ independent of them. In fact, one never exactly knows the border (in the mathematical sense) between two components : for a given accuracy (magnification, discrimination...) there are two portions of space that we call "black" and "white" and between them a narrow strip, let us say $\Delta_1 X$, in which the component cannot be decided. The higher is the accuracy, the narrower the strip becomes. A new $\Delta_2 X$ is substituted for $\Delta_1 X$, more precise in the sense that $\Delta_2 X$ is included in $\Delta_1 X$ ($\Delta_2 X \subset \Delta_1 X$), but with a non-zero thickness anyway. The "border" turns out to be the limit of the intersections of the $\Delta_i X$ included in each other when i → ∞ (Fig. 10).

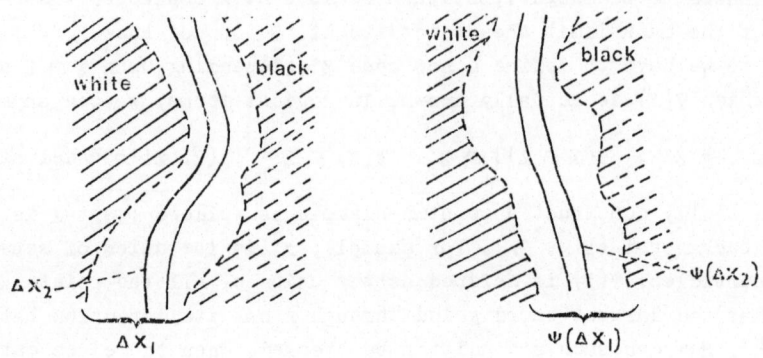

Fig. 10 : Ψ must transform a boundary into another one

Under these conditions, an increasing transformation Ψ is suitable if
and only if it makes one border correspond with another, i.e. :

(4)
$$
\begin{aligned}
&\text{i) } \Delta_i X \downarrow \partial X \\
&\text{ii) } \Delta_i X \subset \Delta_{i+1} X
\end{aligned}
\Rightarrow
\begin{aligned}
&\Psi(\Delta_i X) \downarrow \Psi(\partial X) \\
&\Psi(\Delta_i X) \subset \Psi(\Delta_{i+1} X)
\end{aligned}
$$

which is a methematical upper semi-continuity condition on the Ψ. In
fact, Ψ cannot be increasing, as we assumed for simplicity.(6).
As a counter-example, consider an isolated particle, a pseudo rectangle
X_1 where the two longer sides are slightly enlarged (Fig. 11), and take
for Ψ the union of the maximum number of circles inscribable in X. As
we see on Fig. 11, when X_n reaches its rectangular limit X, the largest
circle jumps suddenly to a long union of circles (the jump is similar
when the limit X is reached by lower values).

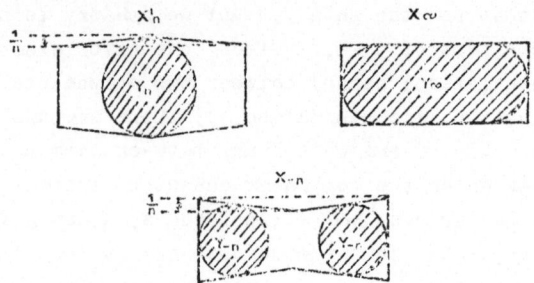

Fig. 11 : example of a discontinuous application Ψ

We have chosen these principles as basic ones, for they seemed to be the smallest group of conditions compulsory for the reproducibility of the measurements. We shall call <u>quantitative</u> any transformation (set → set, as well as set → number) which satisfies them. This axiomatic may be extended also to the functions, considered as families of sets depending upon a parameter (the threshold level) (7) (8) . In the same way, we shall call <u>mathematical morphology</u> the theory of set transformations based on this axiomatic.

<u>QUANTIZATION PRINCIPLES, H.M.T., AND TEXTURE ANALYZER.</u>

Several key theorems associate the quantization principles with the Hit or Miss transformations, and particularly with erosions and dilations. Let us mention the following one, for example :
"Any increasing mapping unvariant under translations is a union of erosions, or an intersection of dilations"(6).
So the erosion and the dilation turn out to be the simplest increasing transformation, or, in other words, the base from which one can construct every increasing mapping. (One will notice the generality of the result, which does not involve any topological requirement at all).
A set transformation $\Psi(X)$ is increasing when $X \subset Y$ implies $\Psi(X) \subset \Psi(Y)$. The morphological operations are not systematically increasing. For example, the one which makes X correspond with its border ∂X obviously does not satisfy this property. However, it is realistic to admit that any transformation met in practice may be written as the difference between two increasing mappings. Under this assumption, the above theorem orients the experiments, and the design of a machine which would be specialised in quantitative image analyses.
The Texture Analyser's family (1965 - 1970 - 1975) of J. Serra and J.C. Klein, currently commercialized by WILD-LEITZ, appears as a direct logical consequence of the above considerations. We designed the Texture Analyser as follows (9) (10) :
i) an image analyser working on the basis of image transformations.
ii) a system using a scanner input, with respect to which it restitutes the transforms in real time.
iii) a system able to iterate the transforms, and more generally to combine them by union, intersection, and difference.
iiii) the transformation carried out at each step is necessarily an erosion by the structuring element, where the pattern is defined by joining up a part of the following cells :

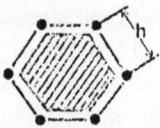

Fig. 12 : basic patterns for str. el. construction used in the texture analyser. The size h is variable. The whole pattern provides 14 binary values for each implantation, according to whether each element (edge, vertex, or full hexagon) is included or not in the set X to be transformed.

So, the texture analyser turns out to be the direct technological issue of mathematical morphology !

PRINCIPLES, CRITERIA AND MODELS

From the standpoint of the principles, every morphological analysis has to elaborate (or choose) the adequate criteria, i.e. the ones suitable for the actual problem, and eventually to fit them with a model. By criterion, we mean a way of image transforming which satisfies, over the principles, certain logical properties of interest. For example, if we wish to perform a size description, we know (·6) that we have to construct a family Ψ_λ of increasing transforms, dependent on the positive parameter λ and such that $\Psi_\lambda(X) \circ \Psi_\mu(X) = \Psi_{Sup(\lambda,\mu)}(X)$, for $\lambda,\mu > 0$. On the other hand, the theory shows that every mapping satisfying the quantization axioms, plus those two granulometric ones, is necessarily a union of morphological openings by convex structuring elements B. If, in addition, we want the transformation to be independent of the orientation of the image, then the B's are obviously the circles. One sees the mechanism which associates any complementary morphological demand with a narrower choice among the transforms : at the limit, only one kind of mapping is focused.

Another important criterion is the one of "connected component". One can decide that the transforms are carried out particle by particle and not point by point, which leads to some restrictions and properties that we shall not present here. It will be noticed that the notion of connected component does not appear at the level of the principles, but only at the level of some of the criteria.

By construction, a criterion is not verifiable. A particular size distribution can be fitted to a model. It is meaningless, and impossible, to test whether the corresponding set was "sizable". On the contrary, the models are mathematical assumptions bearing on the sets we study, (and not on the approach, as the criteria) and thus liable to be verified.

They can be deterministic ("X is the union of separate circles"), or
random ("The particles of X are Poisson implanted").

METHODS AND PROBLEMS

The difference between mathematical morphology and stereology is the
same one which separates a method from a problem. It often happens that
one needs a 3-D characterization from a 2-D analysis (stereology). From
our point of view, such requirements are interpreted as constraints
for the criteria, and may be taken into account in two ways. First, one
can derive significant stereological parameters as mean values, or statistical moments associated with some well chosen criteria (ex. : the
specific area from the number of intercepts ; the star, or mean volume,
from the linear granulometry).
On the other hand, one can make the transformations independent of the
number of dimensions of the space, by defining them point by point
(and not particle by particle), exactly as the H.M.T. does. For example,
the concept of chord distribution of a 3-D specimen is basically the
same as that of its sections (up to correct weights, see R.E. Miles
(11)).
After all, what is mathematical morphology? a mathematical theory, or
a physical one ? Faced with such a choice, we decide in favour of the
second alternative, without any hesitation. In mathematics, a group of
axioms is chosen as the basis of a theory, according to its internal
coherence only. Here, the principles have essentially been picked up
because they transposed into logical words, the a priori conditions
for correct experiments. From the mathematical point of view, the last
principle is the most important one, since it greatly helps the probabilistic version of the method (a semi-continuous mapping is measurable,
thus all the transforms of a random set are also random sets). The
third principle leads to the concept of compact mapping (6) which con-
nects it to the last one. On the contrary, the first two principles,
which deal with some particular properties of the Euclidean space,
would be rejected by a mathematician as a useless restriction of the
generality of the approach.
Moreover, Mathematical Morphology keeps mathematics at a certain distance. According to each special goal, it uses different mathematical
backgrounds : graph theory for the digitalized images, probabilistic
approach by random sets in some cases, deterministic integral geometry
in other cases, etc... Since mathematics appear in our method as

various tools, more or less interchangeable, we can really say that
mathematical morphology is a physical theory.

REFERENCES

1. G. Gamow, One, two, three.... Infinity, Viking Press, N.Y. (1950).
2. S.A. Santalô, Introduction to Integral Geometry, Herman, Paris (1953).
3. G. Matheron, Eléments pour une théorie des Milieux Poreux, Masson, Paris, (1967).
4. J. Serra, Introduction à la Morphologie Mathématique, Ecole des Mines,Paris (1969).
5. K. Preston, Feature extraction by Golay Hexagonal Pattern transform. I.E.E.E Trans. on Computer, Vol. C 20 n° 9, Sept. 1971.
6. G. Matheron, Random Sets and Integral Geometry, Wiley and Sons, New York,(1975).
7. J. Serra, Morphologie de la fonction "à peu près en tout ou rien", Internal report, Paris School of Mines,(1974).
8. F. Meyer, Some Contrast Algorithms, Internal Report, Paris School of Mines, (1977).
9. Texture Analyser patents : n° 1.449.059 (1965) ; 70-21-322 (1970) ; 75-21-925 (1975) ; International patents pending.
10. J.C. Klein, Conception et Réalisation d'une unité logique pour l'analyse quantitative d'images. Thèse, Doct. Ing., Nancy (1975).
11. R.E. Miles, P. Davy, Precise and general conditions for the validity of a comprehensive set of Stereological fundamental formulae, Journal of Micros. Vol. 107, Pt. 3, (1976) , 211-226.
12. R.T. DeHoff, F.N. Rhines, Quantitative Microscopy, McGraw-Hill, New York (1968).
13. E.E. Underwood, Quantitative Stereology, Addison-Wesley (1970).
14. E.R. Weibel, H. Elias, Quantitative Methods in Morphology, Springer-Verlag (1967).

THE FOUNDATION OF COMPUTATIONAL GEOMETRY:
THEORY AND APPLICATION OF THE POINT-LATTICE-CONCEPT
WITHIN MODERN STRUCTURE ANALYSIS

Gustav Bernroider

Zoologisches und Mathematisches Institut der Universität

A-5020   Salzburg,   Austria

SUMMARY

The recent progress in various disciplines of quantitative structure analysis such as Stereology, Image Analysis and Processing, Pattern Recognition and Scene Analysis, opened the question for a common methodological background. This back - ground can be found in probabilistic and computational geometry and is subject to this paper. In using geometrical figures to convey information about structure, the concept of a lattice of points or domains (pixels) becomes decisive. Interaction of structure on one hand and a lattice on the other results in a point-texture which may or not preserve original properties such as the position-invariant, additive and continuous MINKOWSKI-measures known from Integral-Geometry. Some of those measures have a very obvious geometrical meaning such as volume, surface, norm and topological character of the structure but are only defined for certain classes of figures, the most fundamental one being the class of convex-bodies. Obviously, the condition of convexity is a restriction which can hardly ever meet the practical situation found in biological research. As a consequence of this, a new and very general class of figures is introduced for which the Minkowski-functionals can be defined and which can be described quantitatively by a "number of representation" regarding the lattice used to generate the corresponding point-texture. This opens a wide range of practical applications such as the determination of the largest possible grid-constant for Stereology and level of digitalization in Image Processing.

INTRODUCTION

The last years of development in structure analysis brought a gradual but distinct change from qualitative visual observations to quantitative and machine-supported analysis. It is now possible to obtain objective measurements concerning the complexity of structure and to reproduce the results of such investigations. Furthermore, the fact that real-world structure is 3-dimensional and cannot be observed without destruction is about to be overcome by machines which can inter-pret structure from pictures. It is so that the interpretation and analysis of

information contained in pictures becomes the decisive step in processing structure.
A 2-dimensional picture is considered to be a "representative" in either a
probabilistic or deterministic way of a 3-dimensional variety (Miles, 1972a, 1972b,
1976). One possibility to obtain such pictures is to produce thin sections of structure
and to project their contents onto a suitable plane (e.g. a photosensitive layer) by
means of optical imaging. Another possibility would be to produce projections of total
structure by means of x-radiology. The computer can then "reconstruct" an image by
summing all projections (e.g. "computerized tomography", see Hounsfield, 1972).
Applications of quantitative picture analysis cover fields such as Satallite Picture
Processing (SPP, e.g. ERTS-Imaging), Medical Diagnosis and Imaging (Thermography,
Tomography, Radiography, White Blood Cell Differentiation-WBCD), Biology, Metallography,
Simulation and Process Control.

Though the variety and applications are as large as above, the methodological
background can mostly be found within one concept of probabilistic and computational
geometry. In particular, this concept offers a new way of expressing present
stereological and other formulae in a quantitative manner. This means, the expressions
shown here give information about the accuracy of a specific structural estimation
in the sense of approximating integral-geometric relations by numerical methods
(rather than estimating the statistical error). For example, relations are given
for the number of picture points or "pixels" necessary to present the original image
by its digitized point-texture. This will also give rise to a possible connection of
picture processing techniques, originally established in the U.S.A. over 20 years ago
(Kovasznay and Joseph, 1955), with the "European School of Image Analysis and Stereology". This is of particular importance since the directions of structure analysis
in the U.S.A. and Europe are quite different and seem to note very little of each
other. (Rosenfeld, 1972, 1973, 1974, 1977; Jesse, 1974, 1976). Diagram 1 should give
an idea of the present situation regarding various branches in quantitative structure
research. This does of course only reflect a part of the true situation but should
give an impression to the reader about the local distribution and interdisciplinary
connections of the most important developments in this rapidly progressing field.

The "heart" of this body of methods can in many respects be found within
"Integral Geometry" (I.G. in FIG. 1) where the "needle problem", G. BUFFON was
concerned about exactly 200 years ago, is well integrated. The interaction of two
elements, one unknown and one known geometrical object (e.g. needle and lines) can
lead to a fundamental property of this object, such as length. Intergrating over all
possible orientations and translations of an element (e.g. needle) which intersects
lines presents, somehow, the infinite case of measure. In order to actually calculate
those integrals, we must replace the infinite integrals by finite summations. This
decisive step leads to an extension of integral-geometry which we shall call
"computational geometry". (Also see FIG. 1)

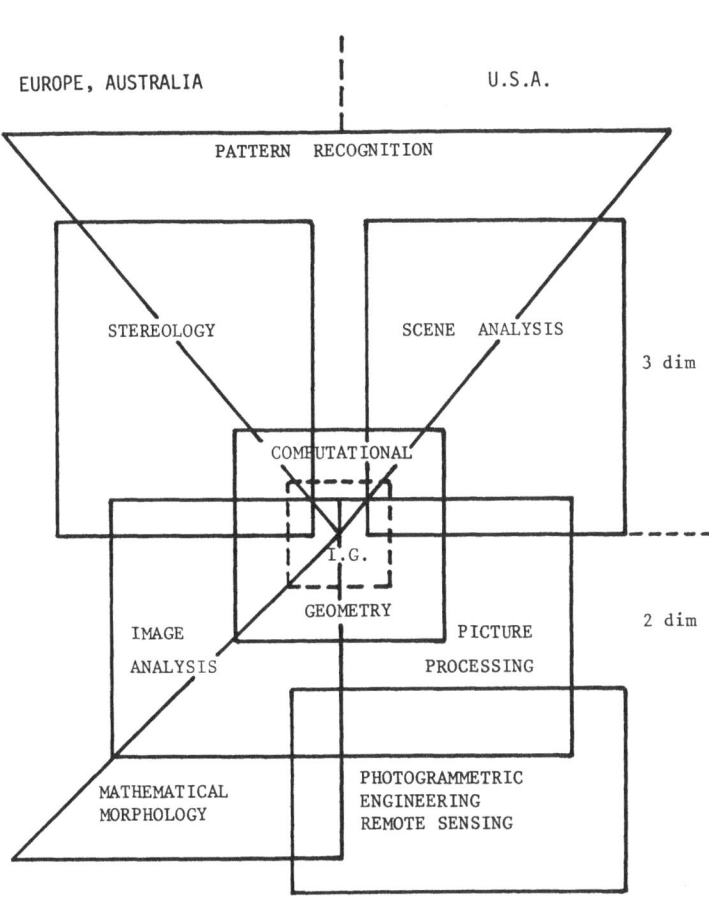

The step from integral to computational geometry is important but not easy at all.
Many new problems arise if an integral is replaced by summation. For instance, the
error which is made by replacing the infinite integral by a finite number of
summations should be known at least roughly. If this is not the case, the results
obtained by different methods or on different structure respectively can hardly be
compared. In this paper I shall therefore give an introduction to "infinite"
geometrical properties first and then gradually extend formulae from pure or
"beautiful" figures (e.g. convex figures) to more realistic and "ugly" ones which
are of particular importance in biology. From the enormous literature in "numerical
integration theory" conditions such as a given "smoothness" of boundaries are well
known. For applied problems in quantitative structure analysis such conditions have
to be expressed in a way so that the degree of "roughness" or "smoothness" can

actually be calculated. To show some relations in this way which can and should be
further extended and improved, is the aim of the present paper.

INFINITE GEOMETRICAL PROPERTIES OF STRUCTURE

Simply speaking the word "geometry" of an object refers to its size, shape, width,
diameter or connectivity. In particular, it does not refer to the "gray-level" of it.
This is how we think about the word "geometry" from our everyday-experience, but
there is much more to say about the geometry of objects than foresaid. Firstly,
infinite geometrical properties are properties of objects which they possess per se.
There is no need to do anything with an object of our physical world that it possesses
some unique "volume" or "norm". Secondly, the class of objects (convex or non-convex
objects, rough or smooth objects, etc.) does not matter as long as we don't "treat"
the objects analytically trying to receive some information like size, shape, width.
If we do so, we have to restrict our considerations to special sets of objects given
by certain conditions like convexity. Clearly, the more general the class of figures,
the more interesting are the results for applied problems. Let us first consider the
class of convex bodies which is the most obvious one from the analytical point of
view. Here we find a basic class of geometrical properties which are the
1. Fundamental properties:

Strictly speaking, this is the only class of pure geometrical properties and all
the rest can either be derived from them or obtained by combination. In the case
of a convex body (i.e. a body where every straight line connecting two points of the
body is totally inside it, also see FIG.3) which is of dimension -n-, there are
exactly n+1 fundamental properties of this body. For example, a convex, 3-dimension.
body posessess 4 such properties. They can be represented by the so-called
"Minkowski - or Quermass- functionals", named after the mathematician Minkowski, who
could first show their unique existence. If we denote the functionals by $W_i(K)$, we
can define them recursively regarding the dimension n of the figure K.
To be specific, for dimension n = 1 we let:

$$W_0(K) = V(K) = 1, \quad \text{a line of length 1}$$
$$W_1(K) = S(K) = w_1 = 2,$$

$$(1)$$

and for n > 1, we define recursively:

$$W_0(K) = V(K),$$
$$W_i(K) = \frac{1}{n \cdot w_{n-1}} \int W'_{i-1}(K,u)\, dw ;$$

$$(2)$$

where w_n denotes the volume of the n-dimensional unit-sphere. To put it in words,
equation (2) states that:

" The i-th Minkowski-functional of a body K equals
the arithmetic mean value of the i-1 functionals
of its orthogonal projections onto planes, with the
exception of a constant factor which only depends
on dimension -n-."

$W'_{i-1}(K,u)$ symbolizes the n-1 dimensional, i-1 th functional of K which appears on a n-1 dimensional plane which is fixed in direction -u-. To illustrate the physical meaning of the functionals, we can write:

$$
\begin{aligned}
W_0(K) &= V(K) &&: \text{volume of K,} \\
n.\ W_1(K) &= S(K) &&: \text{surface of K,} \\
n.\ W_2(K) &= M(K) &&: \text{mean curvature of K,} \\
(1/w_n).W_3(K) &= w_3 = C(K) &&: \text{Euler-characteristic of K;}
\end{aligned}
\tag{3}
$$

The equations (3) are valid for n = 3, thus having i = 0, 1, 2, 3, together 4 functionals which equal the volume, surface, curvature and a topological character of a body K. For the situation n = 2 (the plane) which is of particular importance in Image Analysis, we obtain for the n-1 th functional:

$$ n.\ W_{n-1}(K) = 2.\ W_1(K) = \text{norm of K.} $$

It can be shown (Hadwiger, 1957) that the norm of K is closely related to the "mean-width" $\bar{b}(K)$ of K. We have:

$$ N(K) = \text{norm of K} = (n.\ w_n/2).\ \bar{b}(K); $$

Since w_n equals the contents of the circle for n = 2, namely for r = 1, $w_n = \pi$, we have:

$$ N(K) = \pi.\ \bar{b}(K) \qquad \text{for n = 2;} \tag{4} $$

In this dimension the n-1 functional is identical with the 1 st one, i.e. with the perimeter of the figure. Because the i-1 th functional for i = 1 equals the volume, which again is just the length for n-1, we can see that the mean value of orthogonal projections becomes equal to the mean width $\bar{b}(K)$. The constant factor has reduced to π. FIG. 2 gives an impression of the situation outlined above. The top figures a,b,c illustrate the "fundamental properties" of an idealized cell-image for the 2-dimensional case. Note: while the "mean projection" is a fundamental property, the single projections from which it can be derived by averaging over all possible directions u, are dependent properties. This is important to know, since only the fundamental properties of a figure are *position-invariant, additive* and *continuously* defined for all figures K of the class K(K) (the class of convex bodies).

2. Derived or combined properties:

Here again we have:

        a, shape inducing properties

        b, diameter

        c, geometrical intensity

Shape inducing properties can be derived by simple combination of the former functionals or by derived properties such as width in a given direction. We may for instance classify more circular or streched objects by a shape index combined by two orthogonal widths. (See middle row of FIG.2). A property such as "diameter", though invariant under rigid body motions, is not fundamental because it can in fact be calculated by maximizing the width over all possible orientations. The last group of derived properties is the "intensity" of an object. This is simply the number of objects within a unity of volume. In this case we may think about the "volume" as the body and the objects as its "holes". If N denotes the number of holes and D a domain of unit volume, this intensity can be written as:

$$I(K) = N / |D| ; \tag{5}$$

We have already outlined a simple relationship between fundamental and derived geometrical properties in relation (4) above. This was typical for the continuous and infinite case. How things begin to look for the discrete and finite situation in computational problems, we will discuss in the following.

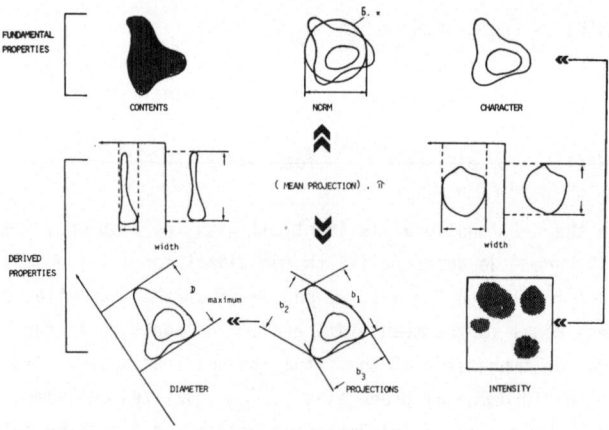

FIG. 2

Fundamental and derived geometrical properties of
an idealized cell image and 2 dimensions.

COMPUTATIONAL GEOMETRY OF NORMAL POINT-TEXTURES

If one wants to receive specific information about the properties discussed above, some interaction between the structure on one hand and a "perceptual tool" on the other must occur. This perceptual tool may be another geometrical object with better or smoother properties than the one to be measured. (Remember just the same situation when approximating a given function by convolution-type integrals !).

To be specific, we are looking for a real-valued function $f(g,K)$ depending on a geometrical figure K and an element g which should simply be a set of points in $\mathbb{R}^2$ (the 2-dimensional real plane). In addition, let m be a function: $\mathbb{R}^2 \to \mathbb{R}^2$ of the type:

$$t(\xi,n)(x,y) = (x+\xi,\ y+n), \qquad (\xi,n) \in \mathbb{R}^2 \tag{6}$$

or

$$\tau_\alpha(x,y) = (x\cos\alpha - y\sin\alpha,\ y\cos\alpha + x\sin\alpha)$$

where τ_α denotes a rotation $0 \leq \alpha < 2\pi$,

We call m a *rigid-body-motion* and let M be the group of all rb-motions. G should further be the collection of all elements g obtained under rb-motions, writing $G = \{ mg_0,\ m \in M \}$. From integral-geometry we know that, for example, by letting $g_0 \in G$ be a straight line in the plane given by

$$x\cos\varphi + y\sin\varphi - p = 0, \qquad (\text{ see FIG. 4 }) \tag{7}$$

the function

$$\mu_G(mg_0) = \int_M dp\ d\varphi \tag{8}$$

has the important invariance property:

$$\mu(m_0 G) = \mu(G) \qquad \forall\ m_0 \in M; \tag{9}$$

Introducing a bounded, closed domain of unit volume, the measure (9) can be turned into a probability measure. For a subset G_0 of G which intersects the domain D and a subset $G \cap K$ (all lines which intersect the figure K) we have:

$$P(G \cap K) = \frac{\mu_G (G \cap K)}{\mu_G(G_0)} \tag{10}$$

Now, if $F(x) = \{ (g: f(g,K) \leq x) \}$ denotes the corresponding probability distribution function, the mean of which is

$$E(X) = \int_\Omega x\ dF(x) = \frac{1}{\mu_G (G_0)} \int_{G_0} f(g,K)\ \mu_G dg \tag{11}$$

the *problem of computational geometry is to determine if and* HOW FAST *the sample mean*

$$\bar{X}_N(g) = \frac{1}{N} \sum_{i=1}^{N} f_i(g,K) \tag{12}$$

converges to the expectation (11).

The random variable $f(g,K)$ has distribution function $F(x,c)$ which highly depends on c, the category of figures to which K belongs. Clearly this category will be decisive for the speed of convergence regarding the mean (12). For example, if K is a simple closed curve (FIG.4), G_0 the set of straight lines intersecting K and $f(g,K)$ the number of intersections, the integral:

$$I(K) = \int_{G_0} f(g,K) \; \mu_G \; dg \tag{13}$$

which is the average of f, will essentially be equal to the perimeter of the convex hull of K, denoted as $\bar{P}(K)$ in FIG. 4. Thus, if -c- is the category of convex bodies, $I(K)$ will be proportional to the perimeter of K itself. Equation (13), in this case, turns to the well known "formula of CROFTON", the derivation of which can easily be seen by Blaschke (1949). At this step it becomes intuitively clear, that the decisive point for the practical evaluation of the above integrals will be a relationship between the number of category c on one hand and some fundamental geometric properties on the other. The reader may get an impression about the dispersity of c for biological structures in the following figure 3. The category c can be expected not to depend purely on the type of structure but also on the magnification and resolution used for the imaging process.

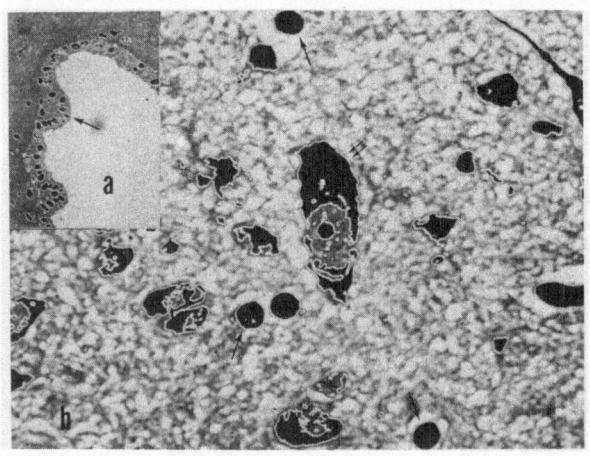

FIG. 3

Bilogical objects electronically extracted from the
the background. Different magnification (a),(b) gives diff. -c- values.

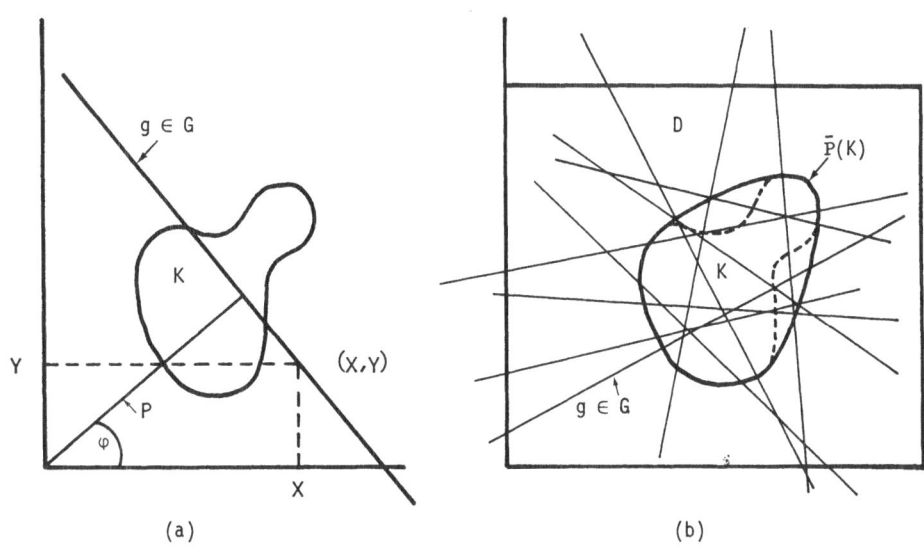

FIG. 4

A simple figure K is given by a "fixed", closed curve
in the plane. A straight line g can be determined by
its normal coordinates p, α for a given point (x,y) of
g. (b) illustrates a set of straight lines g intersecting the figure K and the "mask" D.

We have now briefly outlined the problem of computational geometry. In the heading
of this chapter the term "normal point-texture" has been used. This is to point out
the difference to a "fuzzy point-texture" in the sense of Zadeh (1965) which will be
discussed at another place. To continue with the notation of the "structuring
function" f(g,K), we now generate a lattice of domains which is then to interact
with the figure of interest.

Definition:　By a "lattice of fundamental domains" in the plane we understand
a partition of the space E_2 (2-dim Euclidian space) into a (countable)
number of congruent domains g_i, so that each g_i can be superposed
on a given g_0 by a translating motion.
Each point in the plane should belong to one and only one g_i.

If we further denote the vector which characterizes the lattice by $\bar{h} = (h_1, h_2)$
and the lattice itself by G(t,h), we can define:

Definition:　The section G(t,h) ∩ K (i.e. the points common to the grid and the
figure K) will be called the *point-texture* of K.

The symbol "t" was used above to indicate the translation t of the lattice $G(0,h)$ with $t = (t_1, t_2)$, $0 \leq t < h_i$, $i = 1,2$. Two orthogonal directions of the translating motion will also give rise to different point-textures:

$$G^{(-h_i)} \cap K, \quad i = 1,2 \quad \text{and translation in both direction:} \tag{14}$$

$$G^{(-h_i, -h_j)} \cap K, \quad i,j = 1,2 \quad i \neq j \; ; \tag{15}$$

FIG. 5 illustrates the 4 point-textures resulting from the above translations with respect to a simple non-convex figure K. The value of the "structuring function" $f(g,K)$ now equals the number of semi-open domains g which intersect K. Next, the average of $f(g,K)$ in the sense of (13) is to be defined for the textures:

$$\begin{aligned}
G(t,h) \cap K &\quad : \quad \bar{f}(g,K)_h \\
G^{(-h_i)} \cap K &\quad : \quad \bar{f}(g,K)_{-h_i}, \quad i = 1,2 \\
G^{(-h_i, -h_j)} \cap K &\quad : \quad \bar{f}(g,K)_{-h_i, -h_j} \quad i \neq j = 1,2
\end{aligned} \tag{16}$$

By letting the values of these functions equal N_o, N^1, N^2, N^{12}, the geometrical meaning of the structuring functions becomes obvious if we let:

$$\begin{aligned}
\chi(T) &= N^0 - N^1 - N^2 + N^{12} \\
b_1(T) &= h_1 (N^0 - N^1) \\
b_2(T) &= h_2 (N^0 - N^2) \\
A(T) &= h_1 h_2 N^0
\end{aligned} \tag{17}$$

$\mathbf{N^0} \rightarrow$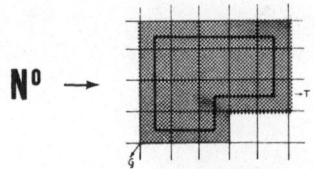

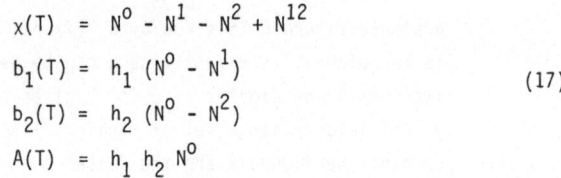

$\mathbf{N^1} \rightarrow$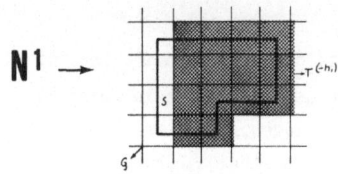

$\mathbf{N^2} \rightarrow$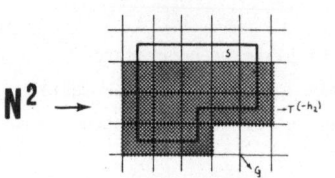

$\mathbf{N^{12}} \rightarrow$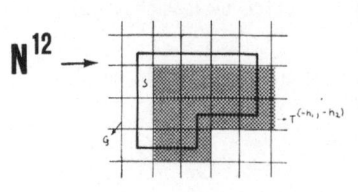

Here "T" was used to denote the point-texture, $\chi(T)$ the *Euler-Characteristic* of T, b_1, b_2 two orthogonal widths of T and $A(T)$ the area of T. Using the example of FIG. 5, the area of T would be $18.h^2$, $b_1(T) = (18-14)$ $h = 14$ h and $b_2(T) = (18-13)$ h = 5 h and finally $\chi(T) = 1$. If we now calculate the mean width (defined by relation (4) above) which is essentially the "norm" of K, equations (17) represent the (n+1) *Minkowski-functionals* of the point-texture T. Obviously, this is not true for the figure K itself. Replacing T by K (which is our aim) causes an error which can be expected to depend on the size of the grid-constant h used to generate the texture T. (Also see Giger, 1975). The error will become smaller as the size of the grid-constant decreases and the point-texture T will become more and more similar to the original structure K.

FIG. 5

For example, the characteristic $\chi(K)$ will become:

$$\chi(K) = N^0 - N^1 - N^2 + N^{12} + O(h_1, h_2)$$

where "$O(h_1, h_2)$" denotes the capital Landau symbol. It is now our intention to find a quantitative expression for this error which should also lead us to the number of category mentioned with the distribution function $F(x,c)$ of the variable $f(g,k)$ before. We give the decisive

Definition: A structure K is called *representable with respect to the lattice* $G(t,h)$ *if*:

$$\max (N^0 - N^{12}) \leq \min (c. N^{12}), \quad c \in \mathbb{R}^+$$

The resulting point-texture $T(K)$ is called *"point-texture of category c"*.

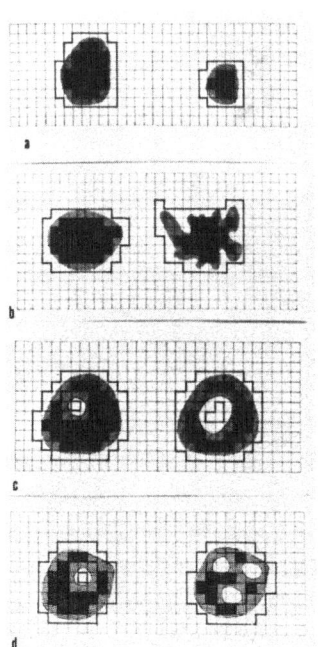

The meaning of the above definition can best be shown graphically. FIG. 6 illustrates the point-textures of different figures K. So that a figure K is "represented" by its point-texture in the sense above, we demand that the maximum of the "boundary-cover" $(N^0 - N^{12})$ over the translations of G within the semi-open interval $[0, h_i)$ is always smaller or equal to the minimum of its "interior" N^{12} times a factor c. Now "c" contains information about:

a, the relative size of K (resolution),
b, the smoothness of the boundary of K,
c, the size of "holes" of K,
d, the topological character of K
   (number of "holes").

The properties a,b,c,d are subsequently shown in FIG. 6. The dark squares correspond to the interior of K and greatly vary with different categories of K .

FIG. 6

Figure K itself is shown grey, its interior dark and the point-texture is given by a black surrounding. In order to satisfy the above definition, the number of white and grey squares would have to be smaller or equal to the number of dark squares times a given factor c. For example, for c = 2 this number has to be smaller or equal the double of dark squares, thus allowing the use of a larger grid-constant.

The key for a practical application of the above considerations can now be expressed by the following theorem:

> *Under the condition* $| N^0 - N^1 - N^2 + N^{12} | = | \chi (K) |$, *a figure K can be represented by a lattice* $G(t,h)$ *with category c, if and only if for* h_1, h_2 *the inequality:*
>
> $$h_1 \cdot h_2 \leq \frac{A\,(c+1)\,(\,2\chi - 2N_1 - 2N_2 + 2N_0\,)}{\bar{N}_0 \cdot N} \tag{18}$$
>
> *holds.*

(For the proof see appendix). "A" was used for total test area, $\bar{N}_0$ for the maximum of N_0, N equals the number of domains which belong to A ($A = h_1 \cdot h_2 \cdot N$) and the numbers $N_1 = (N^0 - N^1)$, $N_2 = (N^0 - N^2)$. If there is equality in the above relation, the largest grid-constant by which K can be represented is determined. This result can be immediately applied to the stereological point-count on one hand and to Image Analysis on the other. On the right hand side of (18) we find numbers which can easily be calculated by applying a "fine" point-grid onto the structure K. For the special case that A = N (the test area is equal to the number of "squares" in the plane), we obtain

$$h \leq \left[\frac{(c+1)\,(2\chi - 2N_1 - 2N_2 + 2N_0)}{N_0} \right]^{1/2} \tag{19}$$

and for the "number of category c" we have:

$$c = \frac{h^2 \cdot N_0}{(2\chi - 2N_1 - 2N_2 + 2N_0)} - 1 \tag{20}$$

These formulae are especially useful for the work with automatic image analysis equipment. In this case the right-hand sides can be calculated by the "infinite fine" point-display which the television monitor supplies. All one has to compute is the area of a figure (N_0), two orthogonal projections (N_1, N_2) and the characteristic χ. The desired category can for example be predetermined and the calculation of (19) finally gives the largest possible value for h. If this value would be 10, every 10th picture point can be used for all subsequent computations and it still guarantees a metrically and topologically correct representation of the picture (e.g. preservation of the Minkowski-measure of the original structure).

Another application is the calculation of -c- by (20) for a given number of "pixels" in picture processing problems. For a predetermined level of digitalization using for instance 500.000 picture points within the area of measurement different objects and magnification and resolution of the imaging system will contribute a different "category -c-".

Note: The number -c- is a "class-property" and expresses the continuity and
smoothness of a picture-function (binary valued) rather than a single geometric
property such as "area". As such it does not vary essentially within a
natural class of bodies (e.g. convex bodies) and may stay constant even if the
relative area (area for a given grid-constant) is increased or decreased
drastically (the body would still be convex if this is done without directional
preference!). On the other hand -c- changes clearly if there is some essential
metric or topological difference or a difference in smoothness respectively.
In this case for example, the area of the figure does not have to change at
all. FIG. 7 gives a graphical presentation of relation (20) using the grid-
constant h and the number c as coordinates. 7 different figures have been
measured in this case giving the change connected by "stretching"a square
into a rectangle under the condition of <u>constant area</u>. Function (1) shows the
square, (2), (3), (4), (5), (6) and (7) rectangles with increasing side-length
but constant area. Taking every 14th picture point from a "square" would give
category c = 100, taking the same from figure (7) gives c = 200. For a low
value of h (e.g. more picture points) the difference in c becomes smaller and
smaller. Due to economic reasons, a high value of h is more interesting but
causes the number c to change dramatically , demonstrating the extreme importance
of its calculation.

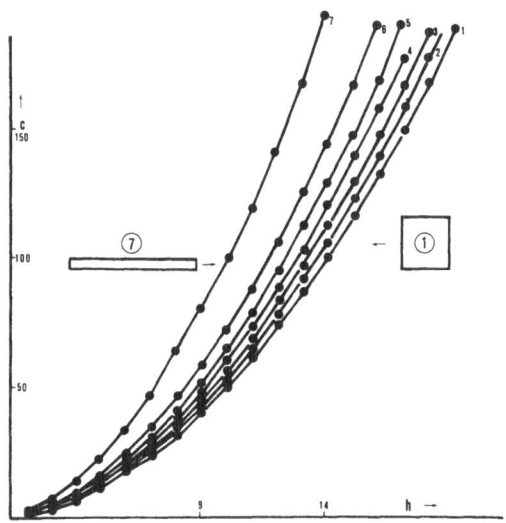

FIG. 7

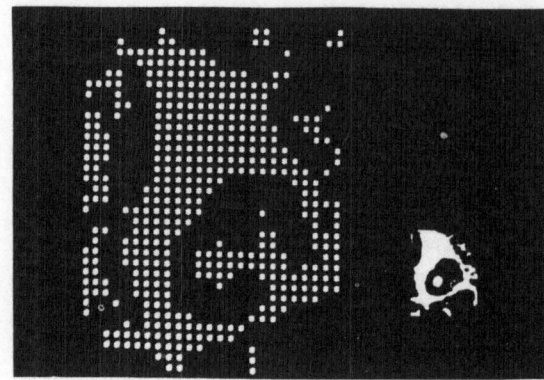

FIG.8

On the right side the binary valued cell-image of a neuron is shown. For this extremely "fine" point-texture the number of category c equals 0.54. On the left side (b) illustrates the same image with c = 10.001 and every 4th picture point for representation.

CONCLUSION

The methods discussed in this paper give the methodological background for the computational analysis of microscopical texture. The relations are derived from Integral Geometry and extended to a new class of figures which we may call "representable by G(t,h) and category c". This extension is based on a quantitative measure for the error which is made if the original point-set K is replaced by its corresponding point-texture T. The problem is then to determine the speed of convergence of a sample mean to its expectation, if the integral-geometric measure is turned into a probability measure by the introduction of a "mask" which contains the point-set of interest. The corresponding distribution function depends on a number c, giving the connection to the numerical condition of "representation" regarding the preservation of the Minkowski-measure of the original point-set. Estimating the upper bound of the grid-constant used to discretize the image, one obtains relations between the lattice G(t,h) on one hand and the category c on the other. This introduction of a very general set of figures covers the great variety of objects which is found when observing biological structure. The work is strongly based on the extensions introduced by Hadwiger (1955, 1957, 1959) and Giger (1975) starting with the class of convex bodies, the convex-ring and finally the class of "normale Körper" and "rasterable point-sets". The present relations seem to be especially useful for the practical application in Image Analysis, Stereology and Picture Processing.

ACKNOWLEDGEMENT

The author is grateful to Prof. Dr. H. Adam and Prof. Dr. P. Zinterhof for their continuing encouregement and support. The study was supported by the Fond zur Förderung der wissenschaftlichen Forschung, Austria.

REFERENCES

Bernroider, G.   Recognition and classification of structure by means of
stereological methods in neurobiology.
J. Microsc. 1o7, 287-295, (1976).

Bernroider, G.   Point-lattice operations in quantitative microscopy.
12. Dortmunder Arbeitsgespräche über quantitative Bildanalyse.
Microscopica Acta, Basel, (1977).

Buffon, G.   Essai d'arithmetique morale.
Suppl. à l'Histoire Naturelle (Paris), 4, (1777).

Crofton, N.W.   Geometrical theorems relating to mean values.
Proc. London Math. Soc., 8, 304-309, (1877).

Giger, H.   Rasterbare Texturen.
Zeitschrift für angewandte Mathematik und Physik (ZAMP),
26, 521-535, (1975).

Hadwiger, H.   Altes und Neues über konvexe Körper.
Birkhäuser Verlag, Basel-Stuttgart, (1955).

Hadwiger, H.   Vorlesungen über Inhalt, Oberfläche und Isoperimetrie.
Springer-Verlag, Berlin-Göttingen-Heidelberg, (1957).

Hadwiger, H.   Normale Körper im euklidischen Raum und ihre topologischen und
metrischen Eigenschaften.
Math. Zeitschrift, 71, 124-140, (1959).

Hounsfield, G.N.   A method of and apparatus for examination of a body by radiation
such as x or gamma radiation.
Patent Spec. 128 3915, London, (1972).

Jesse, A.   Bibliography on automatic image analysis.
Microscope, 22, 89-115, (1974).

Jesse, A.   Bibliography on automatic image analysis.
Microscope, 24, 65-102, (1976).

Kovasznay, L.S.G. & Joseph, H. M.   Image Processing.
Proc. IRE 43, 550-570, (1955).

Miles, R.E.   Multi-dimensional perspectives on stereology.
J. Microsc. 95, 181-195, (1972a).

Miles, R.E.   The random division of space.
Suppl. Adv. Appl. Prob. 242-266, (1972b).

Miles, R.E.   Estimating aggregate and overall characteristics from thick sections
by transmission microscopy.   J. Microsc. 107, 227-233, (1976).

Rosenfeld, A.   Picture Processing 1972.

Computer Graphics and Image Processing, $\underline{1}$, 394-416, (1972).

Rosenfeld, A.   Picture Processing 1973.

Computer Graphics and Image Processing, $\underline{3}$, 178-194, (1974).

Rosenfeld, A.   Picture Processing 1974.

Computer Graphics and Image Processing, $\underline{4}$, 133-155, (1975).

Zadeh, L.A.   Fuzzy sets.

Information and Control, $\underline{8}$, 338-353, (1965).

APPENDIX

Let K be "represented" by $G(t,h)$. Thus we have:

$\max (N^0 - N^{12}) \leq \min (c. N^{12})$ for $c \in \mathbb{R}^+$ which means that:

$1/2 (N_{oi} - N^{12}_i + N_{oj} - N^{12}_j + |N_{oi} - N^{12}_i - N_{oj} + N^{12}_j |) \leq c. (N^{12}_i + N^{12}_j - |N^{12}_i - N^{12}_j|)$

for $\forall$ $i,j = 1,.....,m$ and $0 \leq t_m < h_r$;

We also have the relations: $N^1 = N_o - N_1$, $N^2 = N_o - N_2$ and $N^{12} = N_o - N_1 - N_2 + N_{12}$

Because $| N_{12} | = |\chi(K)|$ was provided to be right, one receives:

$N_{oi} - \chi + N_1 + N_2 - N_o + N_{oj} - \chi + N_2 - N_o + | N_{oi} - \chi + N_1 + N_2 - N_o - N_{oj} - N_1 - N_2 + N_o$

$+ \chi | \leq c. (\chi - N_1 - N_2 + \chi - N_1 - N_2 + N_o - | \chi - N_1 - N_2 + N_o - \chi + N_1 + N_2 - N_o|).$

Now we obtain by some elementary calculations:

$N_{oi} + N_{oj} + | N_{oi} - N_{oj}| \leq 2. \chi (c+1) + 2. N_o (c+1) - 2. N_1 (c+1) - 2. N_2 (c+1)$

If we further use $\bar{N}_o$ for the maximum on the left hand side and remember that the figure K and the lattice $G(t,h)$ were defined within a "mask" D the area of which is $A = h_1. h_2. N$, we finally obtain:

$$\frac{\bar{N}_o . N h_1 h_2}{A} \leq (c+1) (2\chi + 2N_o - 2N_1 - 2N_2) \text{ and thereby equation}$$

(18). ///

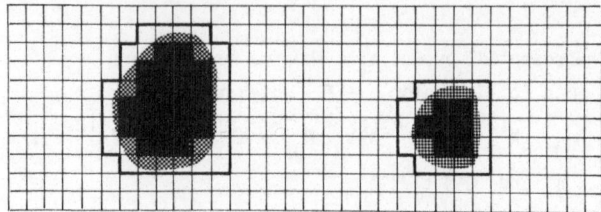

a

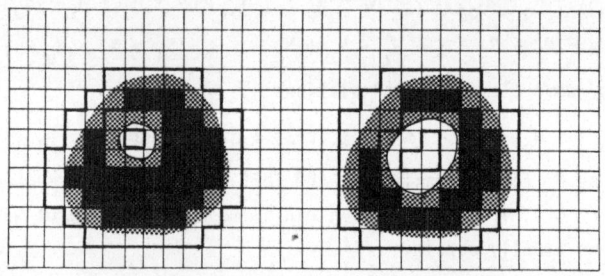

b

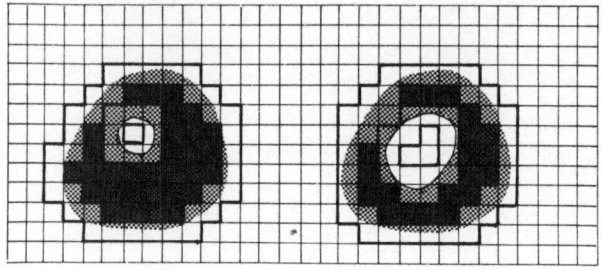

c

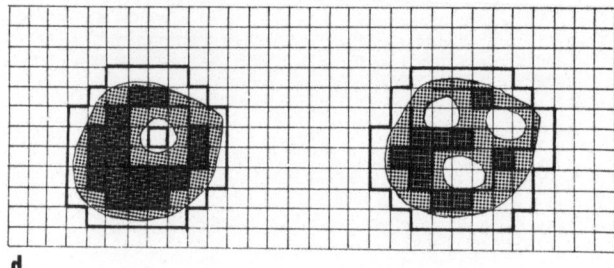

d

THE NON-STATISTICAL NATURE OF BIOLOGICAL STRUCTURE AND ITS IMPLICATIONS ON SAMPLING FOR STEREOLOGY

EWALD R. WEIBEL

Department of Anatomy, University of Berne,
Switzerland

1. Introduction

For any life scientist the talk of random processes in relation to the structure of living systems must at least elicit deep scepticism. Life does not occur by chance; it is rather deeply rooted in a strictly conserved heritage of order that ascends in a continuous line from the most primitive living system - such as viruses or bacteria - to the highest levels of organisation found in mammals and man, allowing for considerable versatility by variations on a common set of themes. Chance interactions of molecules and of cells may certainly occur - but only within the narrow confines imposed by the inherent ordering patterns which are defined by the structure of the biological organism. The existence of such an ordering pattern makes the difference between a living cell and the simple mixture of all its chemical constituents in a test tube.

2. The heritage of biological order

The constancy of biological order is assured by one molecule: DNA, a long chain of 4 different nucleotides which have the interesting property that always 2 can enter into loose binding. A double strand - the famous double helix - is formed by pairing of these nucleotides. If the two strands separate each can serve as a template for the formation of its complement. This is what happens during cell division so that the two daughter cells will each be provided with an exact replica of the genetic code of their parent.

Associated with this self-replication of the genetic code is the principle that a cell can only be produced by a cell: "Omnis cellula e cellula", an axiom first proposed by Raspail in 1825. Cells are therefore likewise self-replicating systems, and this is true for the most primitive as for the highest organisms.

For a higher organism we can by these fundamental laws conclude that it must be made of cells which all have the same genetic potential, because they have all developed from one egg cell. This is true, but then all cells become differentiated to serve specific functions. Indeed, a biological organism is an extraordinary social system in which each member - one of the many cells - serves a certain function in the interest of the entire organism.

Orderly and controlled performance of specific functions and cooperation between all elements are therefore the most important themes underlying biological organisation.

3. Ordering principles in biological structure

The design of a living system follows certain fundamental principles all directed towards establishing order of the elements in view of their cooperative and orderly performance of functions. Most functions are directed processes, and many cells have therefore a distinct polarity. Functions performed by units of a given size are quantitatively reinforced by multiplying the units, so that modular design is frequently encountered. Very often the functional performance of an organ involves a whole sequence of processes, each performed by different cells; the arrangement of these units then follows a serial design, and if each has a distinct directional design or polarity this may result in a hierarchical arrangement of dependent elements. These basic design patterns (fig. 1) are now discussed on some specific examples.

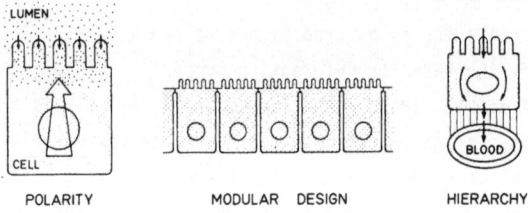

Fig. 1: Ordering principles

3.1. Directional design: polarity

One of the primordial ordering principles is that functional elements must be separated, primarily in the interest of regulating functional processes. Thus all cells are bounded by a <u>membrane</u> partly impermeable to solutes: it establishes the polarity "inside-outside", which allows cells to exist in varied environment; but it also forms the basis for "excitability" by electrical impulses because the enzyme molecules of the membrane can control the directed transit of certain ions across the cell wall.

It is a very common feature of cells that their membrane is different from one place to the other. Figure 2 shows a cell from the lining of the gut with a distinctly polar structure: the membrane facing the intestinal lumen at the "apical pole" of the cell has a morphology clearly different from the other sides and we know that the brush-border formed of finger-like membrane protrusions (fig. 3) is there to actively resorb from the digested food sugars and aminoacids, i.e. some specific nutrients, and to transfer them to the blood.

3.2. Modular design: mosaics, periodicities

This polar or even "vectorial" function of intestinal cells can evidently only be effective if the space containing the digested food, the lumen, and the "internal" space of the organism where blood circulates are clearly separated by a continuous barrier. The capacity of a single cell to "pump" nutrients across this barrier is furthermore limited. As a result a large number of identical polarised cells is arranged in a single layer, an epithelium, in the intestinal wall (fig. 5); they are tightly apposed to each other and the narrow intercellular clefts are sealed to exclude any uncontrolled transit of solutes between the lumen and the internal tissue space.

As a consequence, the intestinal epithelium appears as a mosaic of cells of more or less regular modular design; fig. 4 shows that the finger-like cylindrical protrusions of the brush-border, being precisely identical modules, form a very regular hexagonal lattice and that consequently periodic patterns become particularly conspicuous. The cell bodies are more variable in size and shape, but they are still rather regularly arranged.

This pattern of modular design is typical for a large number of different tissues. More complex mosaics can be formed if cells of different function must cooperate in the performance of the function of the epithelium: in the intestinal wall, for example, some secretory cells are mixed into the lining by resorptive cells.

3.3. Serial design: hierarchy

The separation of functions into different units and their polarised nature may require the establishment of precisely defined sequences of units for the performance of complex functions, where the function performed by one unit may depend on that of one or more other units.

Such a hierarchy is clearly established in the intestinal wall. The brush-border carries the enzymes which are capable of actively pumping the nutrients into the cells; the cell body receives these nutrients, prevents them from flowing back into the intestinal lumen, and directs them into the tissue space, where the blood can pick them up to bring them to the liver and other organs (fig. 1).

One of the typical structures resulting from a combination of serial and modular design are the very frequent tree-like structures. The primordial functional units are present in large number, but they are all connected to one stem by a system of branched tubes. This is the case for most glands: the secretion product is synthesised in specialised cells - all identical - but is then secreted in concentrated fashion through one final port.

Another example of this kind of structure is the lung: in view of an efficient resorption of oxygen air and blood are brought into intimate contact deep in the lung in some 300 million small air chambers (fig. 7) which are connected to cylindrical airway tubes. These tubes are the terminal branches of a complex tree formed from the trachea by about 23 generations of systematic branching. Figure 6 shows that the branches of this tree become progressively smaller as one moves from the trachea through bronchi out to the small alveolated terminals. At each level of this tree the function, and consequently the wall structure of these airways arranged in series is different: a typical example of hierarchical organisation of cooperative units. Clearly, this organ also exhibits the two other ordering principles: its design is polar, and the multiplication of identical units results in a modular design.

4. The design of muscle cells

These basic ordering principles are realised in a nearly perfect and therefore very conspicuous manner in muscle cells. Muscles perform their function by highly directed linear contraction. This contraction is performed by a pair of interconnected threadlike molecules: actin and myosin. Electron micrographs of muscle cells show that these two filamentous molecules are separately arranged in highly regular parallel bundles and that these bundles interpenetrate (fig. 8). One knows today that the muscle contracts or shortens because the actin filaments are pulled into the bundle of myosin filaments, and that the force generated depends on lateral binding between adjoining actin and myosin filaments. This requires a high degree of order. On cross-section one observes a precise arrangement of myosin molecules in a hexagonal lattice with the actin molecules associated in a distinct pattern, each actin filament binding to three neighboring myosin molecules.

On the other hand, all myosin filaments are of identical length, and they are bound together by a central "plate" to form a symmetric "brush" (fig. 8). Likewise, all actin filaments are of identical length and they too are bound together by a central plate, the z-disc, to form a symmetric brush. The serial arrangement of actin (A) and myosin (M) allows these brushes to mutually interpenetrate and to form so-called myofibrils of periodic design:

$$\cdot \, A - M \cdot M - A \cdot A - M \cdot M - A \cdot A - M \cdot M - A \cdot$$

For convenience, the sequence $\cdot \, A - M \cdot M - A \cdot$ is considered the unit repeating period and called a "sarcomere". It is important to note (fig. 9): (1) that myofibrils extend without interruption from one end of the cell to the other and that the potential for length change depends on the length of the myofibril; (2) that each muscle cell contains a number of myofibrils arranged in parallel and that the potential for generating contractive force depends on the number of myofibrils or rather on their total cross-sectional area.

It is furthermore important to note that all myofibrils within one cell or rather their sarcomeres are in register with respect to their actin-myosin periodicity (fig. 9). This is due to the fact that the control of muscle contraction depends on the arrangement between the myofibrils of a complex system of auxiliary structures: mitochondria for the supply of the fuel for contraction work (ATP), and a system of membrane tubules controlling the sequence of events under the influence of nerve impulses. The arrangement of these tubules, which is of considerable biological interest and forms the object of many morphological studies, deserves a brief analysis.

As mentioned above, the excitability of cells, including muscle cells, by electrical nerve impulses, rests on the polarity established by the cell membrane: the cell's interior is, at rest, more negative than the surrounding medium and as a nerve impulse arrives this electrical potential collapses transiently; this so-called action potential, lasting about 3 msec, causes muscle cells to contract. This stimulus needs to be rapidly transmitted into the muscle cell's interior, and this is assured by a network of membrane tubules penetrating deep into the cell from the external cell membrane and surrounding all myofibrils; two such "transverse or T-tubules" are present in each sarcomere (figs. 9 and 10). A complex system of "longitudinal or L-tubules" now surrounds each sarcomere (fig. 10) and is in intimate contact with the T-tubules. The arrival of an excitation potential causes these L-tubules to release Ca ions, and this in turn causes the actin-myosin complex to contract. The control of muscle activity hence depends on the hierarchical series:

cell membrane $\longrightarrow$ T-tubules $\longrightarrow$ L-tubules $\longrightarrow$ sarcomere

The perfect anisotropy of the myofibrils and the close apposition of the L-tubules to the sarcomeres evidently imposes anisotropic properties on the tubules. If these are to be studied by use of stereological methods it is evidently important to know to what degree the tubules are anisotropically arranged. This has been studied by B. Eisenberg (1); she superimposed a set of parallel test lines onto longitudinal sections of muscle cells and recorded the number of intersections between test lines and tubules as a function of the angle between myofibrils and test lines. Figure 11 shows that the tubules clearly exhibit a certain degree of anisotropy, but that they are only "partially" oriented.

5. Conclusions

In summary, we can conclude from this discussion that ordering principles are very prominent in muscle cells. This is basically due to the fact that (a) the contractile material is made of a linear periodic array of strictly identical anisotropic units, the sarcomeres, and that (b) all ancillary components are arranged in close apposition to these basic units to ensure the hierarchy necessary for the control of muscle function. The degree of randomness in muscle structure is extremly limited by these constraints, and it is evident that sampling such cells for a stereological analysis will have to account for these structural properties (1, 2).

In most other tissues structural order may be less prominent. Nevertheless, modular design, directed functions, and hierarchical arrangement of interdependent units, are fundamental principles of construction for almost all biological systems. They must be carefully considered in devising appropriate sampling schemes for the application of stereology in biological morphometry.

References

(1) Eisenberg, B.R., A.M. Kuda, and J.B. Peters (1974): Stereological analysis of mammalian skeletal muscle. J. Cell Biol. 60, 732.

(2) Weibel, E.R. (1972): A stereological method for estimating volume and surface of sarcoplasmic reticulum. J. Microscopy 95, 229 (Stereology 3).

Figure legends
_____

Fig. 2: Intestinal cell with nucleus (N) bounded by cell membrane (arrows).
Membrane facing intestinal lumen (L) is differentiated to brush-border (B).

Fig. 3: Longitudinal section of parallel cylindrical elements of brush-border,
lined by membrane (arrows).

Fig. 4: Cross-section of brush-border shows hexagonal packing of elements.

Fig. 5: Modular design of intestinal epithelium by multiplication of cells like
the one in fig. 1. Blood vessel (V) beneath epithelium.

Fig. 6: Plastic cast of bronchial tree of human lung.

Fig. 7: Terminal ramification of airways (A) in scanning electron micrograph.
Respiratory air chambers form foam-like structure by dense packing of
"modules".

Fig. 8: Complex of actin (A) and myosin (M) filaments forms sarcomere between
Z-discs.

Fig. 9: Serial arrangement of sarcomeres (S) builds myofibrils (F). Membrane tubules (L and T) and mitochondria (C) are periodic, in register with sarcomere period of myofibrils.

Fig. 10: Detail of fig. 9 to show partial orientation of L-tubules.

Fig. 11: Partial anisotropy of L-tubules in muscle cell analysed by intersection
count on oriented test line. Data should follow upper curve for fully
oriented tubules, and horizontal line for random arrangement. (From
Eisenberg et al. {1}).

178

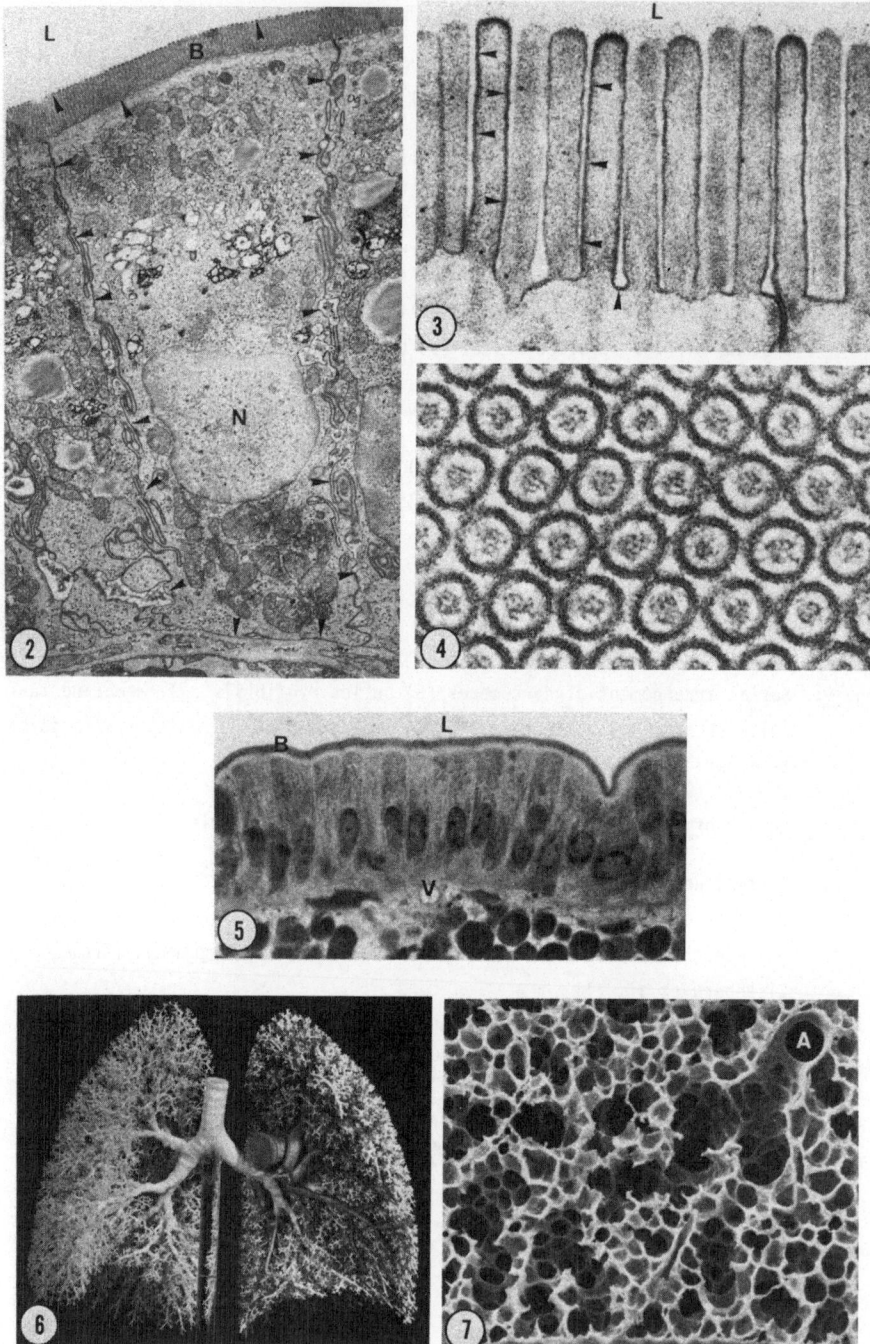

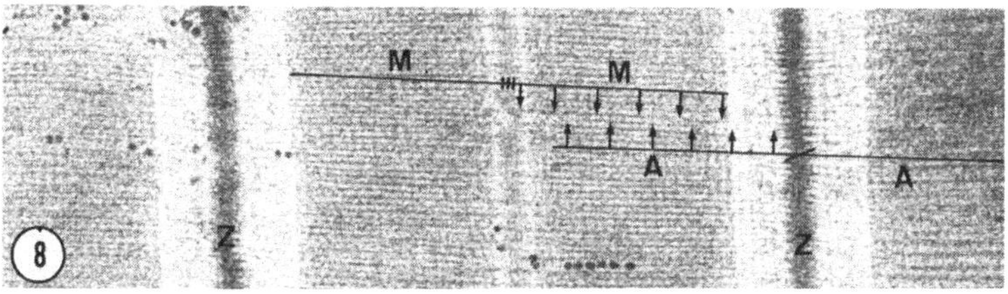

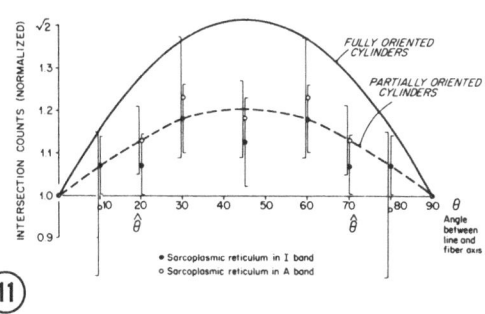

Fig. 8–11

Fig. 8: labels M, M, A, A, Z, Z

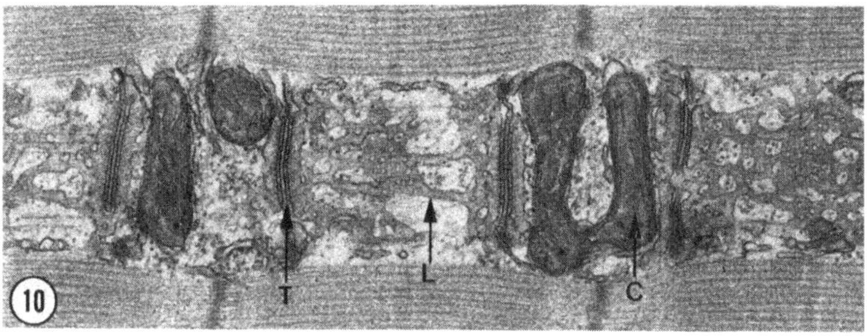

Fig. 9: labels S, F, T, L, C

Fig. 10: labels T, L, C

Fig. 11:

INTERSECTION COUNTS (NORMALIZED)

√2

FULLY ORIENTED CYLINDERS

PARTIALLY ORIENTED CYLINDERS

θ Angle between line and fiber axis

• Sarcoplasmic reticulum in I band
○ Sarcoplasmic reticulum in A band

The non-statistical nature of biological structure and its implica-
tions in sampling for stereology of liver tissue.

Albrecht Reith, Norsk Hydro's Institute for Cancer Research, The Norwegian Radium Hospital, Montebello, Oslo 1, Norway.

One of the tissues most investigated by stereological methods is liver
tissue (see review article (3)). This predeliction for liver is partly explained by the fact that it has been regarded - and treated - as
a tissue in which the structural elements, cells and organelles are
presumed to be randomly distributed and oriented. Consequently, considerations concerning cellular - or subcellular - polarity, aniso-
trophy, etc., which strongly influence the sampling procedure, were
largely disregarded.

Physiological, biochemical and anatomical considerations lead one to
assume that there are structural inhomogeneities (a) in cellular composition and (b) subcellular organisation, i.e. organelle distribution
within the average liver cell. The most conspicuous inhomogeneities
can be assumed to be connected with gradients in oxygen, nutrient supply and bile production. Blood, oxidised and full of nutrients, via
portal vein and hepatic artery reaches liver cells at what is tradi-
tionally called the periphery of the lobule, a small lobe having the
structure of a polygonal prism 1 - 2 mm in diameter, and leaves it,
oxygen-deprived, through the central vein in the centure of the lobule (Fig. 1). On the level of the average liver cell, oxygen and nutrient uptake occurs at one site of the cell, whereas bile is secreted
at another site (Fig. 2).

Fig. 1: Arrangement of liver cells in plates

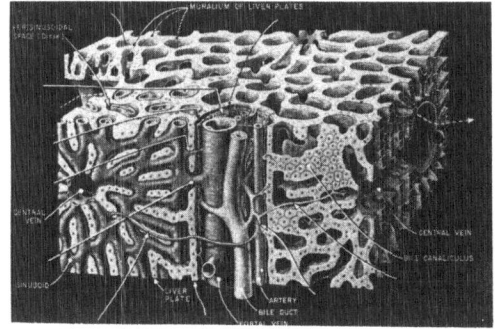

Liver cells are arranged in
interconnecting plates forming
so-called lobules which have a
central vein in their centre.
Blood streams from the artery
and portal at the periphery of
the lobule towards the central
vein. (With kind permission from
Academic Press, New York and
London (1)).

It is only 20 years ago that H. Elias, in his first stereological findings (see (1)), showed that liver cells are arranged in usually one-
cell-thick plates,and not in cords as assumed before, which form a continuous three-dimensional lattice. The cell plates are randomly arranged in all directions within the lobule (Fig. 1). From this follows
that randomly-made tissue sections fulfil the criteria of representative
samples for liver cells.

However, inhomogeneities exist due to the above-mentioned gradients in
oxygen/nutrient supply. Liver cells in the "periphery" have another
biochemical composition, which is also reflected in the fine-structural
composition as seen in Table 1.

Table 1. Fine structure of mitochondria in liver cells in peripheral
and central regions of the lobule. (From (4) .

	Ratio p/c
Average mitochondrion	2.2
Number of mitochondria	0.7
Volume density of mitochondria	1.5
Surface density of cristae membranes	1.3
Enzyme: succinate dehydrogenase	1.4

Most of the studies on liver tissue have been performed without considering the different localisation-dependent composition of liver
cells. In spite of this fact there are not too large variations of
the mean values in the results. For example, for the volume density
of mitochondria in normal rat liver, different working groups (see (3))
found the following values:

$$0.181 \pm 0.005 \text{ (Weibel)}$$
$$0.195 \pm 0.025 \text{ (Weibel \& Bolender)}$$
$$0.181 \pm 0.002 \text{ (Rohr)}$$
$$0.200 \pm 0.001 \text{ (Reith)}$$
$$0.148 \pm 0.006 \text{ (Loud)}$$

On the average, the standard deviation is less that 10% of the mean value, although only 5 blocks per liver with only 6 micrographs were analysed, i.e. a workload which is absolutely tolerable. This result
should indicate that liver cells are mainly homogeneous although structural differences at the extremes, periphery and centre exist, and that
Loud's estimation of structural homogenity in only 80% of all liver

cells (2) is only a rough estimation.

We can conclude from these findings and considerations that in spite of the non-random structure of liver (with respect to the different fine-structure of liver cells at different localisations within the lobules), one can neglect this fact in practical work, since random sampling leads to satisfactory results with a low degree of variation.

But a second inhomogeneity on the subcellular level which has consequences in practical work is observed in every liver cell. This inhomogeneity is connected with the fact that every liver cell has at least two different surfaces (Fig. 2).

Fig. 2: Topography and fine-structure of liver cells

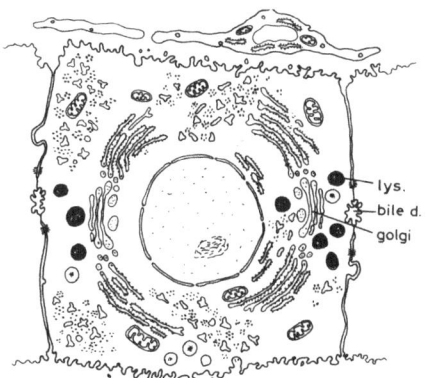

Liver cells are arranged in one-cell plates. Two sites of the surface of the cell face the blood system (top and bottom). Where cells are adjacent to each other a small space is left open which serves as a channel for the bile (canaliculus). In the cytoplasm of the two adjacent cells around the canaliculus, clusters of lysosomes and Golgi apparatus are found.

The predominance of these organelles around the canaliculi results in large cytoplasmic areas which are free of lysosomes and Golgi structures. The non-random distribution of these structures is the reason for the observation that in a large number of micrographs selected randomly for stereological analysis of the organelle composition of liver cells, no hits on lysosomes and Golgi structures are counted at all. This is reflected in the large variation for these structures compared to homogeneously distributed particles such as mitochondria (Table 2)

Table 2: Volumetric subcellular composition of liver cells (mean and standard error of 5 samples).

	m	$\pm$	S.E.(%)[*]
Lysosomes	4.9	$\pm$	1.1 (23)
Golgi apparatus	2.8	$\pm$	0.9 (32)
Mitochondria	318.5	$\pm$	15.0 (4.7)

*S.E. in % of mean

The question arises if it is possible to arrive at better sampling procedures for analysing the organelle composition of the average liver cell, considering the non-random distribution of Golgi structures and lysosomes. This would be a methodological improvement which is necessary if one is especially interested in these structures and on the other hand would like to keep the workload at a practical level. As long as no other sampling procedures than random sampling can be applied in liver cell analysis, the only practical solution left for keeping the variation low is to increase the number of samples.

References

1. Elias, H. & J.S. Sherrick (1969). Morphology of the Liver. Academic Press, New York and London.

2. Loud, A.V. (1968). A quantitative stereological description of the ultrastructure of the normal rat liver parenchymal cells. J. Cell Biol. 37, 27.

3. Reith, A., T. Barnard & H.-P. Rohr (1976). Stereology of cellular reaction patterns. Critical Reviews in Toxicology 4, 219-269.

4. Reith, A. (1972). Intramitochondrial localisation of glycerol phosphate dehydrogenase. A possible marker enzyme for the proliferation of mitochondria. Cytobiologie 5, 384.

SAMPLING PROBLEMS IN THE KIDNEY

R. Østerby and H.J. Gundersen, 2nd University Clinic of Internal
Medicine and University Institute of Pathology, Kommunehospitalet
Århus, Denmark.

The kidney function covers a wide range of different aspects,
some of which are of vital importance for the organism; e.g. the
water and electrolyte balance of the body and the maintainance of
normal blood pressure depend upon normal kidney function. Highly
regulatedmechanisms ensure these equilibria. A precise interaction
between the different parts of the kidney structures is obviously
necessary for this purpose and this is reflected in the very sophisticated organization of the organ. The meaningfulness of some of
the structural details is not yet clear, i.e. the interrelationship
between structure and function is far from being fully clarified.

Kidney structure

The kidney is a bean-shaped organ, in most species with a regular and smooth surface. Only at the concave margin this surface is
discontinuous giving place to the passage of blood vessels and urine
out-flow channels (Fig. 1).

Inspection of any section plane through the kidney shows that
the tissue is layered in fairly distinct zones of different appearance.

Macroscopically we can distinguish two different zones: one
forms a continuous layer immediately beneath the surface and extends
in strands towards the center; this is the cortex, and the remaining
deeper part is the medulla. In the human kidney the medulla forms a
number of conical projections. In the rat and other rodents each
kidney has only one suce medullary cone.

The kidney may be divided into 4 roughly symmetrical parts, one
plane of symmetry containing the long axis and the blunt convex
lateral margin, the other plane perpendicular to the long axis through
the middle of it. The symmetry is not complete since there are differences, for instance in the blood supply between the ventral and
dorsal half and between upper and lower part.

Each kidney is composed of about 1 mill. individual modules
called nephrons (Fig. 2). The different parts of these units may be
seen at light microscopy.

First part of the nephron is the glomerulus which is a tangle

of capillaries enclosed within a capsule. All glomeruli are localized
in the cortex, to some extent aligned in arrays perpendicular to the
kidney surface.

Across the capillary walls of the glomeruli a very great volume
of plasma is filtered (125 l/day in man). The constitution and volume
of the ultra-filtrate is greatly altered while passing through the
very long and complicated tubular system.

The main course which is followed by each tubule is: first a
convoluted segment very much coiled. Then comes a more straight
portion, directed towards the medulla. Within the medulla is a very
regular, thin segment making a hairpin bend. When the tubule is again
back in the cortex it actually makes contact with its parent glomerulus before it once again takes a convoluted course and finally joins
the collecting duct.

These general characteristics hold for all nephrons. But there
are systematic differences as to for instance the length of the different segments. The proximal tubules increase in length with increasing depth within the cortex. The glomeruli close to the medulla are
larger than the superficial ones, and their vascular interconnections
are different (1).

Using electron microscopy we may observe the occurrence of further
non-randomness within the different parts of the nephron. The glomerular capillary tangle is inserted between two somewhat larger vessels.
The breaking up into small capillaries and the reunion takes place
at the vascular pole (Fig. 3). A certain degree of orientation of the
capillaries is therefore present.

In the tubules we see highly polarized cells meant for directional transport systems. Systematic changes between different segments occur, some of which can be distinguished on sections. The
proximal tubule is subdivided into 3 segments. Characteristics which
differ in the 3 regions are for instance the height of the cellular
protrusions towards the lumen, the degree of complexity of basal in-
foldings of the cell membrane, and numerical density and size of certain cell organelles (2).

Scheme for random sampling from the kidney

Our practical problem is now to obtain random sections from
this organ, for instance with the purpose of obtaining an estimate
of some ultrastructural component. Suppose we are in the most favourable situation: an animal experiment where we have at hand a
whole kidney, fixed by intravascular perfusion. Only in this situation there is time enough to follow the somewhat complicated samp-

ling scheme.

In order to obtain random blocks for embedding, the kidney is transsected with parallel, thin (i.e. less than 1 mm) sections e-quidistantly spaced. The thin sections are placed within a 2-dimensional coordinate system, thereby reducing the sampling space to a 2-dimensional one, which makes it immediately accessible for random sampling. Selecting sites with 2 random coordinates we can then cut out random blocks for embedding. If we are interested in for instance glomeruli, we cut out blocks only if the selected site is within the cortex. For our studies we use sections from 3 glomeruli from each kidney and to be sure to obtain that we select one block from each of 7-10 sites. The small blocks are embedded and cut at random orientation.

If glomeruli are present on the random section the glomerular profile which has its center closest to the center of the section is selected, provided the profile is more than one maximal profile- radius from the edge of the block (Fig. 4). This restriction is included, since incomplete glomerular profiles are damaged at the edge and cannot be used. The above restriction with respect to the edge is necessary in order to avoid preferential sampling of smaller profiles. Large profiles near the edge have a higher probability than smaller ones of being cut at the edge.

For the obtaining of micrographs we use a systematical random sampling, moving the specimen stage of the electron microscope at preselected intervals, marked out on the handles. Thus small cross sections are represented with a few, and large cross sections with many micrographs, as they should in this case of sampling from the restricted glomerular space.

Examples of unsolved problems in stereological studies of the kidney

In the following two different stereological questions about kidney structure are outlined, one about glomeruli and one about tubules. In the glomerulus we determine the thickness of the basement membrane in the capillary wall. The structure of interest is the central layer in the wall of the glomerular capillaries (Fig. 5). This extracellular structure is distinctly delineated by cell membranes on both sides.

For what purpose would we like to determine its thickness? In our studies we use the estimate of this basement membrane as a parameter for diabetic vascular disease (3). The basement membrane in the glomerular capillaries thickens steadily over the years of diabetic life - leading eventually to glomerular closure and thereby to kidney insufficiency and death.

Our question therefore usually is: what is precisely the basement membrane thickness in a certain group of diabetic individuals - whether animals or patients - compared to that in a control group? First step is the sampling of kidney tissue. The optimized scheme outlined above is useful in many situations, but not in human patients, where a needle biopsy of the kidney is drawn blindly. This gives a cylinder of about 1-2 cm length and 1.5-2 mm width. But we do not know how deep in the tissue or at which location the specimen was taken. No matter how many glomeruli we find in the biopsy they only represent one small part of the kidney. On the random micrographs from the sample of e.g. 3 glomeruli the basement membrane thickness is measured, placing test-lines at random on the micrographs, and measuring the length of the intercepts with the basement membrane. From these we obtain the harmonic mean membrane thickness and the distribution of membrane thickness (4).

The model which is the basis of this stereological determination of basement membrane thickness assumes random orientation of the membrane. But the capillaries within the tuft are not oriented at random, since within the entire periphery the capillary loops, making a bend, run parallel to the surface. Secondly, there is a certain orientation with respect to the vascular pole, and finally the packing of many channels of this dimension within this limited space means that the direction of individual capillaries is not independent of that of neighbouring ones. These problems of capillary orientation probably cannot be overcome in the sampling procedures.

Another fact is of importance, namely that the membrane thickness does not vary at random at different locations within the tuft. A loop-to-loop variation in thickness is present and the membrane is thicker close to the vascular pole than in the peripheral parts (3). The 3 different cross sections represent only 3 different locations and orientations with respect to the vascular pole.

From each glomerulus we sample about 20 fields covering roughly one quarter of the section and from each field about 5 intercepts are measured.

The above restrictions posed on the degree of variation of both orientation and thickness mean that 300 intercepts measured on 3 glomerular cross sections do not represent 300 degrees of freedom.

To minimize these problems one obvious suggestion would be to section a greater number of glomeruli per individual. However, for the statistical evaluation of our results we reduce as much as possible the variation due to interindividual differences, i.e. the number of individuals included in each series should not be too small. Therefore, to make the projects practicable within a reasonable time we usually have to be content with about 3 glomeruli per individual.

So, we are left with at least two major problems: the number of degrees of freedom is unknown, somewhere between 3 and 300. Therefore, the number of intercepts necessary for obtaining a certain low variation of the estimate must be determined empirically rather than analytically. The other problem is that the non-random sampling of glomeruli in patients gives rise to an estimate with great variation due to differences in glomerular characteristics at different sites of the kidney. This variation cannot be reduced by measuring more glomeruli from the same biopsy. Its magnitude is unknown.

Finally, a sampling problem concerning the tubular system.

The question which is raised is: what is the surface with respect to cell volume of the apical cell membrane in the first compared with the second segment of the proximal tubule?

The two segments are both located within the same part of the cortex. They both belong to the convoluted part and the transition between them is gradual.

The long and slender protrusions from the cell surface are highly oriented, perpendicular to the cell surface. This brush border is higher in the first than in the second segment - and this is one of the criteria for the distinction. The great problem in sampling from the two segments is due to the fact that only when both the basal and the luminal part of the cell is in the field of the section plane can the tubule in question be categorized as belonging to 1st or 2nd segment. This means that the definition of the objects to be measured requires a certain and presumably undeterminable degree of orientation.

Would the best solution be to consider only tubular profiles which seem to be cut at a plane closely perpendicular to the long axis of the cylindrical tubule, as estimated by the shape of the tubular profile - and then accept the degree of uncertainty with which this can be done?

References

1. Barger, A.C. and Herd, J.A.: Renal vascular anatomy and distri-
    bution of blood flow. Handbook of Physiology. Renal Physiology.
    Washington, D.C.: Am. Physiol. Soc., 1973, sect. 8, pp. 249-313.

2. Maunsbach, A.B.: Observations on the segmentation of the proximal
    tubule in the rat kidney. Comparison of results from phase con-
    trast, fluorescence and electron microscopy. J. Ultrastruct. Res.
    16, 239-258, 1966.

3. Østerby, R.: Early phases in the development of diabetic glome-
    rulopathy. A quantitative electron microscopic study. Acta med.
    Scand. suppl. 574, 1975.

4. Gundersen, H.J.G., Jensen, T.B. and Østerby, R.: Distributions
    of membrane thickness determined by lineal analysis. J. Microsc.,
    to be published.

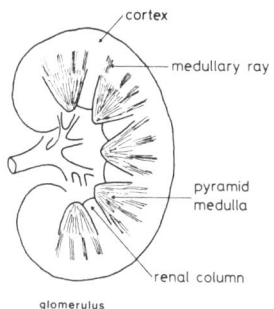

Fig. 1.   Sketch of a human kidney

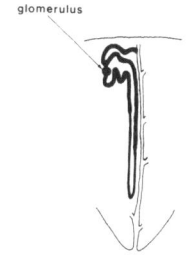

Fig. 2.   Sketch of one kidney module: a nephron
           consisting of glomerulus and the tu-
           bular system.

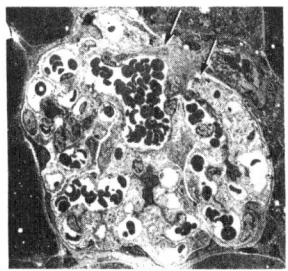

Fig. 3.   Low magnification electron micrograph
           of glomerular tuft. The vascular
           pole is at the arrows.

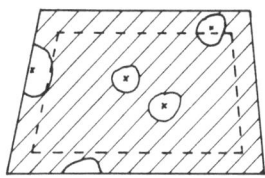

Fig. 4.   Semithin section from which to select
           one glomerular profile for electron
           microscopy.
           Centers of glomerular profiles are
           marked with crosses. Only glomeruli
           within the dotted line are considered
           for selection.

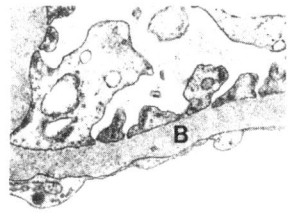

Fig. 5.   Glomerular capillary wall. The peri-
           pheral basement membrane (B) between
           the endothelial and epithelial cells.

CLUSTERING AND LAYERING OF NEURONS IN THE CENTRAL NERVOUS SYSTEM

H. HAUG

Abtlg. für Anatomie, Medizinische Hochschule Lübeck, Germany

SUMMARY:

In many biological tissues, especially in the nervous system, structures are not randomly distributed. Therefore stereological evaluation needs guide-lines that indicates how an optimal systematic sampling can be done. For the cortex cerebri a first solution is mentioned.

During the last years stereology has developed into a science which is capable of solving many quantitative questions in the evaluation of flat planes in sectioned tissues [4, 11, 13]. However, all of these procedures presuppose that the structures are randomly distributed in the investigated specimen. In many cases, especially in the biological field, structures are actually arranged systematically and not randomly [2, 5, 6]. This may be explained by a metaphor.

Buffon [3] introduced the geometrical probability theory which has often been compared with the problem of the distribution of needles in a haystack. The present development in stereology might be described in modern context of the haystack-problem.

A farmer has a field with long grass that he mows twice and a field with clover that he mows four times a year. The grass and the clover are stored in the same silo. In preparation for winter, he wants to know the composition of the silage that he has put away. He forgot to keep careful records of when and what he placed in the silo. However, being an University educated farmer he will use the sampling and evaluation techniques of stereology to study the composition of the silage. He is particularily interested in the quantity and quality of the two, not randomly but periodically distributed hay types.

Thus, we can translate the silage problem into the modern question of a stereological solution in the analysis of heterogeneously constructed tissues.

First it is necessary to give a short sketch of the brain's construction. The brains of men and animals are highly complex organs and are composed basically of two kinds of tissue. One has a grey colour and consists of formations of neurons and their connections. This part is called "grey matter" and can be compared with the central processing unit of a computer. The other part is called "white matter" and contains the myelinated fibres, which connect the different grey structures.

The grey and white matters of the central nervous system are arranged in a strange manner. The neuronal construction is called cytoarchitectonics. Knowledge of the specific cytoarchitectonic composition is important in understanding the brain's activity.

The individual grey matters differ depending on their various functions in size and arrangement of their neurons. Each neuron consists of at least three parts: the perikarya and two kinds of processes. The perikaryon generates the electrical activity. Its architectonic is wellknown in light microscopy because it can easily be stained. Actually the volumepart of the neuronal processes is much larger than that of the perikarya.

One kind of process is the dendrite which samples the input from other neurons with thousands of synapses. Another one is the axon or neurite which transfers the activities of one perikaryon to others, sometimes distant neurons. Therefore, the neuron is a highly polarized structure. The dendrites and neurites are intermingled in a highly complex net-work called neuropil which is situated between the perikarya. This composition of the brain is such that the structures are not randomly distributed.

The strangely arranged structural composition of the brain tissue can be evaluated neither by morphometric nor by stereologic procedures with the usual random sampling techniques. Thus it becomes necessary to work with the aid of a systematic sampling and such a procedure will be described later on. Fig. 1 demonstrates first typical brain compositions and second the possibility of obtaining stereological results which have, however, a probabilistic basis.

A medium power light micrograph of the hypothalamus shows in the middle part darkly stained large nerve-cells which are arranged in mosaic-like clusters of 3 to 10 cells. The individual cells are not randomly distributed, however, the clusters may be that way. Such an arrangement can be observed nearly in all greys of the brain.

The stereological evaluation of such a structural composition depends on the desired results. If the investigator is only interested in knowing the amount of nerve-cells in a certain grey matter, he may evaluate in counting-fields, which are slightly larger than the greatest distance from one cluster to an other [10, 12]. However, two problems cannot be avoided.

1.) Any information on the phenomenon of clustering will be lost.
2.) In some cases errors arise due to the fact that the size of the structures which must be counted becomes too small in comparison with the resolving power of the test lattice.

While the first point can be accepted by the investigator, the second is important and sometimes unknown.

Another construction of even greater interest in the grey matter of the brain is the layered cortex. Fig. 2, a longitudinal section through the cortex of a small ape (Saimiri), shows cytoarchitectonics which are quite different. On the right side especially in the region KS we observe a well-pronounced layering with higher and lower densities of smaller nerve-cells. This area is responsible for somatesthetic function. On the left side (Gig) an area having large neurons but little distinct layering can be seen. This area is responsible for the conscious motor function of the body.

Fig. 3 demonstrates that the architectonics of various structural parts in one area is highly different. On the left, you see the construction of the nerve-cells; in the middle, of the large nerve-fibres, and on the right, of the vessels. Stereology of such layered tissues is only possible if we compromise and use a procedure of systematic evaluation in those areas which have similiar architectonics.

Fig. 4 demonstrates the results obtained after an evaluation of the cortex field by field [8]. In this manner one is able to use that magnification which optimally resolves the structure. The lower right graph demonstrates that the nerve-cell densities in the individual fields are very irregularily distributed. The results in the form of the sums of individual rows lying perpendicular to the surface can be seen in the lower left graph, they are normally distributed. The upper right graph demonstrates the sums of all those fields lying along a line having the same distance from the surface. These values correspond to the structural layering. Within the layers we see again a normal distribution. Such an evaluation provides new and important information on the construction of the layers.

Fig. 4 demonstrates as well the principle of evaluation which has now been facilitated by a computer program. The desired stereological parameters are measured from

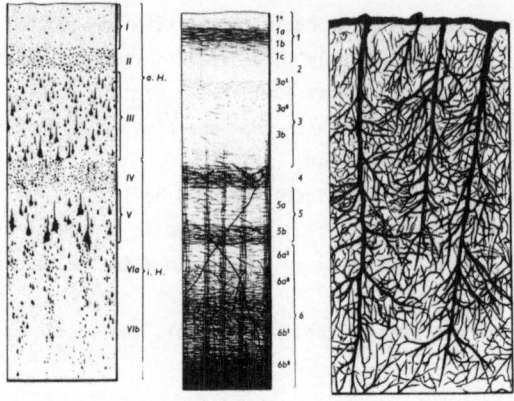

Fig. 3: Three different views of the same human cerebral area:
left cytoarchitectonics, middle myeloarchitectonics,
right architectonics of vessels.

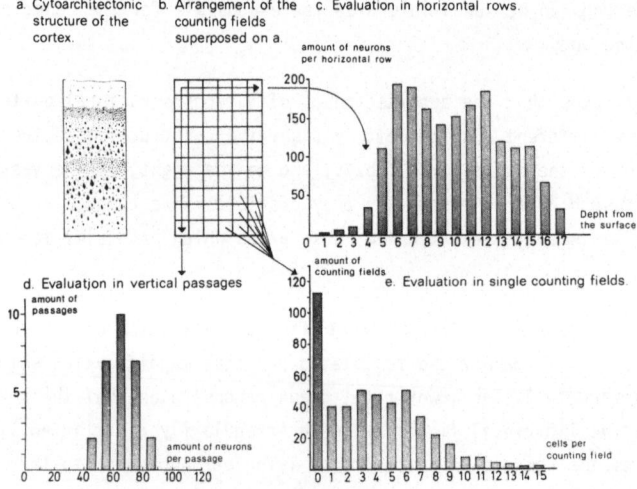

Fig. 4: Scheme of evaluation of the cerebral cortex. Each square
represents one evaluation-field (for more details see text).

one microscopical field to the next one. The fields are arranged in rows perpendicular to the surface of the cortex. The distance of every field to the surface as well as the measured values are stored automatically by a magnetic type recorder [7].

An evaluation of one cortical area needs about 15 to 30 rows, each 15 to 50 fields long. The number of fields in one row depends on the thickness of the cortex. With a program developed in our institute we can compute results for different stereological assumptions; such as numbers per volume, surface areas, lenghts, volumeparts and the size-distribution of various structures. An important further advantage of our program is the possibility of selecting rectangular samples in the perpendicular or horizontal direction. The calculation in the horizontal direction can be done by rotating the matrix of values. The stored original values make it possible to study distinct details in our material without additional measurements.

Fig. 5 presents a graph with results obtained from such a rotation of the value matrix by our computer program. It shows the different densities of the nerve and glia cells in the human visual cortex. The "Zeile" indicates the distance from the cortical surface. The scale of one "Zeile" is the size of one counting field, the number corresponds to the amount of counting fields which lie between the surface and the investigated field. For each "Zeile" the average actual values in 1 mm^3 and their confidence limits are plotted. To avoid irregularities in our samples the program makes it possible to smooth the values by averaging over one, two or three fields.

The same procedure can be used in electron microscopy after making a few changes. Such an electron-microscopical investigation has been completed on the development of myelinated fibres in the visual cortex of the cat. Fig. 6 demonstrates the results on 15 cats ranging in age from 36 days after birth to the adult, younger stages have no myelinization in the visual cortex. For this study 43 000 fibres in 820 electron micrographs at a magnification of 5 400 were evaluated [9]. The possibility of excellent findings in respect to the single layers can be demonstrated in the two parts of fig. 6. In the upper diagram the total fibre length in the layers is presented. However, the boundaries of the layers cannot be located exactly, therefore only the direction from layer I at the surface to layer VI at the base of the cortex is designated. The upper part shows that the amount of fibres in proximity to the surface is high, then decreases to relatively low values. A great amount lies in layer I and a low one in layers II and III. A second maximum can also be seen in layer IV.

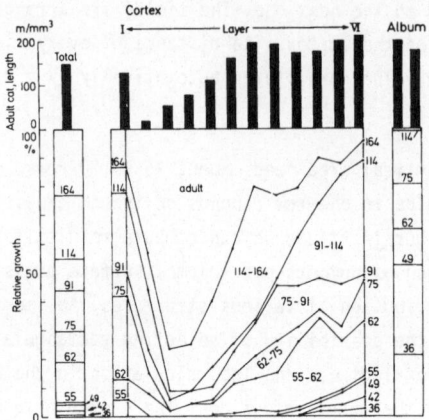

Fig. 6: The length-development of the myelinated fibres in the layers of the
cat's visual cortex (for more details see text).

The lower diagram shows the relative development of the fibres. The numbers on
the lines represent the age in days after birth. In this way it is possible to
follow the growth rate. It is interesting that the fibres do not develop continuously, rather the development appears in a kind of growing pushes. Furthermore,
these growth pushes are different in every layer with regard to age. A third unexpected fact has been that the late development in the layers II and III continues even after the 160th day. The age of a 160 day old kitten is equivalent
to the middle human adolescence.

At the end of my report on evaluations in inhomogeneously structured tissues I
would like to add the following remarks that investigations are actually much
more complicated as demonstrated above. Fig. 7 shows a low power micrograph of a
section through the convoluted human visual cortex. The above-mentioned evaluation procedure presupposes that all of the layers have nearly the same thickness.
This can only be realized in small regions of the cortex. Generally, the thickness
of the cortex and/or the layers changes: The relation of the layers is very
different especially at the bottom, in the walls, and on the top of the sulci and
gyri. This fact can be seen in fig. 7 [1].

At present an evaluation with stereological procedures in such curved layered
tissues remains an unsolved problem. However, I hope that a close collaboration
with mathematicians will provide adequate solution to this and other stereological
problems. Such a solution should facilitate a rapid and accurate evaluation

procedure especially in an appropriate selection of representative fields.

The need for stereological procedures is of great importance in the evaluation of brain tissue because the complicated composition of its structure can only then provide actual values with respect to the third dimension. Stereological thinking facilitates our ability to understand the construction of the brain.

REFERENCES

1. BOK, S.T.: A quantitative analysis of the structure of the cerebral cortex. Verhandl. K. Akad. v. wetensch., Amsterdam, 2 de. sect. 35, 1-55 (1936).
2. BOK, S.T.: Quantitative analysis of the morphological elements of the cerebral cortex. In: S.T. BOK (Ed.), Structure and function of the cerebral cortex, pp 7-17. Amsterdam: Elsevier Publ. Comp. 1960.
3. BUFFON, G.L.L.: Essai d'arithmétique morale. Suppl. a l'Histoire Nature IV. Paris 1777.
4. ELIAS, H., HENNIG, A., SCHWARTZ, D.E.: Stereology: Applications to biomedical research. Physiol. Rev. 51, 158-2oo (1971).
5. HAUG, H.: Quantitative Untersuchungen an der Sehrinde. Stuttgart: Georg Thieme 1958.
6. HAUG, H.: Stereological methods in the analysis of neuronal parameters in the central nervous system. J. Microsc. 95, 165-180 (1972).
7. HAUG, H.: Experiences with optomanual automated evaluation-systems in biological research, especially in neuromorphology. National Bureau of Standards Special Publication 431. Proceedings of the Fourth International Congress for Stereology held at NBS, Gaithersburg, Md., Sept. 4-9, 1975.
8. HAUG, H., KEBBEL, J., WIEDEMEYER, G.-L.: Die Messung der mittleren Zelldichte und ihre Verteilung in Geweben mit erheblichen Zelldichteunterschieden. Auswertung am Cortex cerebri als Beispiel. Microsc. Acta 71, 121-128 (1971).
9. HAUG, H., KÖNIG, M., RAST, A.: The postnatal development of the myelinated fibres in cat's visual cortex. A stereological and electron microscopical investigation. Cell Tiss. Res. 167, 265-288 (1976).
10. LEDER, O.: Die ungerichtete, zweidimensionale,systematische Stichprobe als stereologisches Punktnetz. (Personal communication 1976).
11. UNDERWOOD, E.E.: The stereology of projected images. J. Microsc. 95, 25-44 (1972).
12. WEIBEL, E.R.: Selection of the best method in stereology. J. Microsc. 100, 261-269 (1974).
13. WEIBEL, E.R., ELIAS, H.: Introduction to stereology and morphometry. In: E.R. WEIBEL, H. ELIAS (Eds.), Quantitative methods in morphology, pp. 3-19. Berlin, Heidelberg, New York: Springer 1967.

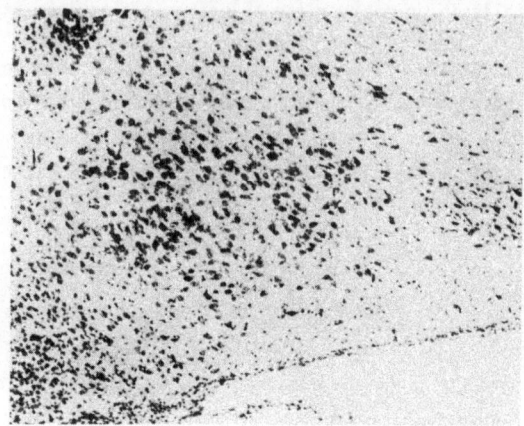

Fig. 1: A grey nucleus in the hypothalamus of a guinea-pig
with clustered large nerve-cells.

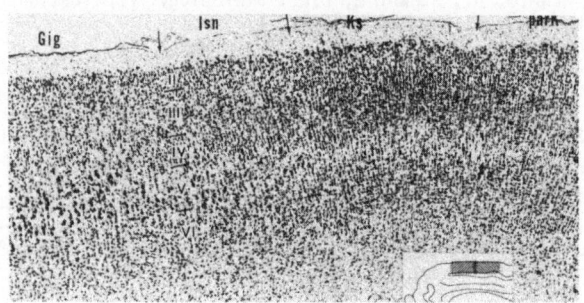

Fig. 2: Four different cytoarchitectonic areas of the cortex
cerebri of a small ape (Saimiri sciurius).

SEHRINDE MENSCH 76 J.

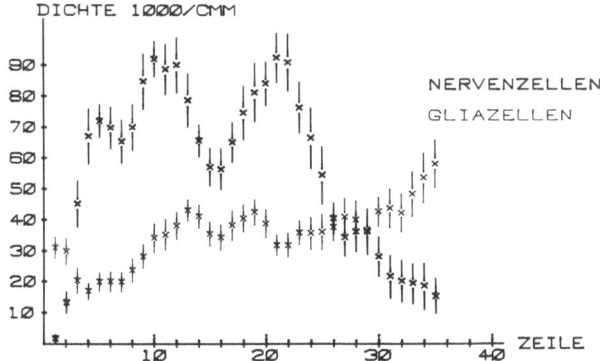

Fig. 5: A plot-diagram of the neuronal densities and their confidence-
limits obtained by an evaluation of the human visual cortex.

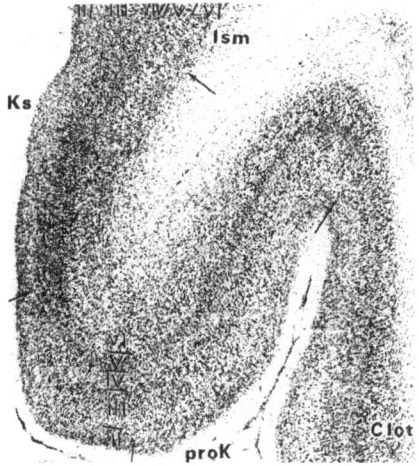

Fig. 7: The changing cytoarchitectonics of the cortex and its layers
in different locations of the convolution (Saimiri sciurius).
Arrows indicate borders between cortical areas.

ANISOTROPY AND DISTANCE FROM THE CENTRE.PROBABILISTIC MODELS

AND STATISTICAL ANALYSIS.

By

Franz STREIT

University of Geneva, Switzerland.

1. Introduction

Geometrical stochastics is the branch of mathematics which is concerned with the probabilistic and statistical analysis of randomly generated sets in space. Sets encountered in every day life can often
reasonably be interpreted as realizations of random sets. This may
for instance be the case for the system of holes in a porous material (Mathéron (1967)) or for the set of sites on an area of ground
where individual plantes of a particular species are growing (Pielou
(1969, chapter 9 and 10)).

For the derivation of many results of geometrical stochastics some
kind of assumption of invariance with respect to a group of transformations is of crucial importance (for examples see e.g.Kendall
and Moran (1963) and Santaló (1976)). Very often it is supposed that
the random set is composed of rigid, undeformable objects which are
randomly laid out in an Euclidean space and that the probability law
of the position of each object is invariant with respect to any motion acting on this object. Intuitively this means that every position in space of such an object is equally likely.

In concrete situations it is not always possible to justify this
assumption of complete invariance. It may for instance be that the
randomly laid out object in space prefers certain orientations or
that it prefers certain regions in space (for instance certain dis-

tances from the centre). Streets interpreted as random lines may tend
to agglomerate in certain parts of the countryside, a needle falling
randomly on a table may show the effect of magnetism and the cells of
a muscle may point mainly in one prefered direction. Under such conditions the use of a model allowing only partial invariance (for instance with respect to the subgroup of translations or with respect to
the subgroup of rotations rather than with respect to the full group
of motions) will be more appropriate.

Two types of problems arise in this context :

1) The construction of various stochastic models for the considered
problem and the probabilistic solution of the problem for these models.

2) The construction of optimal statistical procedures for the evaluation of the different stochastic models taken into consideration
(typically problems of this second type do not arise in the classical
part of geometrical stochastics commonly called 'geometrical probability' because the invariant probability measure is essentially unique).

In the following we illustrate these ideas in treating a problem inspired by the classical work of Buffon (1777). Similar investigations
can be made for the well-known problem of Bertrand. It turns out that
the statistical tests for the comparison of the important stochastic
models depend on the randomly laid out objects (straight lines) only
in terms of their distances from a centre (which may be chosen at Z)
(Streit (1976)). In contrast to this the angular distribution is the
relevant distribution for the statistical analysis presented in section 3.

2. Probabilistic Analysis

2.1. A Buffon Problem Allowing Anisotropy and its General Solution.

Suppose that the Euclidean plane E^2 with origin Z and a system of Cartesian x-y-coordinates is subdivided by the grid K consisting of the parallel lines $x \equiv k$ ($k = 0, \pm 1, \pm 2, \ldots$). Denote by $S_1(x,y,\varphi)$ the segment of lenght 1 whose midpoint is located in the point (x,y) of E^2 and whose angle formed by the corresponding straight line and the x-axis is φ ($x,y \in (-\infty, \infty)$, $0 \leqslant \varphi < \pi$). Let $S_1(X,Y,\Phi)$ be an analogous segment in random position (X,Y,Φ) in E^2. In this paper we suppose that $S_1(X,Y,\Phi)$ is of moderate lenght, i.e. that $0 \leqslant 1 \leqslant 1$. Furthermore we assume that the (improper) probability measure of the position variables is induced by the differential form dx dy dG(φ) where G designates a distribution function which concentrates all the probability mass on the interval $[0,\pi)$. We thus admit that the segment may tend to take prefered directions.

We want to determine the probability $Pr(S_1(X,Y,\Phi) \cap K \neq \emptyset : G)$ that the random segment intersects a line of the system K. As indicated, this probability depends on the choice of G. Taking into account that for reasons of symmetry this probability is equal to the corresponding probability evaluated under the restrictions $-\frac{1}{2} \leqslant x < \frac{1}{2}$ and $y = 0$ and using $\int^*$ to denote integration over those points $x \in [-\frac{1}{2}, \frac{1}{2})$ for which $|x| \leqslant \frac{1}{2} 1 \cdot \{|\cos(\varphi)|\}$ we find

$$Pr(S_1(X,Y,\Phi) \cap K \neq \emptyset : G) = \{ \int_0^\pi \int^* dx \, dG(\varphi) \} \cdot (\int_{-\frac{1}{2}}^{+\frac{1}{2}} dx)^{-1} \text{ and}$$

thus

$$Pr(S_1(X,Y,\Phi) \cap K \neq \emptyset : G) = 1 \int_0^\pi |\cos(\varphi)| \, dG(\varphi). \qquad (2.1)$$

This is the general solution of a Buffon-type problem allowing anisotropy. It may be seen that the probability (2.1) is the product of

the lenght l of the segment and the expectation of the absolute value of cos(φ) evaluated by means of the distribution function G.

2.2 Special Cases.

Lets now consider a few specific stochastic models by making special
choices on the angular distribution. We suppose that this distribution
is absolutely continuous and denote the density function corresponding to G by g.

Case I

$$g_1(\varphi) : = \pi^{-1} \qquad (0 \leqslant \varphi < \pi).$$

In this stochastic model the angular distribution is the uniform
distribution over $[0,\pi)$. The probability law of the positions is
induced by the so-called integral geometric kinematic density of
$S_1(X,Y,\Phi)$ in E^2 (see for instance Santaló (1976), pp.89,92). Our
problem coincides with the original needle problem of Buffon, treating the case in which every orientation of the segment is equally
likely. Its solution is well-known. We obtain from (2.1)

$$P_1 : = P(S_1(X,Y,\Phi) \cap K \neq \emptyset : G_1) = 2 \, l/\pi = 0,637 \, 1 \, . \quad (2.2.1)$$

Case II

$$g_2(\varphi) : = \{ \pi \, I_0(\mathcal{K}) \}^{-1} \exp \, \{\mathcal{K} \cos(2\varphi) \} \qquad (\mathcal{K} \geqslant 0)$$
$$(0 \leqslant \varphi < \pi)$$

Here I_0 denotes the modified Bessel function of the first kind of
order 0, i.e.

$$I_0(\mathcal{K}) : = \sum_{r=0}^{\infty} \{ (\tfrac{1}{2}\mathcal{K})^{2r} \Big/ (r \, !)^2 \} \, .$$

For this stochastic model the angular distribution is the circular
von Mises distribution rescaled to $[0,\pi)$ (see Mardia (1972), p.70).

The distribution, which depends on the additional parameter $\mathcal{K}$, may be considered as one possible circular analogon to the normal distribution on the real line. Applying this model means that we believe that the segment prefers the angles around 0 and π and this to a greater extent the greater $\mathcal{K}$ is. Since $g_2(\varphi : 0) \equiv g_1(\varphi)$, the particular choice $\mathcal{K} = 0$ leads back to the case I. From (2.1) we obtain

$$P_2(\mathcal{K}) : = Pr(S_1(X,Y,\Phi) \cap K \neq \emptyset : G_2, \mathcal{K})$$

$$= 1 . \{ \pi I_0(\mathcal{K}) \}^{-1} (J_1 + J_2)$$

where $J_1 : = \int_0^{\frac{\pi}{2}} \cos(\varphi) \exp[\mathcal{K}\cos(2\varphi)] \; d\varphi$

$$= \tfrac{1}{2} \int_0^\pi [\tfrac{1}{2} . \{ 1 + \cos(\$)\}]^{\frac{1}{2}} \exp\{\mathcal{K}\cos(\$)\} \; d\$$$

$$= \tfrac{1}{2} \int_{-1}^{+1} \{ 2(1 - z)\}^{-\frac{1}{2}} \exp(\mathcal{K}z) \; dz$$

$$= \{ 2(2\mathcal{K})^{\frac{1}{2}} \}^{-1} e^{\mathcal{K}} \int_0^{2\mathcal{K}} u^{-\frac{1}{2}} e^{-u} \; du$$

and $J_2 = J_1$, as may be easily verified.

Introducing the incomplete γ-integral, defined by the relation

$$\gamma(a,x) : = \int_0^x t^{a-1} e^{-t} \; dt \qquad\qquad (a, \; x > 0)$$

we find

$$P_2(\mathcal{K}) = 1 \; e^{\mathcal{K}} \{ \pi I_0(\mathcal{K}) (2\mathcal{K})^{\frac{1}{2}} \}^{-1} \quad \gamma(\tfrac{1}{2}, \; 2\mathcal{K}) \qquad (2.2.1a)$$

As may be seen from the alternative representation

$$P_2(\mathcal{K}) = c_2(\mathcal{K}) . P_1$$

with $c_2(\mathcal{K}) : = \{ 2 I_0(\mathcal{K}) (2\mathcal{K})^{\frac{1}{2}}\}^{-1} e^{\mathcal{K}} \; \gamma(\tfrac{1}{2}, \; 2\mathcal{K}),$ \qquad (2.2.1b)

$P_2(\mathcal{K})$ may be obtained from the solution of the original problem P_1

by multiplying with the correction factor $c_2(\mathcal{K})$. Numerical values of $c_2(\mathcal{K})$ and $P_2(\mathcal{K})$ are indicated in the following table

$\mathcal{K}$	$c_2(\mathcal{K})$	$P_2(\mathcal{K})$
0	1	0,637 1
1	1,284	0,817 1
2	1,430	0,910 1
5	1,527	0,972 1
10	1,550	0,987 1
∞	1,571	1

In interpreting this data the supposition made that $0 \leqslant 1 \leqslant 1$ has to be kept in mind.

Case III

$$g_3(\varphi) := \{ \pi \, I_0 \, (\mathcal{K}) \}^{-1} \; \exp[\mathcal{K} \cos(\varphi)] \qquad (\mathcal{K} \geqslant 0)$$
$$(0 \leqslant \varphi < \pi)$$

For this stochastic model the angular distribution is the circular von Mises distribution with mean direction 0 truncated at π (Mardia (1972), p.57). Applying this model means that we believe that the segment prefers the angles around $\varphi = 0$ and shuns the angles around $\varphi = \pi$, and this to a greater extent the greater $\mathcal{K}$ is. For $\mathcal{K} = 0$ the model is again the one described under case I.

From (2.1) we obtain

$$P_3(\mathcal{K}) := \Pr(S_1(X,Y,\Phi) \cap K \neq \emptyset : G_3, \mathcal{K})$$
$$= 1 \, \{\pi \, I_0(\mathcal{K})\}^{-1} \, \{ M(\cdot) + M(-\mathcal{K}) \} \, ,$$
where $M(\mathcal{K}) := \frac{1}{2} \sum_{\nu=0}^{\infty} \{ \mathcal{K}^{\nu} B(\frac{\nu}{2} + 1, \frac{1}{2}) \, / \nu! \} \, ,$

and B denotes the Beta function. Thus

$$P_3(\mathcal{K}) = 1 \{ \pi \ I_0(\mathcal{K}) \}^{-1} \ \sum_{\nu=0}^{\infty} [\mathcal{K}^{2\nu} \ B(\nu + 1, \tfrac{1}{2}) \ / \ \{ (2\nu) \ ! \} \ ] \qquad (2.2.2a)$$

Another representation, which is more appropriate for numerical evaluation, is

$$P_3(\mathcal{K}) = 1 \{ I_0(\mathcal{K}) \}^{-1} \{ L_1(\mathcal{K}) + 2/\pi \} , \qquad (2.2.2b)$$

where L_1 denotes the modified Struve function of order 1, i.e.

$$L_1(z) : = (\tfrac{z}{2})^2 \ \sum_{\nu=0}^{\infty} [\ (\tfrac{z}{2})^{2\nu} \ \{ \Gamma \ (\ \nu + \tfrac{3}{2})\Gamma(\nu + \tfrac{5}{2}) \ \}^{-1}]$$

(see Abramowitz and Stegun (1970), p.498 and p. 501). For making comparisons with the classical result P_1 we may use the relation

$$P_3(\mathcal{K}) = c_3(\mathcal{K}) \ P_1 , \qquad (2.2.2c)$$

where the multiplicative correction factor is given by

$$c_3(\mathcal{K}) = \{ I_0(\mathcal{K}) \}^{-1} . \{ 1 + \pi \ L_1(\mathcal{K}) \ / \ 2 \ \} .$$

The following table gives some numerical indications related to the formula (2.2.2c).

$\mathcal{K}$	$c_3(\mathcal{K})$	$P_3(\mathcal{K})$
0	1	0,637 1
1	1,071	0,682 1
2	1,199	0,763 1
5	1,405	0,894 1
10	1,490	0,949 1
∞	1,571	1

Case IV

$$g_4(\varphi : \tilde{\alpha}, K) = \pi^{-1} [\ 1 + K \cos\{ 2(\varphi - \tilde{\alpha}) \}] \ (0 \leqslant \tilde{\alpha} < \pi, -1 \leqslant K \leqslant +1)$$
$$(0 \leqslant \varphi < \pi)$$

For this stochastic model the angular distribution is the circular cardioid distribution rescaled to $[0,\pi)$ (Mardia (1972), p.51) . This distribution has two parameters $\widetilde{\alpha}$ and K. For K = 0 the model discussed in case I is recovered. Applying this model means that we believe that the segment prefers the angles around $\varphi = \widetilde{\alpha}$ for K > 0 and around the direction which differs from $\widetilde{\alpha}$ by $\pi/2$ for K < 0.

From (2.1) we obtain a result which was previously discovered by Marriott (1971), namely

$$P_4(\widetilde{\alpha}, K) : = Pr(S_1(X,Y,\Phi) \cap K \neq \emptyset : G_4, \widetilde{\alpha}, K)$$

$$= [2 \, 1 \, \{ 1 + K \cos(2\widetilde{\alpha})/3 \}]/\pi \qquad\qquad (2.2.3a)$$

This may be also expressed by the relation

$$P_4(\widetilde{\alpha}, K) = c_4(\widetilde{\alpha}, K) \cdot P_1$$

with

$$c_4(\widetilde{\alpha}, K) : = 1 + K \cos(2\widetilde{\alpha}) /3$$

Numerical values for $c_4(\widetilde{\alpha}, K)$ and $P_4(\widetilde{\alpha}, K)$ are thus easy to determine. It is of interest to note that

$$P_4(\pi/4, K) \equiv P_4((3\pi) /4, K) \equiv P_1 \quad .$$

3. Statistical Analysis

Suppose that we want to judge on the appropriateness of the stochastic
models discussed in section 1 and that we are considering to this end
a random sample $S_1(X_1,Y_1,\Phi_1)$, ..., $S_1(X_n,Y_n,\Phi_n)$ of segments. In this
paper we make only some comments on the problem of comparing the ani-
sotropical models with the isotropical models, since this seems to
be the question most likely raised in this context.

For the comparison of case I with case II or case III the problem
consists in finding an optimal test for the hypotheses $H_0 : \mathcal{K} = 0$
and $H_1 : \mathcal{K} > 0$. From considerations developed in the book of
Mardia ((1972), p.132) it may be seen that the uniformly most powerful test for comparing H_0 with H_1 at the level of significance α
depends on the segments only in terms of their angles and may be
based on the test statistics

$$\overline{C} = \begin{cases} \overline{C}_{II}(\underset{\sim}{\Phi}) = \frac{1}{n} \sum_{i=1}^{n} \cos(2\,\Phi_i) & \text{alternative models of case II} \\ \overline{C}_{III}(\underset{\sim}{\Phi}) = \frac{1}{n} \sum_{i=1}^{n} \cos(\Phi_i) & \text{alternative models of case III} \end{cases}$$

and the decision rules

reject H_0 if $\overline{C}(\underset{\sim}{\varphi}) > c(1-\alpha)$
accept H_0 if $\overline{C}(\underset{\sim}{\varphi}) \leqslant c(1-\alpha)$ ' .

A table of the critical values $c(1-\alpha)$ exists and may be found in
the quoted book or in Stephens (1969). For the comparison of case I
with case IV it may be useful to look for a most powerful test for the
hypotheses $H_0 : K = 0$ and $H_1 : K = K_1 \neq 0$ and $\tilde{\alpha} = \tilde{\alpha}_1$. Such a test
may be based on the quantity

$$\Delta_n := \sum_{i=1}^{n} (\ln \, [1 + K_1 \cos \{ \, 2(\Phi_1 - \tilde{\alpha}_1) \, \} \,] \,)$$

and a critical region rejecting large realized values of this statistic. Advantage may be taken of the fact that Δ_n follows under the null hypothesis asymptotically a normal distribution, i.e. that for $|K_1| < 1$

$$\sigma_n^{-1} (\Delta_n - \mu_n) \sim N(0,1) \quad \text{for } n \to \infty \, ,$$

where $\mu_n = -n \sum_{s=1}^{\infty} \{ (2 \pi s)^{-1} \, (2K_1)^{2s} \, B(s + \tfrac{1}{2}, \, s + \tfrac{1}{2}) \}$

and $\sigma_n^{2} = n \sum_{s=1}^{\infty} \{ (\pi s)^{-1} \, (2K_1)^{2s} \, B(s + \tfrac{1}{2}, \, s + \tfrac{1}{2}) \, (\sum_{v=1}^{2s-1} v^{-1}) \} - n\mu_1^2 .$

REFERENCES

ABRAMOWITZ, M. and STEGUN, I.A. - (1970) Handbook of Mathematical Functions. New York : Dover.

BUFFON, G.L.L. (1777). Essai d'arithmétique morale, Supplément à l'histoire naturelle, $\underline{4}$.

KENDALL, M.G. and MORAN, P.A.P. (1963). - Geometrical Probability. London : Griffin.

MARDIA, K.V. (1972). Statistics of Directional Data. London : Academic Press.

MARRIOTT, F.H.C. (1971). Buffon's problem for non-random directions. Biometrics $\underline{27}$, 233-235.

MATHERON, G. (1967). Eléments pour une théorie des milieux poreux. Paris : Masson.

PIELOU, E.C. (1969). An Introduction to Mathematical Ecology. New York : Wiley Interscience.

SANTALO, L.A. (1976). Integral Geometry and Geometric Probability. Reading : Addison Wesley.

STEPHENS, M.A. (1969). Tests of randomness of directions against two circular alternatives. J.Amer.Statist.Soc. $\underline{64}$, 280-289.

STREIT, F. (1976). On Methods and Problems of Geometrical Stochastics. Bulletin of the ISI $\underline{46}$(2), 600-605.

CHARACTERISTIC STATISTICAL PROBLEMS

OF STOCHASTIC GEOMETRY

by

G. S. Watson
Princeton University

Technical Report No. 130, Series 2
Department of Statistics
Princeton University
August 1977

This work was partially supported by the
Office of Naval Research contract N00014-
75-C-0453 awarded to the Department of
Statistics, Princeton University, Princeton, New Jersey, and written while the
author was a Guggenheim Fellow.

ABSTRACT

The Buffon needle problem and some variations
are used to illustrate classical statistical methods
of estimation and to lead into, and contrast with,
the problems which arise when a sample of some random
structure is the data. The flavor of these problems
is conveyed largely by discussion of the simplest,
and most described, case, that of point processes.

Note: This technical report was originally delivered
as a lecture at the Buffon Bicentenary Symposium on
Geometrical Probability, Image Analysis, Mathematical
Stereology and their relevance to the determination of
Biological Structure, Paris, June 20-24, 1977.

1. INTRODUCTION

The Buffon needle problem has its origins in gambling. But, unlike card and dice games, it requires some geometry and calculus. Thus it was a bold step in generalizing the idea of mathematical probability. The needle problem and minor variants have been studied by many authors up to the present day as a means of estimating π statistically. More sensibly, however, they provide a means of inferring the ratio of the length of the needle to the scale of the regular network onto which it is thrown. And of course this is the origin of modern methods for the sampling study of geometric bodies. Thus it seems most appropriate at this Symposium to begin my talk by going over this work which used only quite standard statistical ideas.

However when the needle is replaced by a more general probe or sampling window -- in other words, some set -- and the regular network is replaced by a random structure, we must face statistical problems of quite a different character. They are akin to those met in time series analysis which has a long history by statistical standards but in which there is still some confusion between the exploratory, modelling and confirmatory aspects and a lack of communication between probabilists and statisticians. The spatial problems are decidedly more difficult. There are not so many explicit probability models to get statistical experience with. One must consider the shape as well as the size of sampling windows. The data has a more awkward form. The second part of the lecture will therefore merely give some idea of these problems and their literature. Hopefully other speakers will address them in detail. But it is clear that statistical geometry or morphology is just beginning.

2. BUFFON PROBLEMS

The classical problem considers a parallel grid with spacing a
and a needle of length ℓ where $\ell \leq a$. If the needle is tossed so
that its position and orientation are random, the probability p that
it cuts a grid line is given by

$$p = \frac{2}{\pi} \frac{\ell}{a} \quad . \tag{1}$$

Uspensky (1937), Kendall and Moran (1963), for example, give proofs of
this and most of the results used below.

Define $\phi = \frac{1}{\pi}$, $r = \frac{\ell}{a}$ and suppose that n independent trials
(tosses) yield C cuts. Then $\mathcal{L}(C)$ = Binomial (n,p) where $p = 2\phi r$.
The statistical problems that arise are

       (i)   estimate ϕ , i.e., estimate π

      (ii)   estimate r, i.e., estimate a if ℓ is known

     (iii)   test that $r = r_0$ (known).

The most complete reference on (i) in this case and those given below
is Perlman and Wichura (1975). Oddly no one seems to have considered
(ii) and (iii). Trivial though they are, they are the prototypes of
the real problems.

The likelihood of the data when c cuts are observed is

$\binom{n}{c} p^c (1-p)^{n-c}$ and the number of cuts is a complete sufficient statistic for p. Thus we may assert: among all functions f(c) such that
Ef(C) = p, for all p in (0,1), it is true that

$$\text{var } f(C) \geq \text{var } \frac{C}{n} = n \, p(1-p) \quad . \tag{2}$$

Thus the obvious estimator of p, $\hat{p}$ = C/n is, uniformly in p, the minimum variance unbiased estimator of p.

If one knows r and wishes to estimate π, the same is true of

$$\hat{\phi} = \frac{c}{2rn} \;,\; \text{var}\,(\hat{\phi}) = \frac{\phi^2}{n}\left(\frac{1}{p} - 1\right)\;. \tag{3}$$

However $\hat{\pi}_1 = 1/\hat{\phi}$ is biased. If n is large, we may argue that

$$E\hat{\pi}_1 = E\frac{1}{\phi}\left[1 + \frac{\hat{\phi}-\phi}{\phi}\right]^{-1}$$

$$\approx \frac{E}{\phi}\left\{1 - \frac{\hat{\phi}-\phi}{\phi} + \frac{(\hat{\phi}-\phi)^2}{\phi^2}\right\}$$

$$= \pi\left\{1 - 0 + \frac{1}{n}\left(\frac{1}{p} - 1\right)\right\}$$

$$\rightarrow \pi \quad \text{as} \quad n \rightarrow \infty\;.$$

Further it follows from (3) that

$$\text{var}\,\hat{\pi}_1 \approx \frac{1}{n\phi^2}\left(\frac{\pi a}{2\ell} - 1\right)\;. \tag{4}$$

Thus if one is set on estimating π this way and can "design the experiment" one should take $\ell = a$ since this choice minimizes (4). In this case $\text{var}\,\hat{\pi}_1 \approx 5.63/n$. Lazzerini (1901) conducted such an experiment with $n = 3408$ and found $\hat{\pi}_1 - \pi = 3 \times 10^{-7}$. As Kendall and Moran (ibid.) suggest, he must have stopped when he noticed the remarkable and fortuitous accuracy!

In fact we know ϕ and are more likely to want to know a. By the same arguments used above, $\hat{a} = \frac{2\ell\phi n}{c}$, and we will have $E\hat{a} \rightarrow a$ as $n \rightarrow \infty$ and

$$\text{var}\,(\hat{a}) \approx \frac{a^3}{n}\left(\frac{\pi}{2\ell} - \frac{1}{a}\right)\;. \tag{5}$$

Again if we can choose the needle size we should try to make it near to but less than a. Thus the effort to optimize may lead to bias since (1) is false when $\ell > a$. Trivially if n is large, $\hat{a}$ is Gaussian so tests are easy to make.

Instead of a parallel grid, Laplace considered a rectangular grid, the A lines being a apart, the B lines being b apart. He showed that the probability that the needle cuts at least one line is

$$\frac{2\ell(a+b) - \ell^2}{\pi ab} , \tag{6}$$

a fascinating formula whose direct derivation is tricky so that it seems easier to get it from a more general Crofton argument. To illustrate my points here set $a = b$ and

$$\frac{\ell}{a} = \frac{\ell}{b} = r , \quad \phi = \frac{1}{\pi} .$$

Introducing the notation

$$P_{\overline{A}B} = \text{Prob(needle cuts a B line but not an A line)},$$

$$P_{\overline{AB}} = \text{Prob(needle cuts neither an A nor a B line)},$$

etc., formula (6) is clearly $1 - P_{\overline{AB}}$. Further if

$$P_A = \text{Prob(needle cuts an A line)}$$

$$P_B = \text{Prob(needle cuts a B line)}$$

then

$$P_A = P_{AB} + P_{A\overline{B}} = 2r\phi ,$$

$$P_B = P_{AB} + P_{\overline{A}B} = 2r\phi ,$$

and

$$P_{AB} + P_{A\overline{B}} + P_{\overline{A}B} + P_{\overline{AB}} = 1.$$

These three equations plus (6) for $1 - P_{\overline{AB}}$ yield

$$P_{AB} = r^2\phi \ , \ P_{\overline{AB}} = P_{A\overline{B}} = r(2-r)\phi \tag{7}$$

and for brevity we follow Perlman and Wichura in writing

$$P_{\overline{AB}} = 1 - (4r-r^2)\phi = 1 - m\phi \ . \tag{8}$$

Let n trials yield results (in an obvious notation)
$\underline{N} = \{n_{AB}, n_{A\overline{B}}, n_{\overline{A}B}, n_{\overline{AB}}\}$. Then

$$\mathcal{L}(N) = 4\text{-nomial } (n; P_{AB}, P_{A\overline{B}}, P_{\overline{A}B}, P_{\overline{AB}}) \ .$$

Thus the likelihood of the data is proportional to

$$L = P_{AB}{}^{n_{AB}} \ P_{A\overline{B}}{}^{n_{A\overline{B}}} \ P_{\overline{A}B}{}^{n_{\overline{A}B}} \ P_{\overline{AB}}{}^{n_{\overline{AB}}} \ .$$

Defining

$$N_0 = n_{\overline{AB}} = \# \text{ no cuts}$$

$$N_1 = n_{A\overline{B}} + n_{\overline{A}B} = \# \text{ 1 cuts}$$

$$N_2 = n_{AB} = \# \text{ 2 cuts}$$

$$n = N_0 + N_1 + N_2$$

and using (7) and (8), L may be written as

$$L = (1-m\phi)^{N_0} \ \phi^{N_1+N_2} \ r^{2N_2+N_1} \ (2-r)^{N_1}. \tag{9}$$

Thus if r is known, N_0 or $N_1 + N_2$ is a complete sufficient statistic for ϕ and $L(N_1+N_2) = \text{Binomial}(n, m\phi)$ so the story of the estimator of π follows the previous pattern. The resulting estimator $\hat{\pi}_2$ has, for n large, a variance equal to $0.47/n$ so that the extra work in using a square grid yields an estimator which is 12 times as efficient as that for the parallel grid.

However to estimate r knowing π is quite different. The practical method is to choose r to maximize the likelihood (9). Setting $\partial \log L/\partial r$ equal to zero leads to the equation

$$\frac{N_0}{1 - \frac{4r-r^2}{\pi}} \frac{2r-4}{\pi} + \frac{2N_2+N_1}{r} - \frac{N_1}{2-r} = 0 \qquad (10)$$

which must be solved iteratively to yield $\hat{r}$. The standard theory of maximum likelihood estimation gives us an asymptotic formula for $\hat{r}$,

$$\text{var } \hat{r} \sim 1 \bigg/ - \left(\frac{\partial^2 \log L}{\partial r^2} \right)_{\hat{r}} \qquad (11)$$

and asymptotic normality of $\hat{r}$ so that tests can be made.

The rectangular grid follows the same pattern with $r_1 = \frac{\ell}{a}$, $r_2 = \frac{\ell}{b}$, L is a function of r_1 and r_2 which are esti- mated by solving $\partial \log L/\partial r_1 = 0$, $\partial \log L/\partial r_2 = 0$. Other regular networks do not introduce the need for further techniques.

Above we considered only the case of $\ell \leq a = b$. The case when ℓ is <u>much</u> greater than a is simple and instructive to consider. Let then this long needle intersect the B lines at an angle θ. Then θ is uniformly distributed on $(0,\pi/2)$. If we define

$$\left. \begin{array}{l} N_A = \# \text{ A lines cut} \approx r \cos \theta \ , \\ N_B = \# \text{ B lines cut} \approx r \sin \theta \ , \\ N = \# \text{ lines cut} \approx r(\cos \theta + \sin \theta) \ , \end{array} \right\} \qquad (12)$$

we have

$$EN_A = \frac{2}{\pi} \int_0^{\pi/2} r \cos \theta \, d\theta = \frac{2r}{\pi} = E(N_B) \ ,$$

$$EN = \frac{4r}{\pi} \ ,$$

$$EN^2 = E(r^2 + 2r^2 \cos \theta \sin \theta) \ ,$$

$$= r^2(1 + \frac{2}{\pi}) \ ,$$

so

$$\text{var } N = r^2\left(1 + \frac{2}{\pi} - \left(\frac{4}{\pi}\right)^2\right) \ .$$

Hence if we make n throws and find $\bar{c}$ as the average number of cuts, it will be an unbiased estimator of $E(N)$. Hence the estimator of π that is suggested, following our ealier work, is

$$\hat{\pi}_3 = \frac{4r}{\bar{c}}$$

and an easy calculation shows that

$$\text{var } \hat{\pi}_3 \approx \frac{\pi^4}{16r^2} \ \frac{\text{var } N}{n}$$

$$\approx \frac{0.0095}{n} \ .$$

While this seems a great improvement, we will show below that one can do better still with this experiment.

To estimate r from $\bar{c}$, the "natural" method is to set $r = \frac{\pi}{4}\bar{c}$ with $\text{var } \hat{r} \approx \frac{r^2}{n} \frac{\pi^2}{4^2}\left(1 + \frac{2}{\pi} - \frac{16}{\pi^2}\right)$. This estimate too can be improved because neither is the maximum likelihood estimator, as was true of our first three examples.

From (12), $N = \sqrt{2}\ r \cos\left(\theta - \frac{\pi}{4}\right)$ so that

Prob$(N \leq k)$ = Prob$(\cos \psi \leq k/r\sqrt{2})$ where $\mathcal{L}(\psi)$ is uniform on $(-\pi/4,\ \pi/4)$. Let $\psi_0 = \cos^{-1} \dfrac{k}{r\sqrt{2}}$. Then

$$\text{Prob}(N \leq k) = 2\ \text{Prob}(\psi > \psi_0)\ ,$$

$$= \frac{4}{\pi}\left(\frac{\pi}{4} - \psi_0\right)\ ,$$

(13)

so that the probability density of N at k is the partial derivative of (13) with respect to k, namely

$$\frac{4}{\pi}\ \frac{1}{\sqrt{2r^2 - k^2}}\ .$$

Thus given counts $k_1,\ldots,k_n$ in n trials, their likelihood is

$$\prod_1^n \frac{4}{\pi}\ \frac{1}{\sqrt{2r^2 - k_i^2}}$$

provided $r \leq$ all the k_i's $\leq \sqrt{2}\ r$, and zero otherwise. Thus the maximum likelihood estimate of r is

$$r^\ast = \frac{1}{\sqrt{2}}\ \max(k_1,\ldots,k_n)\ ,$$

(14)

not $\hat{r} = \pi\bar{c}/4$. It may be shown that the variance of $r^\ast$ is of order n^{-2}, not order n^{-1} like that for $\hat{r}$. Thus for large n, $r^\ast$ is a very much better estimator than $\hat{r}$. This shows dramatically that the usual practice in geometrical statistics of obtaining estimators by equating theoretical and observed means may be very inefficient. So much for the "long needle." Other details may be found in Diaconis (1976) and in a forthcoming monograph by H. Solomon.

Buffon's needle may be used to obtain a connection with a quite different aspect of geometrical statistics. Let us analyze the tossing of the needle. Suppose now that an origin is marked on one of the lines of the parallel grid and that the needle is thrown so that its center rests on the plane a distance X from the marked line. Let $\mathcal{L}(X)$ = Gaussian $(0,\sigma^2)$ so it has probability density

$$f(x) = (\sigma\sqrt{2\pi})^{-1}\exp(-x^2/2\sigma^2) \ .$$

It is then clear that if Y is the distance from the center of the needle to the nearest line below it,

$$\text{Prob}(x < Y \leq x + dx) = \sum_{\nu=-\infty}^{\infty} f(x - \nu a)\,dx$$

$$= g(x)dx \ , \ \text{say}. \tag{15}$$

The density $g(x)$ is concentrated on $(0,a)$ and $g(x)$ is periodic, period a. Thus we may write

$$g(x) = \sum_{j=-\infty}^{\infty} g_j \exp(-2\pi i j x/a) \ . \tag{16}$$

It is shown that (see Hartman and Watson (1974)) this density can be very well approximated by

$$(2\pi I_0(\kappa))^{-1}\exp \kappa \cos (2\pi x/a) \tag{17}$$

where κ is a suitably chosen function of σ and that by certain randomizing an exact result may be obtained. One of the commonest distributions for describing non-uniformly distributed angles (which we would need if we wished to give the needle a preferential orientation) is the von Mises distribution

$$(2\pi I_0(\kappa))^{-1} \exp \kappa \cos \theta \, . \tag{18}$$

Our final two examples lead into random structures. Suppose that the spacings of the parallel grid are identically and independently distributed (I.I.D.) with some density function $h(a)$ which is zero when $a < a_0$. Consider a needle of length $\ell \leq a_0$ tossed at random. Then

$$\text{Prob}\left(\begin{array}{c}\text{center of the needle}\\ \text{falls in a space,}\\ a < \text{space} < a + da\end{array}\right) = \frac{ah(a)da}{\displaystyle\int_{a_0}^{\infty} ah(a)da}$$

$$= \frac{ah(a)da}{E(a)} \, .$$

Since

$$\text{Prob}(\text{cut}|\text{space } a) = \frac{2}{\pi}\frac{\ell}{a} \, ,$$

$$\text{Prob}(\text{cut}) = \int_{a_0}^{\infty}\frac{2}{\pi}\frac{\ell}{a} \cdot \frac{ah(a)da}{E(a)}$$

$$= \frac{2}{\pi}\frac{\ell}{E(a)} \, .$$

Thus by repeating this experiment the only thing we can learn about the spacings is $E(a)$. There is, for example, no way one can check whether they are I.I.D. This would require a long needle.

Thus let us consider an infinitely long needle and suppose that we could know, after it is tossed at random onto an arbitrary parallel grid, the sequence of spaces on the needle between line crossings, $\{s_i\}$. If the needle makes an angle θ (which we do not know) and the grid spacings are a_i then $a_i = s_i \sin \theta$, for all positive and negative integers i. From only the sequence $\{s_i\}$ we can check all properties of the $\{a_i\}$ sequence that do not depend upon scale, e.g., that it is I.I.D., stationary, etc.

The cut points on the needle form a <u>Point Process</u> in one dimension. The discussion of Point processes in space is the largest aspect of geometric probability and statistics in the modern sense, the topic to which we now turn.

3. <u>WHAT IS STATISTICAL GEOMETRY?</u>

Everyone is fairly clear what is meant by geometry but statistics is less well defined. It has a number of facets - (i) exploring data for regularities, i.e., patterns; (ii) estimating "population" characteristics from a sample; (iii) testing hypotheses; (iv) designing sampling plans to be effective and efficient (usually by including a random element).

Probability models enter (i) to (iv) in several ways:

    a) by a scientific mechanism or model,

    b) by assumption,

    c) via a random sampling plan,

and to different degrees. In (i) they may not enter explicitly at all.

In statistical geometry our data will be a sample from some geometrical "population."

Such definitions do not convey much, so we now give some examples of problems and the groups that pursue them. (A) Grenander's books on "Pattern Synthesis" (1976) cover a vast area in a novel way not represented at all at this Symposium, and I think they are of basic

importance. He has developed an abstract way of generating and distort-
ing patterns and then restoring them. The latter is of course statistical. He gives a wealth of diverse examples; one of the simplest
is discussed below. (B) Classical problems such as may be found in
the Kendall and Moran (1963) book. (C) The publications of the
Fontainebleau School of Mathematical Morphology represent a different
line again. Their major achievement seems to me to be the wedding of
the image analyzer and mathematical description of the objects
scanned. While much practical work is done, the publications deal
more with the mathematical theory than with the statistical aspects.
(D) The Stochastic Geometry pursued in Cambridge by D. G. Kendall
(see, e.g., Harding and Kendall, 1974) and others overlaps theoretically with the French School but has, it seems, purely mathematical
motivations. Like so much of this literature, it has not been reduced to a level of mathematical simplicity for practical statistical
use. (E) The Point Process literature, stemming from Bartlett (see,
e.g., 1963, 1964, 1976) originated in practical statistical problems
and mainly in one dimension. It is now pursued at a highly mathematical level by Krickeberg (1977) and other Europeans in many dimensions. Earlier practical work in Forestry, especially Matern's (1960)
has led to many papers -- see, e.g., the issues of Biometrika.
Ripley's recent papers (1977a, 1977b) have a combination of theory
and practice and extensive bibliographies.

While there are many mechanisms for generating point processes
in time, the few that do so in space are summarized by Ripley (1977a),
for example. The main emphasis, in line with second order stationary
processes, is the definition and estimation of functions that control

the enhancement or inhibition of neighboring points. Here, as in time series analysis, there is a large exploratory element. Even if there were parametric models, it would rarely be possible to write down the likelihood of the data so that the time-honored statistical methods illustrated earlier cannot be used. Computers are essential for almost all calculations, unlike the Buffon problems, e.g., variances must usually be found by simulating.

In practice we will often want random sets, rather than the Poisson fields of points, lines, flats, etc. that are most often discussed. In his 1967 book Matheron made one of the early models that can be dealt with easily -- the Boolean scheme. Here I.I.D. copies of a random set K_i are attached to a Poisson field of points $\{x_i | x_i \epsilon X\}$ in a vector space to obtain the random set

$A = \bigcup_{x_i \epsilon X} (K_i + x_i)$. Such a set is intuitively stationary, i.e.,

spatially homogeneous though not necessarily isotropic. If we know, for any fixed set B,

$$q(B) = \text{Prob}(K \cap B = \emptyset)$$

then

$$\text{Prob}(A \cap B = \emptyset) = \exp\{-\lambda \int_{R^p} (1-q(B+\xi)) d\xi\} \tag{19}$$

where λ is the intensity of the Poisson process. In this Symposium Coleman went further in this construction than (19) which is the zero term of a Poisson distribution.

Time series analysis is about 100 years old. It began as a practical endeavor, became very mathematical and it is only recently that practical books and programs have been readily available. The time lag for this subject could be greatly shortened if theoreticians would make the effort to write for practical users and not only for other mathematicians.

We conclude with the simplest instance of problems in Grenander's book (ibid.). It illustrates (i) how a finite window is different from a finite sample of I.I.D. observations, (ii) the use of Fourier analysis. It is the restoration of a linear lattice whose points have been independently displaced. The complete set of points is

$$X_\nu = a + \nu\xi + n_\nu \qquad (\nu = \ldots, -1, 0, 1, \ldots)$$

where a is a phase, ξ = the lattice spacing (unknown) and n_ν is the noise. The window is the interval $(0,L)$. When the noise is small with respect to L, almost all the points that should be in $(0,L)$ will be there and no two points will have their true order inverted. Then we have an ordinary regression problem in estimating a and ξ. When the noise is not small, successive X points may not have successive indices and the "wrong" points may be in the window -- this illustrates point (i). (This model is essentially the same as that set up by D. G. Kendall (1974) to detect a unit of measurement in an archeological site.)

Here one automatically thinks of Fourier analysis. To save time, set $a = 0$ and define,

$$\phi(\omega) = \frac{1}{L} \sum_{X_\nu \epsilon(0,L)} \exp i \, \omega X_\nu, \tag{20}$$

$$m(\omega) = E\phi(\omega) ,$$

$$= \frac{1}{L} \int_0^L \exp(i\omega x)p(x)dx,$$

where

$$p(x) = \sum_{\nu} f(x - \nu\xi),$$

$$f(x) = \text{density of the noise} \quad n_\nu$$

and

$$p(x) = \sum_{k} p_k \exp(-2\pi i k/\xi).$$

Then

$$m(\omega) = \sum_{k} p_k \frac{\exp\left[\{\omega - \frac{2\pi k}{\xi}\}iL\right] - 1}{iL\{\omega - \frac{2\pi k}{\xi}\}}.$$

As $L \to \infty$,

$$m(\omega) \to 0, \quad \omega \neq \text{multiple of} \quad 2\pi k/\xi$$

$$\to p_k, \quad \omega = \text{multiple of} \quad 2\pi k/\xi.$$

Thus we would hope to see a pattern of peaks near the points $2\pi k/\xi$ from which we would first try to see if there is a pattern, and if so to estimate ξ.

$$\text{var } \phi(\omega) \sim \frac{1}{L} \frac{1}{\xi} \left[1 - |f^*(\omega)|^2\right]$$

where f^* is the Fourier transform of f so that

$$p_k = \frac{1}{\xi} f^*\left(\frac{2\pi k}{\xi}\right).$$

Thus $\text{var}(\phi(\omega))$ may also help us learn about the noise since this is governed by f.

The same computation with a Poisson process leads to

$$E(\phi) = \lambda \frac{\exp(iL\omega) - 1}{iL\omega}, \quad \text{var } \phi = \frac{\lambda}{L}, \quad \text{a very different picture also}$$

seen with all renewal processes. Thus ϕ does not differentiate between stationary point processes but one hopes that the variance might. If we define

$$N(x) = \# \text{ points in } (0,x) \; ,$$

we have

$$\phi(\omega) = \frac{1}{L} \int_0^L \exp(i\omega x)\,dN(x) \; ,$$

$$|\phi(\omega)|^2 = \frac{1}{L^2} \int_0^L\!\!\int \exp\{i\omega(x-y)\}\,dN(x)\,dN(y) \; .$$

Assume with Bartlett (1976) that

$$E\left(dN(x)\right)^2 = \lambda\,dx$$
$$E\,dN(x)\,dN(y) = \{\lambda^2 + W(x-y)\}dx\,dy \; . \tag{21}$$

If $W(\cdot)$ in (21) is identically zero, the points are Poisson. Let $W(v) = W(-v)$. It is clear that if $W(v)$ is positive, a point at y means that there is, relative to the Poisson process, more chance of having a point at $y + v$, i.e., enhancement. Negative values mean inhibition. Now

$$E|\phi|^2 \sim \frac{\lambda}{L} + \lambda^2 + \frac{1}{L}\int_{-\infty}^{\infty} W(v)\exp(i\omega v)\,dv \tag{22}$$

which verifies our notion that knowledge of $|\phi|^2$ should yield information about the function $W(v)$.

In this use of Fourier analysis one should note that the F.F.T. cannot be used; it is hard to adjust for bias and finite L and hard to find the variance of $|\phi|^2$, even in R'. The vagueness in these last two paragraphs is to some extent unavoidable. When dealing with unknown functions, one simply has to use judgment, try various tricks with the computer -- there cannot be any simple and apparently clear cut methods of the t-test type.

REFERENCES

BARTLETT, M. S. (1963). *The spectral analysis of point processes.* J. Roy. Stat. Soc. B, <u>25</u>, 264-296.

_____ (1964). *The spectral analysis of two-dimensional point processes.* Biometrika, <u>51</u>, 299-311.

_____ (1976). *The Statistical Analysis of Spatial Pattern.* Chapman & Hall, London.

COX, A. R. & LEWIS, P. A. M. (1966). *The Statistical Analysis of Series of Events.* Methuen, London.

DIACONIS P. (1976). Unpublished.

GRENANDER, U. (1976). *Pattern Synthesis: Lectures in Pattern Theory, Volume 1.* Springer-Verlag, New York.

HARDING, E. F. & KENDALL, D. G. (1974). *Stochastic Geometry.* John Wiley & Sons, New York.

HARTMAN, P. & WATSON, G. S. (1974). *"Normal" distributions on spheres and the modified Bessel function.* Ann. Prob., <u>2</u>, No.4, 593-607.

KENDALL, D. G. (1974). *Hunting quanta.* Phil.Trans. Roy. Soc. London, <u>A</u>, <u>276</u>, 231-266.

KENDALL, M. G. & MORAN, A. P. (1963). *Geometrical Probability.* Griffin, London.

KRICKEBERG, K. (1977). "STATISTICAL PROBLEMS ON POINT PROCESSES". Conférences au Centre Banach, Varsovie, Sept. 1976.

LAZZERINI, M. (1901). <u>Periodico di Mathematica</u>, <u>4</u>, 140.

MATERN, B. (1960). *Spatial variation.* Meddelanden fran Statens Skogsforskringsinstitut, <u>45</u>, No. 5.

MATHERON, G. (1967). *Eléments pour une theorie des milieux poreux.* Masson et Cie., Paris.

PERLMAN, M. D. & WICHURA, M. J. (1975). *Sharpening Buffon's needle.* The Amer. Stat., <u>29</u>, No. 4, 157-163.

RIPLEY, B. D. (1977a). *Modelling spatial patterns.* To appear in J. Roy. Stat. Soc. A.

_____ (1977b). *Spectral analysis and the analysis of pattern.* Unpublished.

UPSENSKY, J. V. (1937). *Introduction to Mathematical Probability.* McGraw-Hill, New York and London.

THE FRACTAL GEOMETRY OF TREES
AND OTHER NATURAL PHENOMENA

BENOIT B. MANDELBROT

IBM Thomas J. Watson Research Center
POBox 218, Yorktown Heights, New York 10598

Before we can tackle some specific new technical tidbits, which this paper hopes to contribute to the study of the geometry of plants, we must deal with the first term in the title. You are not expected to know it, because I coined it only recently. Before I define it, I beg you to examine Figure 1, which is the bottom half of the combined Figures 1 and 2.

My point is that Figure 1 DOES NOT represent what I hope you think it might represent. With due apologies to Mr. Baedeker and his heirs and competitors, this is NOT a landscape on the Earth, the Moon, or any other planet, but an artificial surface generated deliberately to mimic a landscape. Of course it is computer-generated and -plotted. The program, due to Richard F. Voss, is based upon the algorithm advocated in my mathematical model of the Earth's relief. The two most significant parameters are a real number D lying between 2 and 3 and the seed of a pseudo random subroutine. In this talk I shall have a lot to tell about this parameter D, to be called *fractal dimension* of the surface. In Figure 1, in order to insure the resemblance that I hope you have perceived, the suitable value had to be D=2.2500.

Had we picked a different D, the resulting surface would have been "different in form". But what is really meant by *form*? The question is vital in many sciences, and particularly so in biology, but it is far from having been answered satisfactorily. If we were to trust our mathematical friends blindly, we would interpret *form* as that which is studied by topology, but in the present case we would be sorely disappointed. Or perhaps delighted not to have to learn topology. Indeed, at least in principle, the surface you see can be obtained from a square without a tear, using a one-to-one continuous transformation, and this property *defines* it as being topologically a square. The surfaces corresponding to the same algorithm but different values of D are also squares from the topological viewpoint; nevertheless their form depends greatly upon the value of D. When D is closer to 2, they are smoother in detail and less flat overall. When D is closer to 3, they are much less smooth in detail and much flatter overall – as seen on pages 210 to 215 in my book, *Fractals: Form, Chance, and Dimension*. (The back of the book's jacket reproduces the present combined Figures 1 and 2.)

For a further example of the limitations of topology, think of a membrane lining the air ducts of the lungs. It too could be stretched without a tear into a square, at least in principle. Yet it is completely different in form from the surface shown on Figure 1.

For yet another example, kindly examine Figure 2, which sits on top of Figure 1. Again, its point is that it DOES NOT represent what I hope you think it might represent. With due apologies to the crew of Apollo XI, it is NOT a line version of the photograph of a planet, but the projection (upon a tangential plane) of the surface of a sphere on which a curve was drawn by a computer, instructed again by Voss, using an algorithm I supplied. Topologically, this curve can be obtained by deforming a collection of a few circles by a one-to-one continuous transformation. Therefore this drawing is topologically a collection of circles, but you must agree that its topology can account for only a small part of the whole truth. There ought to exist some mathematical way of expressing that between a collection of circles and the curve you see here there is, again, a profound *difference in form*.

As a matter of fact, I hope that a second and more careful look at Figure 2 leads you to conclude that it is *not quite* Earth-like. The curves you see are markedly too wiggly, too complicated to represent the contours of Earth's continents. The primeval continent Pangaea, which we are told existed before the continents split up, is also reported to have had a more complicated coastline, but I would not be so sure: those who remember Pangaea are a bit old, so their memories may play tricks.

The fact that very irregular shapes are often encountered in Nature requires little elaboration. Does it follow from the inappropriateness of topology that the degree of wiggliness must remain an intuitive notion, that is, a notion inaccessible to mathematical description? The answer is a resounding NO. It turns out that mathematicians had evolved long ago a notion that serves to measure the degree of *local* irregularity of a surface or of any other set; it is called *Hausdorff-Besicovitch dimension*, and its value will be denoted by D. However, this notion had few applications in mathematics, so it remained classical but obscure, and it failed completely to draw the attention of scientists. Applications, and the very idea that (in the case of certain geometric shapes) D can also be used to quantify an aspect of "form", did not come out until my papers that eventually led to the book, *Fractals*.

Originally, the above D, was geared towards the study of certain artificially designed standards of irregularity, most notably of sets due to George Cantor and Giuseppe Peano. I hope that those among you who are aware of the reputations of Cantor and Peano are surprised at hearing their names in the present context, so let me elaborate. The great mathematicians active during the period 1875-1922, which was very critical in pure

FIGURES 1 AND 2

mathematics, introduced many sets intended solely to prove that in comparison to the concepts of the old mathematics, those of modern mathematics were indeed of increased generality. Such sets, however, were never meant to be applied in science. In fact, both the creators and their followers were practically unanimous in considering them as pathological monsters. But I think they were mistaken. My claim is that these specific sets also have another and very different sort of use, as "standard" models of the irregularity and fragmentation constantly observed in Nature. Far from being pathological, they possess features that turn out to be by far closer to certain aspects of Nature than the geometry on which all of us have been brought up, that of Euclid.

Their usefulness is linked to the fact that from the viewpoint of irregularity they have something very deep in common with a curve that is on the contrary quite familiar to everyone concerned with Nature, namely Brownian motion. Together, these various sets are examples of a class of sets that I have proposed to call *fractals*, and they are the topic of the book to which I have already alluded, *Fractals*, as well as of its earlier version, *Les objets fractals*, and of a paper I have recently contributed to *La Recherche*.

To attempt to summarize these books here would be foolish, but I would like to take this opportunity to put on record a few fresh illustrations and applications. The more important ones will refer to branching structures such as trees. (Related discussions of direct interest to biologists can be found in the section of *Fractals* devoted to the vasculature, and in the sections of my *La Recherche* paper devoted to the lung.)

While the accompanying text cannot avoid being sketchy and allusive, I hope that the graphics – especially Figures 5, 6 and 7 – are evocative. It is useful, however, for some readers' convenience, to begin by restating the construction of some basic fractals and by commenting upon the notion of fractal structure, which makes it possible to refer to Hausdorff-Besicovitch dimension as being a *fractal dimension*.

SOME BASIC FRACTALS

The largest and most complicated of the five diagrams that make up Figure 3 is the composite of two wondrous and very-many-sided polygons, devoid of self contact. They are drawn by a computer instructed by Sigmund W. Handelman. I am tempted to call them *teragons*, to take advantage of the original basic meanings of the Greek word *teras*, *teratos*, "a wonder or a monster", and of the use of *tera* in the metric system as the prefix to designate the very large number 10^{12}. The first teragon is so violently folded upon itself as to give the impression that it attempts a monstrous task for a curve: to fill the interior of the second teragon. This impression was intended and is quite justified, since indeed the first teragon is an advanced stage of the construction of a *space-filling* curve. By way of contrast, the second curve can be called a *wrapping*.

The first space-filling curve was discovered by Giuseppe Peano in 1890 and many others followed in the next 25 years. Then the search for new ones went out of fashion until 1961, when J.E. Heightway discovered the *dragon curve*, and until 1976, when R. W. Gosper discovered the *flowsnake*. Each of these new curves is the limit of tera-

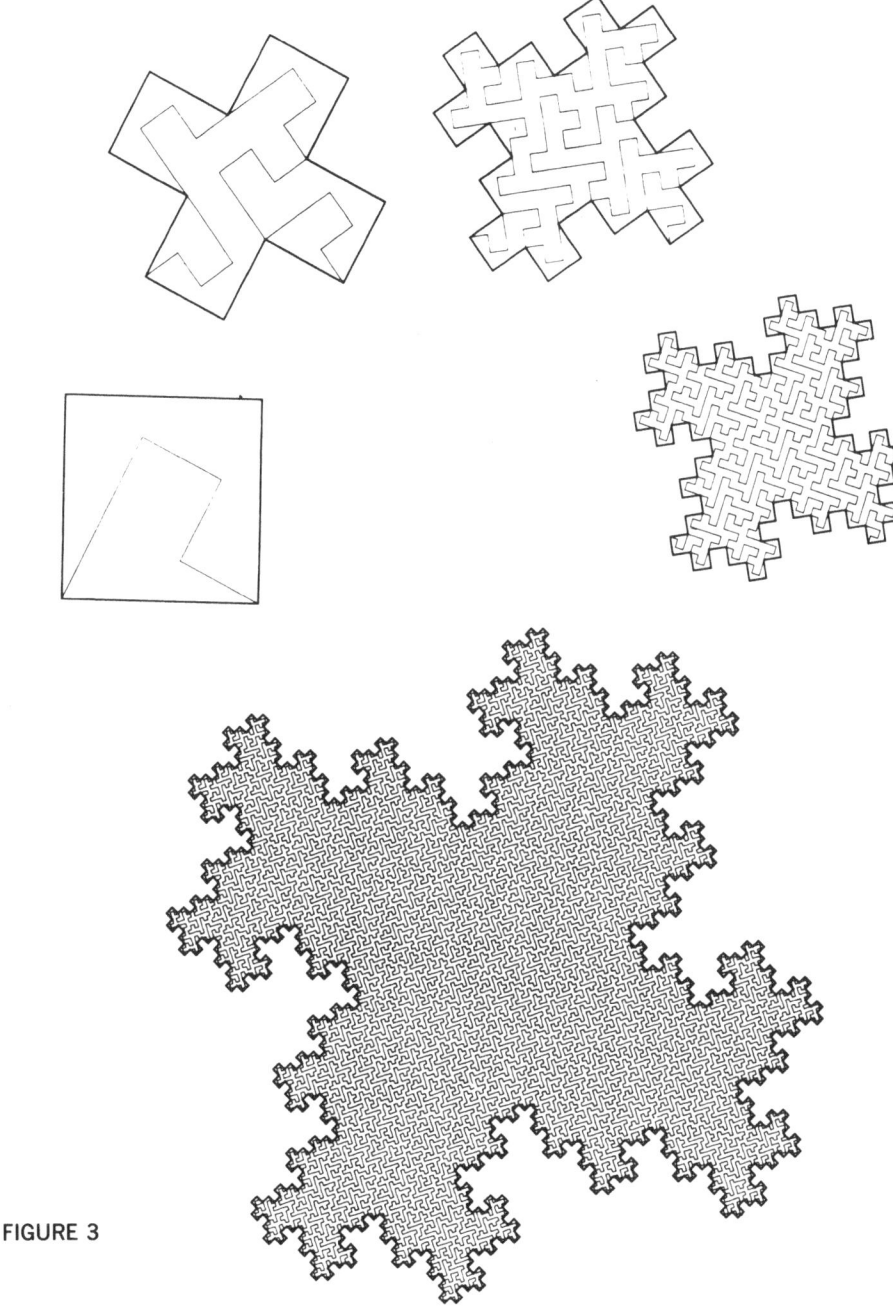

FIGURE 3

gons, the second one being devoid of self contact and both curves are constructed by a *Koch cascade*. (Figure 3 constitutes a variant of Gosper's construction.)

Koch cascades are named after the Swedish mathematician, Helge von Koch, who had the pioneering idea (first implemented in his *snowflake* curve) to seek shapes with the property of being precisely as complicated in the small as in the large. His motivation was purely mathematical, but in effect he wanted the fine details seen under the micro-scope, whatever the magnification, to be the same (scale aside) as the gross features seen by the naked eye. To implement this goal, the best is to work step by step. Point of departure: one selects an initial set that is either an interval or an equal-sided polygon, and a standard polygon. The first construction stage replaces each side of the initial polygon by an appropriately rescaled and displaced version of the standard polygon. Then a second stage repeats the same construction with the polygon obtained at the first stage, and so on and on ad infinitum.

The early stages of the constructions shown on Figure 3 are illustrated by the four small diagrams of this Figure, to be followed clockwise from left center, in order of increasing complication. The initial polygons are a unit square for the wrapping, and for the filling it is an irregular open equal-sided pentagon (the first "leg" from the left is really two sides). This pentagon does its best to fill the square. (Indeed, one perceives an underlying square lattice of lines $1/\sqrt{5}$ apart, and our original filling passes through every lattice vertex contained in in the original wrapping.) As for the filling, the initial and standard polygons are identical. In other words, in the next stage of the construction, each side of the pentagon is replaced by an image of its whole reduced in the ratio of $1/\sqrt{5}$. The result no longer fits within the square, but it fills uniformly a modified version of the wrapping, to wit, the crosslike shape obtained by replacing each side of the square by a polygon made of $N=3$ sides of length $r=1/\sqrt{5}$. The same two constructions are then repeated ad infinitum in parallel. A designer who zooms in as the construction proceeds will see the same degree of filling, but one who stays put sees a curve that fills increas-ingly uniformly a varying wrapping that becomes itself of increasing complexity. The wrapping's limit having four corners, it may be called *quartet*. (But it may equally well be called *quintet* , because it is the sum of five reduced versions of itself; the fifth one hides in the center.)

(Observe, in a digression, that a so-called plane-filling curve is not meant to fill the whole plane, rather a domain thereof. This feature was not apparent with the Peano curves which mathematicians designed during the heroic period up to 1922, because they all filled a square or a triangle, but the dragon and the Gosper curve each involve a more imaginative boundary and so does the present construct.)

Figure 4 carries the construction of the two curves of Figure 3 one step further, and presents the result in a different light. The filling is now interpreted as the cumulative shoreline of several juxtaposed river networks, and the wrapping as the combination of a drainage divide surrounding these networks and of a portion of seashore. (To build up the network, one proceeds step by step: 1) Each deadend square in the basic underlying lattice – defined as such that three sides belong to the filling teragon – is replaced by a short stream from its center beyond the open side. After that the deadends are cut off.

FIGURE 4

2) Then one proceeds in the same fashion with the straightened teragon. 3) And so on until the filling is exhausted.) In this light, the wrapping becomes reinterpreted as the juxtaposed network's external drainage divide. To use an old sophomoric line, after you think of it imaginatively, carefully and at great length, it becomes quite obvious that a river network's shore gives an idea of the structure of a plane-filling curve. And the converse is also true, except of course that the present network is much too squarish to be realistic. Much better-looking ones are given in my book, but the basic idea is present here. You may say that the present network resembles a bonsai tree.

To sum up, the mathematicians who tell us that Peano curves are totally non-intuitive are mistaken. Or perhaps, to be kinder, let us students of Nature accuse them of having tried to prevent us from knowing of a beautiful new tool, and applaud their skill in succeeding in holding to it exclusively for so long.

Further comments on the topic of Peano curves and of their intrinsic trees, with additional illustrations, are to be found in my *La Recherche* paper.

THE NOTION OF FRACTAL DIMENSION

It is easy to see that each stage of a Koch construction multiplies a polygon's length by a fixed factor $Nr > 1$; hence the limit curves obtained by pursuing the constructions of Figure 3 ad infinitum are of infinite length. But it is obvious, so to speak, that the filling is "much more infinite" than its wrapping. In particular, its length tends to infinity more rapidly. This intuitive feeling is expressed mathematically by the notion of *fractal dimension*, to which we have alluded repeatedly; the notion is due to Hausdorff and was perfected by Besicovitch. The explanation of the underlying idea begins with the very simplest shapes: line segments, rectangles in the plane, and the like.

Because a straight line's Euclidean dimension is 1, it follows for every integer γ that the "whole" made up of the segment of straight line $0 \leq x < X$ may be "paved over" (each point being covered once and only once) by $N = \gamma$ "parts". These "parts" are segments of the form $(k-1)X/\gamma \leq x < kX/\gamma$, where k goes from 1 to γ. Each part can be deduced from the whole by a similarity of ratio $r(N) = 1/N$. Likewise, because a plane's Euclidean dimension is 2, it follows that, whatever the value of γ, the "whole" made up of a rectangle $0 \leq x < X$; $0 \leq y < Y$ can be "paved over" exactly by $N = \gamma^2$ parts. These parts are rectangles defined by $(k-1)X/\gamma \leq x < kX/\gamma$ and $(h-1)X/\gamma \leq y < hY/\gamma$, wherein k and h go from 1 to γ. Each part can now be deduced from the whole by a similarity of ratio $r(N) = 1/\gamma = 1/N^{1/2}$. For a parallelepiped, the same argument gives us $r(N) = 1/N^{1/3}$. Finally, we know that there is no serious problem in defining spaces whose Euclidean dimension is $E > 3$. In that case, D-dimensional parallelepipeds can be defined for any $D \leq E$, and they satisfy $r(N) = 1/N^{1/D}$, that is,

$$\log r(N) = \log (1/N^{1/D}) = -(\log N)/D.$$

Finally, we see that in all the classical cases, the dimension satisfies

$$D = -\log N/\log r(N) = \log N/\log (1/r).$$

It js the latter equality which will now be generalized. In order to do so, note that the exponent of self similarity continues to have formal meaning for some shapes which are neither a segment nor a square. The main requirement is that we have the following property of "scaling": the whole may be split up into N parts deducible from it by self similarity having the ratio r (followed by displacement or by symmetry). Such is precisely the case with the limits of our teragons. For the wrapping, we see that $N=3$ and $r=1/\sqrt{5}$, hence

$$D = \log 3 / \log\sqrt{5} = \log 9 / \log 5 = 1.3652$$

For the filling, we see that $N=5$ and $r=1/\sqrt{5}$, hence

$$D = \log 5 / \log\sqrt{5} = 2.$$

Thus, the impression that the quartet-quintet's filling is more infinite than its wrapping is confirmed and quantified by the inequality between their dimensions. The impression that the filling really fills a plane domain is confirmed and quantified by its dimension being $D=2$.

The preceding argument may seem overly special, so it may be comforting to know A) that fractal dimension can be defined in other ways in full generality and rigor, and B) that the result behaves like the old-fashioned one in many other ways. For example, consider the notion of *measure*. If a set is self similar and measure is taken properly, then the portion of this set that is contained in a sphere of radius R is of measure R^D.

(You may have forgotten that I promised a definition. *I define fractal as a set such that the above D – or one obtained by more intrinsic methods due to Hausdorff and Besicovitch – is greater than the topological dimension – or the intuitive dimension.* The wrapping is a Gaussian curve of topological dimension 1; hence it is a fractal curve. The filling is a more involved matter we shall not discuss further at this point.)

SOME FRACTAL TREES IN THE PLANE

We are now ready to sketch briefly a few new applications. The marvelous *Jardin des Plantes* behind this hall's windows mandates that we put stress on various kinds of trees.

In addition to examples given in *Fractals*, one tree has already been featured in Figure 5 above. Had it been extended ad infinitum, it would have filled the plane, which is why one may have expected that its fractal dimension is $D=2$. Now we would like to broaden the range of dimensions one can expect in a tree; to do this, it will be convenient to begin by taking one step backwards. Consider therefore the trees represented on Figure 5; it covers two pages and is to be viewed sideways and examined clockwise from bottom left. These trees are less structured than those of Figure 4, hence they are definitely duller, but the algorithm used to draw them boasts of a fully adjustable dimension. All these trees are seen to be regular, in the sense that a) the ratio between branch and trunk length is the same at every branching point, equal to some prescribed r, and b) the angle θ between branches is also constant throughout. It can be shown that it follows in every case that the set of branch tips is self similar with the dimension $D = \log 2 / \log(1/r)$.

(As an aside, which may be skipped by the reader, let us mention that the dimension of the whole tree is a more complicated matter. The whole tree is the sum of two reduced size trees *plus* a remainder: the trunk. Hence, self similarity fails by a "detail" that may seem small but in certain cases turns out to be absolutely non-negligible, so that one cannot apply the notion of dimension until after it has been properly generalized. When $D > 1$, the generalized concept shows that the whole tree is also of dimension D, but when $D \leq 1$, the tree is of dimension 1. The fact that its dimension may not go below 1 could have been expected a priori, because a tree includes branches which are of dimension 1, and because it is true – as it should be – that a shape's fractal dimension can in no case be less than the fractal dimension of a shape's part.)

Furthermore, the trees of Figure 5 are designed to be maximally tight, in the sense that the algorithm used in this Figure attributes to r (and hence to D) the largest value compatible with each successive value of θ (the gaps you see here between the branches would decrease to zero if the construction could be continued ad infinitum).

The construction is extremely plain, nevertheless it is seen to generate an astonishing wealth of very different looking structures.

We begin at bottom left with a dry-looking plant, a kind of broom, of which we discover here that its fractal dimension is close to 1.

Then we proceed upwards to tastier plants, namely to several varieties of the species *brassica oleracea bothrytis*, and we are thus made aware of the fact that from the purely geometric viewpoint the main difference between cauliflower and broccoli lies in their respective fractal dimensions.

As dimension continues to increase, quantitative change ends up by generating a fresh qualitative difference. Some will recognize in one of the drawings of Figure 5 a sketch of yet another fortification by the Marquis de Vauban, Marechal de France.

As the value of θ increases still further and reaches $\theta = \pi/2$, r attains its maximum value $1/\sqrt{2}$ and D reaches its maximum $D = 2$. At this point, we deal with a plane-filling tree. It is most uninspiring, compared to the tree shown on Figure 4; nevertheless it has the advantage of demonstrating that the plane can be covered by a tree in more ways than one.

As θ increases beyond $\pi/2$, r must decrease, and D falls below 2. The resulting trees no longer hark to botany; their contortions are worthy of the dancers on classical Indian sculptures.

Those familiar with d'Arcy Thompson's classic *On Growth and Form* will recall the striking illustrations which show how different species or varieties of fish can be related to each other by transformations known to elementary Euclidean geometry. The transformations that relate the various trees of Figure 5 partake of the same overall inspiration (and in particular they leave the topological structures untouched). However, they belong to an entirely different chapter of mathematics, which I would call fractal geometry. I cannot

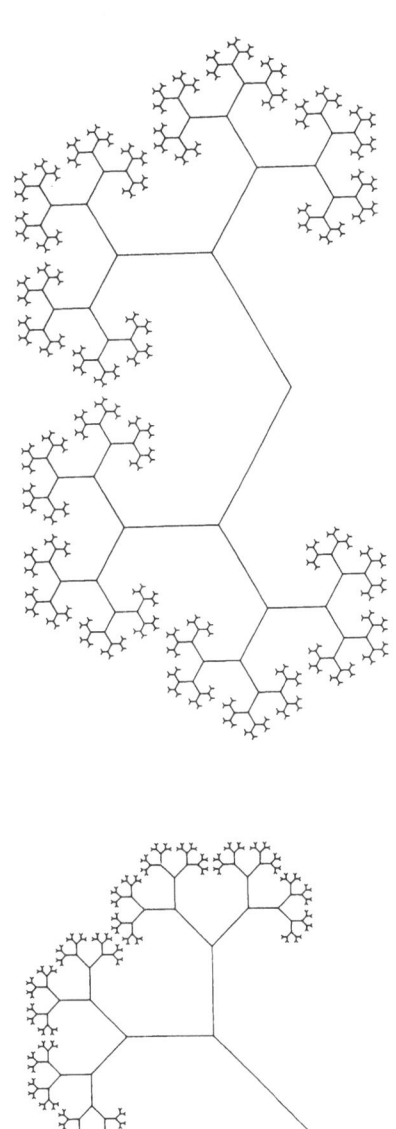

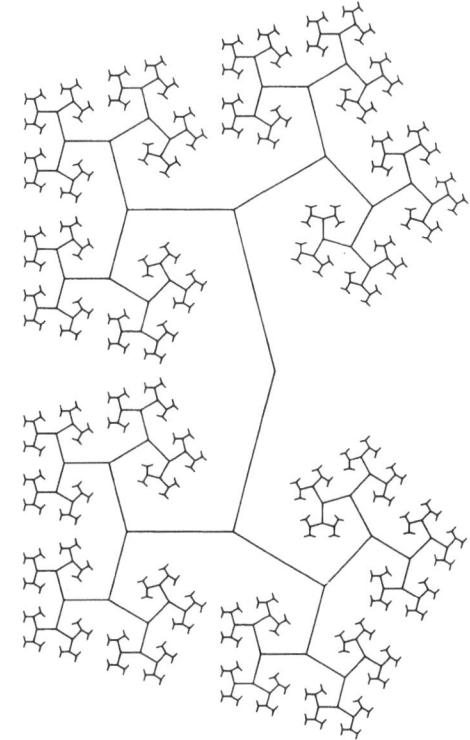

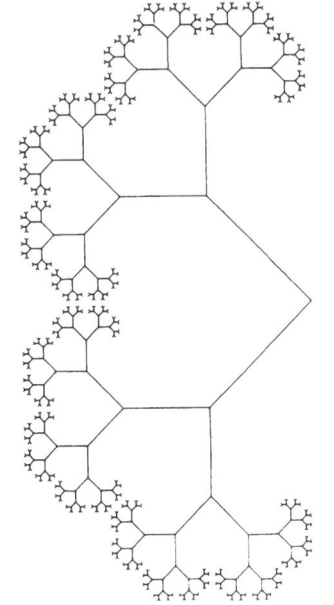

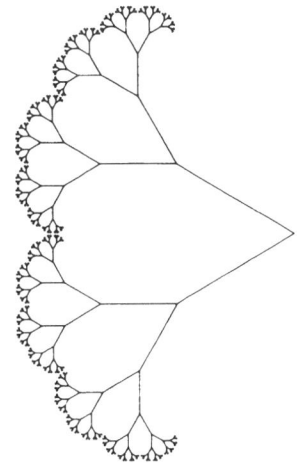

FIGURE 5 (TOP PORTION)

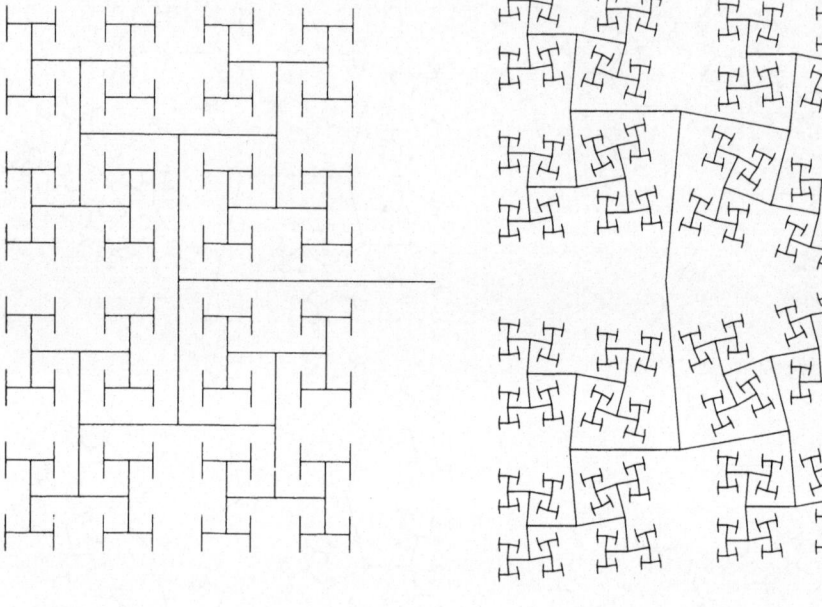

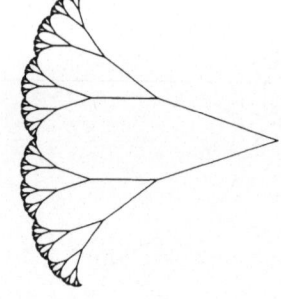

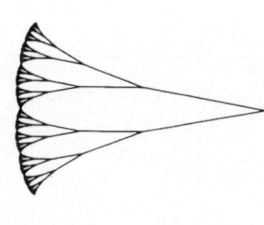

FIGURE 5 (BOTTOM PORTION)

imagine it could fail to be of the most direct relevance to those sharing Thompson's consuming interest in the geometric form of plants and animals.

THICKENED TREES IN THE PLANE

Of course, the skeletons of cruciform plants that we have drawn thus far are not recommended by any ethnic tradition of cooking. Hence, our next task is to thicken the branches. Sticking to the plane, two closely related methods for achieving thicker branches are illustrated on Figures 6 and 7. (It may be noted that the results are no longer "trees" in the sense given to this word by topologists.) Figure 6 corresponds to the fourth diagram of Figure 5 (clockwise from bottom left), but now each branch is replaced by a kind of bud or vase made of two triangles sharing a side and symmetric of each other along the branch's skeleton. This configuration is hard to figure out, but its outline is quite exciting in its form. You will note in particular the empty spaces circumscribed by buds. (Could it be that the air occlusions in cauliflowers are ordained by geometry?)

To make the illustrations less impenetrable, the cauliflower or broccoli buds must be separated a bit. On Figure 7, this is done starting from the third diagram of Figure 5 (clockwise from bottom left). To improve legibility further, the tree's skeleton is opened up a bit, and is just short of being maximally tight.

SPATIAL TREES

Actual plants are not flat but three-dimensional objects. We shall not attempt to provide model drawings for them, but it is useful to refer to the late portions of Chapter II of *Fractals* and to add some remarks concerning wooden trees. Arguing teleologically, a tree's goal – insofar as allowed by many external constraints – is to maximize the area of contact between its leaves and air and sunshine. In the absence of interference from neighbors, such a tree would attempt to fill as thoroughly as possible as large a volume as it can mechanically support. Thus, one may argue that in an ideal situation the surface of a tree's bark and leaves should be nearly a space-filling surface – a three-dimensional version of the "bark" of the tree of Figure 5.

A readily-seen consequence is that when a tree is self similar and nearly space-filling, the quantities r_1 and r_2, defined as ratios of the lengths of the two branches and the trunk, should satisfy the relationship $r_1^D + r_2^D = 1$ with the exponent $D \sim 3$. When r_1 and r_2 are approximately equal, call their common value r, one would have $r \sim 2^{-1/3}$. This ratio is well known (and constantly rederived in new and unexpected ways). It appears that it was first considered in *Growth and Form*. D'Arcy Thompson's derivation invoked specific arguments of mechanics, but we see that the $r \sim 2^{-1/3}$ law is really a matter of elementary geometry of space packing.

Now what about branch diameters and their effects upon a tree's overall equilibrium? Under the assumption of self similarity, the diameters d_1 and d_2 of the two branches and the diameter d_0 of the trunk would all be proportional to the corresponding lengths, hence they would satisfy the relationship $d_1^D + d_2^D = d^D$. As a corollary, the sum of the cross sections of the two branches originating at a branching point would be equal to the trunk's cross section multiplied by the factor $2 \times (2^{-1/3})^2 = 2^{1/3}$. This factor is greater than 1, however, which means that in a self similar wooden tree the total cross section and the corresponding weight would increase as one goes along the branches. This is clearly a mechanical impossibility. Hence a wooden tree cannot conceivably be self similar. In the different context of bronchial trees, the situation is actively different: the above-written relationship involves no impossibility whatsoever, because the lung is supported as a whole from the outside. Therefore, as argued in my *La Recherche* paper, the lung *can* be self similar, and indeed *it is*. But in the case of trees, the situation is different. The best one could achieve by deliberate design is to make the sum of branch cross sections equal to the trunk cross section, thus satisfying the relationship $d_1^2 + d_2^2 = d^2$.

A corollary of the above elementary geometric constraints is that the diameter of a branch situated at a distance δ from the branch tips would be proportional to $\delta^{1.5}$. This last relation happens to be empirically correct. Furthermore, McMahon & Kronauer, from whom I heard of its existence, deduced it from a specific dynamic aspect of self similarity. Now we see that, like Thompson's ratio $2^{-1/3}$, it may also be merely of geometric origin.

* * * * *

Finally, let us proceed to a different application by moving down from trees to the ground on which they grow. In this way, we shall be able to come back to Figure 1, which we used for initial effect but to which we devoted less attention than it deserves. This comment will be even sketchier than the preceding ones. We have already alluded to the fact that Brownian motion in a plane or a higher space is a prototypical fractal: indeed it is topologically a simple curve of dimension 1, but it is fractally of dimension 2. Its coordinate functions, such as X(t), are curves of fractal dimension 3/2. And Paul Levy has generalized the Brownian function to functions from a multi-dimensional "time", say (x,y), to a scalar Z. If Z is interpreted as an altitude, such a function represents a landscape; it is of fractal dimension D=5/2. Finally, one can modify it by performing a certain operation called Riemann-Liouville fractional integration. If the order of this integration is 1/4, one obtains the surface shown on Figure 1, which in technical terms is described as a fractional Brownian surface of dimension D=2.2500. It brings us back to our point of departure, and I shall stop here for today.

REFERENCES

Mandelbrot, B. B. *Les objets fractals: forme, hasard et dimension.* Paris and Montreal, Flammarion, 1975.

Mandelbrot, B. B. *Fractals: Form, Chance, and Dimension.* San Francisco and London: W. H. Freeman and Company, 1977.

Mandelbrot, B. B. Des monstres de Cantor et Peano à la géométrie des rivières et des poumons. *La Recherche,* January 1978.

McMahon, T.A. & Kronauer, R.E. Tree Structures: Deducing the Principle of Mechanical Design. *Journal of Theoretical Biology* 1976.

Thompson, d'A.W. *On Growth and Form.* Cambridge University Press. 1917-1942-1961. The dates refer to the first, second and abridged editions.

FIGURE 6

FIGURE 7

THE INFINITISIMAL EROSIONS

G. MATHERON
CENTRE DE MORPHOLOGIE MATHÉMATIQUE
E.M.P. 35 Rue St-Honoré 77305 FONTAINEBLEAU

On the space $C(\mathcal{K})$ of the compact convex sets in R^n, the erosion by the homothetic sets ρK of a fixed $K \in C(\mathcal{K})$ constitutes a semi-group, the generator of which is defined by the relationship $K(A) = \lim (A \ominus \check{A}_\rho)/\rho$, when $\rho \downarrow 0$ $(A = A \ominus \rho \cdot K)$. This generator is called "infinitesimal erosion". $K(A)$ depends only on the support S_A of the surface measure associated with A only. More precisely : $K(A)$ is the largest convex on S_A. As an application of this theorem, one solves the equation $X \ominus K = A$ (A, K known, $X \in C(\mathcal{K})$ unknown).

La morphologie mathématique utilise et combine entre elles des transformations d'ensemble de types extrêmement variés. Je souhaite qu'un algébriste, un jour, nous donne une théorie synthétique permettant de comprendre pourquoi tel ou tel types d'opérations réussissent dans tel ou tel cas. Dans cet article, j'examine un cas particulièrement simple : celui des _érosions_ $A \to A \ominus \rho \check{K}$ par les homothétiques ρK d'un compact convexe K, qui constituent un _demi-groupe_ d'opérateurs sur l'espace $C(\mathcal{K})$ des compacts convexes de l'espace euclidien $\mathbb{R}^n$. Mon but est d'essayer de caractériser ce demi-groupe par son générateur infinitésimal, et cela me conduit à la notion d'_érosion infinitésimale_. Dans cette théorie, la mesure de surface G_A associée au convexe $A \in C(\mathcal{K})$, et son support S_A, jouent un rôle très important. On voit s'introduire de manière très naturelle des notions qui s'apparentent à la théorie du potentiel - le "potentiel" associé à la mesure de surface G_A n'étant autre que la fonction d'appui φ_A du compact convexe A. Mais (sauf dans l'espace à 2 dimensions) la correspondance $G_A \to \varphi_A$ n'est _pas_ linéaire. D'une manière générale, le cas $n = 2$, où cette correspondance est linéaire, présente des particularités remarquables qui ne subsistent pas pour $n \geq 3$. Par exemple, c'est seulement dans l'espace à 2 dimensions que l'ensemble A initial peut être reconstitué par intégration de ses érosions infinitésimales.

1. RAPPELS ET NOTATIONS.

Dans ce qui suit, A, K etc... désignent des éléments de $C_o(\mathcal{K})$, c'est-à-dire des compacts convexes de $\mathbb{R}^n$ contenant l'origine 0 ; φ_A , φ_K ... désignent leurs fonctions d'appui, et G_A , G_K ... leurs mesures de surface sur la sphère unité S_o. Par définition, pour tout $u \in S_o$:

$$\varphi_A(u) = \text{Sup } \{ <u,x> , x \in A\}$$

Vis-à-vis de l'addition de Minkowski, on a la règle $\varphi_{A\oplus K} = \varphi_A + \varphi_K$. Vis-à-vis de l'_érosion_ $A \ominus \check{K}$ (où $\check{K} = \{-x, x \in K\}$ est le symétrique de K par rapport a l'origine) on trouve seulement $\varphi_{A\ominus\check{K}} \leq \varphi_A - \varphi_K$. Plus précisément : l'érodé $A \ominus \check{K}$ est le plus grand des compacts convexes C tels que $\varphi_C \leq \varphi_A - \varphi_K$.

A tout $A \in C_o(\mathcal{K})$ est associée une mesure positive G_A sur la _sphère unité_ S_o, appelée _mesure de surface_ de A, caractérisée par la relation:

$$(1-1) \qquad W_1(A,K) = \frac{1}{n} \int_{S_o} \varphi_K(u) \; G_A(du) \qquad (K \in C_o(\mathcal{K}))$$

(n est le nombre des dimensions de l'espace euclidien ; W_1 la première fonctionnelle mixte de la géométrie intégrale, égale à la dérivée à droite en $\rho = 0$ du volume de la somme $A \oplus \rho K$, ou, aussi bien, de l'érosion $A \ominus \rho \check{K}$). La correspondance $A \rightarrow G_A$ est _continue_ (pour la convergence faible des mesures sur S_o), invariante par translations, et dé_croissante_ pour les érosions [1] :

$$(1-2) \qquad\qquad G_{A\ominus\check{K}} \leq G_A$$

Le résultat suivant, qui remonte à Minkowski [2], montre que la donnée de G_A détermine A à une translation près :

Soit G une mesure positive sur la sphère unité, _non concentrée sur un grand cercle_. Alors G est la mesure de surface G_A associée à un compact convexe A, _nécessairement unique_ à une translation près et de volume $V(A) > 0$, si et seulement si elle vérifie la _condition barycentrique_.

$$(1-3) \qquad \int u \, G(du) = 0$$

Si G est concentrée sur un grand cercle, elle est la mesure de surface d'un compact convexe A (non unique) de volume nécessairement nul si et seulement si elle est symétrique et concentrée sur deux points diamétralement opposés de la sphère unité.

Dans le cas $n = 2$, la restriction relative au cas dégénéré disparait : car les grands cercles de la sphère unité de $\mathbb{R}^2$ sont, justement, constitués de deux points diamétralement opposés : toute mesure G sur le cercle unité vérifiant $\int \cos \theta \, G(d\theta) = \int \sin \theta \, G(d\theta) = 0$ est la mesure périmétrique d'un compact convexe de $\mathbb{R}^2$ unique à une translation près. Dans $\mathbb{R}^2$, la relation entre fonction d'appui φ_A et mesure périmétrique G_A est <u>linéaire</u> et peut s'écrire :

$$(1-4) \qquad G_A = \varphi_A + \varphi_A''$$

(avec une dérivée seconde qui doit, en général, être prise au sens des distributions). Inversement, en intégrant $(1-4)$, il vient :

$$(1-5) \qquad \varphi_A(\theta) = a \cos(\theta - \theta_o) + \int_o^\theta \sin(\theta - \varphi) G_A(d\varphi)$$

Les deux constantes arbitraires a et θ_o expriment que G_A détermine A à une translation près seulement. On notera aussi le <u>caractère local</u> de la relation $(1-4)$: la donnée de la fonction d'appui φ_A sur un ouvert σ du cercle unité détermine la mesure G_A sur ce même ouvert. Nous verrons que le caractère local de la correspondance $\varphi_A \to G_A$ (mais non la linéarité) subsiste pour $n \geq 3$.

De ces relations résulte aussitôt que l'on a dans $\mathbb{R}^2$ (mais non dans $\mathbb{R}^n$, $n \geq 3$)

$$(1-6) \qquad G_{A \oplus K} = G_A + G_K$$

Par suite aussi, toujours dans $\mathbb{R}^2$, A est ouvert selon K (i.e. $A_K = A$ avec $A_K = (A \ominus \check{K}) \oplus K$) si et seulement si $G_A \geq G_K$.

En effet, si A est ouvert suivant K, il est de la forme $A = C \oplus K$ et (1-6) donne $G_A \geq G_K$. Inversement, si $G_C = G_A - G_K \geq 0$, cette mesure positive vérifie la condition barycentrique et par suite est la mesure périmètrique d'un compact convexe C. La relation (1-5) donne alors $A = C \oplus K$ à une translation près.

Plus généralement, si A et K sont compacts convexes dans $\mathbb{R}^n$ et $K \subset A$, posons $A_1 = A \ominus \check{K}$; $K_1 = A \ominus \check{A}_1 = A \ominus (\check{A} \ominus K)$.

On peut écrire $K_1 = (\check{A}_1)^c \ominus A = (K \oplus \check{A}^c) \ominus A^c = K^{A^c}$ (A^c est le complémentaire de A). Donc K_1 est <u>la fermeture de K suivant A^c</u>, c'est-à-dire <u>l'intersection de tous les translatés de A qui contiennent K.</u>

Si l'on réitère l'érosion, on trouve $A \ominus \check{K}_1 = A_1$. Car $A \ominus \check{K}_1 = A \ominus (\check{A} \ominus A_1)$ est la fermeture de A_1 selon A^c. Mais $A_1 = A \ominus \check{K}$ est déjà fermé selon A^c, en tant qu'intersection de translatés de A, et donc $(A_1)^{A^c} = A_1$. Dans $\mathbb{R}^n$, on a l'inclusion $A \supset A_1 \oplus K_1$, mais non en général l'égalité. Par contre, dans <u>l'espace à 2 dimensions</u>, on a toujours :

$$(1-7) \qquad\qquad A = A_1 \oplus K_1$$

Autrement dit, si n = 2, <u>A est ouvert selon tout érodé $A \ominus \check{K}$ non vide.</u> En effet, on a $G_{A_1} \leq G_A$ (puisque la correspondance $A \rightarrow G_A$ est décroissante pour les érosions) donc A est ouvert selon A_1, soit $A = A_1 \oplus C$. On en déduit $C = A \ominus \check{A}_1 = K_1$ et $A = A_1 \oplus K_1$.

De même, dans $\mathbb{R}^2$, on vérifie sans peine que l'on a $K = K_1$ (K fermé selon A^c) si et seulement si $A = A_K$ (A ouvert selon K). Dans $\mathbb{R}^n$, $n \geq 3$, $A = A_K$ entraîne encore $K = K_1$, mais la réciproque n'est plus vraie.

2. L'EROSION INFINITESIMALE.

Venons-en maintenant à la définition de l'érosion infinitésimale. Soient A et K compacts convexes dans $\mathbb{R}^n$ et contenant l'origine O. Nous supposons dans tout ce qui suit que A a un <u>intérieur non vide</u> ($V(A) > 0$), de sorte que l'érodé $A_\rho = A \ominus \rho \check{K}$ est non vide pour $\rho > 0$ assez

petit. Nous désignerons par $R = \text{Sup } \{\rho : A_\rho \neq \emptyset\}$ <u>le module de l'éro-</u>
<u>sion ultime</u>. Il est facile de voir que A_R est non vide mais d'intérieur
vide (de volume nul) et que l'application $\rho \to A_\rho$ est continue sur l'intervalle fermé $(0,R)$. De fait, cette application $\rho \to A_\rho$ est <u>concave</u> :
pour ρ, $\rho' \in (0,R)$ et $0 \leq \alpha \leq 1$, on a :

$$(2\text{-}1) \qquad \alpha A_\rho \oplus (1-\alpha) A_{\rho'} \subset A_{\alpha\rho+(1-\alpha)\rho'}$$

Cela résulte du calcul élémentaire suivant : $A_{\alpha\rho+(1-\alpha)\rho'} =$
$[(\alpha A \oplus (1-\alpha) A) \ominus \alpha \rho \stackrel{\vee}{K}] \ominus (1-\alpha) \rho' \stackrel{\vee}{K} \supset [\alpha A_\rho \oplus (1-\alpha) A] \ominus (1-\alpha)\rho' \stackrel{\vee}{K}$
$\supset \alpha A_\rho \oplus (1-\alpha) A_{\rho'}$. Pour $\rho \in (0,R)$, nous poserons ensuite $K_\rho = A \ominus \stackrel{\vee}{A_\rho}$.
K_ρ est donc la fermeture de ρK selon A^c. D'après ce qui précède, on a
$A \ominus \stackrel{\vee}{K_\rho} = A_\rho$ et $A_{\rho K} = A_\rho \oplus \rho K \subset A_{K_\rho} = A_\rho \oplus K_\rho \subset A$, avec d'ailleurs l'é-
galité $A_\rho \oplus K_\rho = A$ dans le cas $n = 2$. Notons aussi <u>que</u> K_ρ <u>est le plus</u>
<u>grand compact convexe C tel que</u> $A_\rho = A \ominus \stackrel{\vee}{C}$. Car $A_\rho = A \ominus \stackrel{\vee}{C}$ entraine
$K_\rho = A \ominus \stackrel{\vee}{A_\rho} = A \ominus (\stackrel{\vee}{A} \ominus C) = C^{A^c} \supset C$. Inversement, d'ailleurs, $K_\rho = C^{A^c}$
entraine $A_\rho = A \ominus \stackrel{\vee}{C}$, de sorte que <u>l'on a</u> $A_\rho = A \ominus \stackrel{\vee}{C}$ <u>si et seulement si</u>
$\underline{C^{A^c} = K_\rho}$.

D'après la relation de concavité $(2\text{-}1)$, la fonction d'appui φ_{A_ρ} admet
en ρ une dérivée à droite (pour $0 \leq \rho < R$) et une dérivée à gauche
(pour $0 < \rho \leq R$). Toutefois ces dérivées ne sont pas elles-mêmes, en
général, des fonctions d'appui. Néanmoins, pour $0 \leq \rho < R$ et $0 < \varepsilon < \alpha$,
on a

$$\varphi_{A_\rho \ominus \stackrel{\vee}{A}_{\rho+\varepsilon}} \leq \varphi_{A_\rho} - \varphi_{A_{\rho+\varepsilon}} \leq \frac{\varepsilon}{\alpha} (\varphi_{A_\rho} - \varphi_{A_{\rho+\alpha}})$$

à cause de la concavité de $\rho \to \varphi_{A_\rho}$. Comme $A_\rho \ominus \stackrel{\vee}{A}_{\rho+\alpha}$ est le plus grand
convexe C tel que $\varphi_C \leq \varphi_{A_\rho} - \varphi_{A_{\rho+\alpha}}$, il en résulte : $\frac{1}{\varepsilon} (A_\rho \ominus A_{\rho+\varepsilon}) \subset$
$\frac{1}{\alpha} (A_\rho \ominus \stackrel{\vee}{A}_{\rho+\alpha})$ pour $\alpha \geq \varepsilon$. D'où l'existence de la <u>limite décroissante</u> :

$$K'_\rho = \lim_{\varepsilon \downarrow 0} \downarrow \frac{A_\rho \ominus \stackrel{\vee}{A}_{\rho+\varepsilon}}{\varepsilon}$$

ou érosion infinitésimale. La notation K'_ρ (ou, plus explicitement,
$K'_\rho(A)$) est justifiée par la relation : $A_\rho \ominus \stackrel{\vee}{A}_{\rho+\varepsilon} = K_{\rho+\varepsilon} \ominus K_\rho$ (en

effet : $A_\rho \ominus \check{A}_{\rho+\varepsilon} = A \ominus (\check{K}_\rho \oplus \check{A}_{\rho+\varepsilon}) = (A \ominus \check{A}_{\rho+\varepsilon}) \ominus K_\rho = K_{\rho+\varepsilon} \ominus \check{K}_\rho)$.

On montrerait de la même façon l'existence d'une dérivée à gauche en tout ρ tel que $0 < \rho \leq R$), limite croissante des $(A_{\rho-\varepsilon} \ominus \check{A}_\rho)/\varepsilon$. Pour $\rho = R$, cette limite est, en général, un fermé convexe non nécessairement compact.

L'intégrale de K'_ρ. La famille croissante K'_ρ , $\rho \in (0,R)$ est évidemment intégrable (au sens d'une intégrale de Riemann-Minkowski). Mais en général, $\underline{K_\rho}$ n'est pas égal à l'intégrale de sa dérivée. Car l'inégalité

$$\varphi_{K_{\rho+\varepsilon}} \ominus \check{K}_\rho \leq \varphi_{K_{\rho+\varepsilon}} - \varphi_{K_\rho} \quad \text{donne seulement} \quad \varphi_{K'_\rho} \leq \underline{\lim} \frac{\varphi_{K_{\rho+\varepsilon}} - \varphi_{K_\rho}}{\varepsilon} \quad .$$

Comme $K_\rho \supset K$, on trouve toutefois : $\rho_0 K \subset \int_0^{\rho_0} K'_\rho \, d\rho \subset K_{\rho_0}$ et par suite (puisque $A_{\rho_0} = A \ominus \rho_0 \check{K} = A \ominus \check{K}_{\rho_0}$) :

$$A_{\rho_0} = A \ominus \int_0^\rho \check{K}'_\rho \, d\rho$$

Mais, dans l'espace à deux dimensions, on a toujours l'égalité

$$(2\text{-}2) \qquad K_{\rho_0} = \int_0^{\rho_0} K'_\rho \, d\rho$$

En effet, $A_{\rho+\varepsilon} = A_\rho \ominus \varepsilon \check{K}$ est l'érodé de A_ρ par εK, ce qui implique, comme on l'a vu, pour $n = 2$, que A_ρ est ouvert selon $A_{\rho+\varepsilon}$. Par suite :

$$\varphi_{A_\rho \ominus \check{A}_{\rho+\varepsilon}} = \varphi_{A_\rho} - \varphi_{A_{\rho+\varepsilon}} \quad \text{et} : \quad \varphi_{K'_\rho} = -\frac{d}{d\rho} \varphi_{A_\rho} \quad . \quad \text{En intégrant, il vient}$$

alors :

$$\int_0^{\rho_0} \varphi_{K'_\rho} \, d\rho = \varphi_A - \varphi_{A_{\rho_0}} = \varphi_{K_{\rho_0}}$$

c'est-à-dire $(2\text{-}2)$. Mais la relation $(2\text{-}2)$ est fausse, en général, si $n \geq 3$.

3. CARACTERISATION DE L'EROSION INFINITESIMALE K'_0

Nous allons maintenant étudier plus en détail la structure de la dérivée à droite en $\rho = 0$, soit : $K'_\rho = \lim_{\rho \downarrow 0} \downarrow \frac{K_\rho}{\rho}$. Il suffit de remplacer l'ensemble initial A par A_ρ pour voir que les propriétés de K_o' se transposent d'elles-mêmes à K_ρ' $(0 \leq \rho < R)$. Notons d'abord un premier résultat :

LEMME 3-1 - On a $\varphi_{K_o'} = \varphi_K$ sur le support S_A de la mesure de surface G_A, et $\varphi_{K_o'} \geq \varphi_K$ ailleurs. Plus généralement, pour $0 \leq \rho < R$, on a $\varphi_{K_\rho} = \rho \, \varphi_K = \rho \, \varphi_{K'}$ sur le support S_{A_ρ} de G_{A_ρ}.

En effet, on a $\rho \, K \subset \rho \, K_o' \subset K_\rho$ pour $0 < \rho \leq R$. Comme $A_\rho = A \ominus \check{K} = A \ominus \check{K}_\rho$, il en résulte aussi $A_\rho = A \ominus \rho \, \check{K}_o'$ et $V(A_\rho) = V(A \ominus \rho \, K_o')$. En prenant la dérivée à droite en $\rho = 0$ du volume de cette érosion, on trouve donc : $\int \varphi_K(u) \, G_A(du) = \int \varphi_{K_o'}(u) \, G_A(du)$. Il en résulte que $\varphi_K = \varphi_{K_o'}$ sur S_A, puisque $\varphi_{K_o'} \geq \varphi_K$ et que ces fonctions sont continues. D'après la concavité de A_ρ, on a $(A \ominus \check{A}_\rho)/\rho \subset (A_\rho \ominus \check{A}_{\rho+\varepsilon})/\varepsilon$ et par suite $K_\rho \subset \rho \, K_\rho'$, et $\rho \, \varphi_K \leq \varphi_{K_\rho} \leq \rho \, \varphi_{K'}$. Mais $\varphi_K = \varphi_{K'}$ sur S_{A_ρ}, d'après la première partie de la démonstration, donc aussi $\varphi_{K_\rho} = \rho \, \varphi_K$ sur S_{A_ρ} ■.

Pour aller plus loin, nous utiliserons le résultat suivant, qui évoque la théorie du potentiel :

THEOREME 3-1 - Soient C et D dans $C_o(\mathcal{K})$, et C d'intérieur non vide. Si l'on a $\varphi_D \leq \varphi_C$ sur le support S_C de la mesure de surface G_C associée à C, alors $\varphi_D \leq \varphi_C$ sur la sphère unité S_o, c'est-à-dire $D \subset C$.

En effet, notons d'abord que $K \supset C$ et $W_1(C,K) = V(C)$ entraine $K = C$. Cela résulte des inégalités de Brunn-Minkowski :

$$W_1(C,K) \geq (V(C))^{\frac{n-1}{n}} \, (V(K))^{1/n} \geq V(C)$$

Pour que la première inégalité soit une égalité, il faut que l'on ait $K = \lambda \, C$ à une translation près. Comme $K \supset C$ par hypothèse, la seconde inégalité devient une égalité si et seulement si $\lambda = 1$ et $K = C$.

Ainsi, $\varphi_C \leq \varphi_K$ et $\varphi_C = \varphi_K$ sur S_C implique $\varphi_C = \varphi_K$. Désignons par D_o le plus grand compact convexe dont la fonction d'appui minore φ_C sur S_C : il vérifie $\varphi_{D_o} \geq \varphi_C$ et $\varphi_{D_o} = \varphi_C$ sur S_C . Donc, $\varphi_{D_o} = \varphi_C$ sur la sphère unité. ■

Voici maintenant la caractérisation cherchée de K_o' .

THEOREME 3-2 - K_o' est le plus grand des compacts convexes C tels que l'on ait $\varphi_C \leq \varphi_K$ sur le support S_A de la mesure de surface G_A .

D'après le Lemme 3-1, on a bien $\varphi_{K_o'} \leq \varphi_K$ sur S_A . Désignons par C_o le plus grand compact convexe C tel que $\varphi_C \leq \varphi_K$ sur S_A . On a, évidemment, $C_o \supset K$, et $\varphi_{C_o} = \varphi_K$ sur S_A . On a donc d'une part $A \ominus \rho \check{C}_o \subset A \ominus \rho \check{K}$ $(0 \leq \rho < R)$ et d'autre part $\varphi_{A \ominus \rho \check{K}} \leq \varphi_A - \rho \varphi_K = \varphi_A - \rho \varphi_{C_o}$ sur S_A , soit encore : $\varphi_{(A \ominus \rho \check{K}) \oplus \rho C_o} \leq \varphi_A$ sur S_A . D'après le théorème 3-1, cela implique $(A \ominus \rho \check{K}) \oplus \rho C_o \subset A$, donc $A \ominus \rho \check{K} \subset A \ominus \rho \check{C}_o$, et finalement $A_\rho = A \ominus \rho \check{K} = A \ominus \rho \check{C}_o$. Mais on a vu que K_ρ est le plus grand compact convexe D tel que $A_\rho = A \ominus \rho \check{D}$. On en déduit $C_o \subset (1/\rho) K_\rho$ et, pour $\rho \downarrow 0$, $C_o \subset K_o'$. ■

4. LE SUPPORT S_A DE LA MESURE DE SURFACE G_A.

A étant toujours supposé d'intérieur non vide, le théorème 3-2 montre que l'érosion infinitésimale K_o' ne dépend, en réalité, de A que par l'intermédiaire du support S_A de sa mesure de surface. Nous verrons que les directions $u \notin S_A$ qui ne chargent pas la mesure G_A sont associées aux éléments anguleux de la frontière ∂A de ce convexe. Si l'on convient de dire, dans un langage imagé, que S_A est le spectre de A, on voit que la fonction K_o' réalise, en quelque sorte, l'analyse spectrale de A et de ses érodés A_ρ, et doit permettre de mettre en évidence les propriétés d'angulosité de A et de ses érodés. On verra qu'il existe également, au moins pour n = 2, une synthèse spectrale, permettant de reconstituer A à partir de la donnée des seuls <u>supports</u> S_{A_ρ} de ses érodés successifs.

D'une manière générale, pour $u \in S_0$ (sphère unité) et $a \geq 0$, désignons par $E_{a,u}$ le demi-espace défini par : $E_{a,u} = \{x, < u \; x > \; \leq a\}$. D'après le théorème 3-2 ci-dessus, nous pouvons écrire :

$$(4\text{-}1) \qquad\qquad K_0' = \bigcap_{u \in S_A} E_{u, \varphi_K(u)}$$

et cette relation montre bien que K_0' ne dépend que du support S_A, et non de A lui-même.

On peut d'ailleurs généraliser. Si S est un sous-ensemble quelconque de la sphère unité S_0, on peut considérer le fermé convexe $K(S)$ défini par :

$$(4\text{-}2) \qquad\qquad K(S) = \bigcap_{u \in S} E_{u, \varphi_K(u)}$$

C'est le plus grand fermé convexe dont la fonction d'appui minore φ_K sur S. Comme φ_K est continue sur la sphère unité, on a toujours $K(\overline{S}) = K(S)$, $\overline{S}$ désignant l'adhérence de S, de sorte que l'on peut se limiter au cas où S est un sous-ensemble fermé de S_0. En général, $K(S)$ est fermé, mais non compact. Pour qu'il soit compact, il faut et il suffit que l'origine 0 soit un point intérieur de l'enveloppe convexe $C(S)$ de S.

Supposons remplie cette condition (0 appartient à l'intérieur de $C(S)$). Alors, d'après le théorème de Minkowski rappelé au paragraphe 1, on peut trouver un $A \in C_0(\mathcal{K})$ d'intérieur non vide dont la mesure de surface G_A admette le support $S_A = S$. On a alors $K(S) = K_0' = \lim (1/\rho)$ $A \ominus \check{A}_\rho$. Mais le support de la mesure de surface de $A \ominus \check{A}_\rho$ est contenu dans $S_A = S$, puisque $A \to G_A$ est décroissante pour les érosions. Comme d'autre part l'application $G \to \text{Supp } G$ est semi-continue inférieurement, on en déduit $S_{K_0'} \subset S$. Ainsi, le $K(S)$ défini en (4-2) vérifie la condition :

$$(4-3) \qquad S_{K(S)} \subset S$$

On en déduit un résultat intéressant concernant le support des mesures de surface.

THÉORÈME 4-1 - Soit A un compact convexe d'intérieur non vide dans $\mathbb{R}^n$, et S_A le support de sa mesure de surface G_A . Alors S_A est le plus petit sous-ensemble fermé S de la sphère unité S_0 tel que l'on ait : $A = \bigcap\limits_{u \in S} E_{u, \varphi_A(u)}$. De plus, on a $A(S) = A$ si et seulement si le fermé S contient S_A .

En effet, avec les notations définies ci-dessus, $A(S) = A$ entraine $S_A \subset S$, d'après (4-3). Il reste à montrer $A(S_A) = A$. En prenant $K = A$, on trouve $A_\rho = (1-\rho)A$, $A \ominus \overset{\vee}{A}_\rho = \rho\, A$ et $K(S_A) = K'_0 = A$. Mais cela signifie $A(S_A) = A$. Si maintenant $S \supset S_A$, on trouve $A \subset A(S) \subset A(S_A) = A$, et donc $A(S) = A$. ∎

Ce théorème conduit à une caractérisation très géométrique du support S_A de la mesure de surface G_A. Comme $\overset{\circ}{A}$ est non vide, nous pouvons toujours (quitte à effectuer une translation) supposer $0 \in \overset{\circ}{A}$. Dans ces conditions, le dual $A^* = \{y : \underset{x \in A}{\text{Sup}} < x,y > \leq 1\}$ de A est lui-même un compact convexe admettant 0 comme point intérieur.

On sait, [3], que les points de la frontière ∂A^* de ce dual sont alors les points de la forme $x = u/\varphi_A(u)$, $u \in S_0$. A l'intersection $A(S) = \bigcap\limits_{u \in S} E_{u, \varphi_A(u)}$ est associée, par dualité, le compact convexe $A(S)^*$, enveloppe convexe de $A^* \cap \widetilde{S}$ (où $\widetilde{S}$ désigne le cône engendré par les demi-droites de directions $u \in S$).

Mais on sait que l'enveloppe convexe fermée d'un sous-ensemble $D \subset A^*$ coïncide avec A^* lui-même si et seulement si l'adhérence $\overline{D}$ de D contient tous les points extrêmaux de A^*. On aura donc $A^*(S) = A^*$, c'est-à-dire $A(S) = A$, si et seulement si le fermé S contient la direction $u = x/|x|$ pour tout point x extrêmal sur ∂A^*. Convenons de dire que ces directions sont les directions extrêmales pour A. On peut voir facilement, par dualité, que ces directions extrêmales sont, effectivement, celles des génératrices extrêmales des cônes C(y) des normales aux différents points y de ∂A. Par conséquent, nous concluons de ce qui précède que l'on a A(S) = A si et seulement si le fermé S contient toutes les directions extrêmales pour A. Autrement dit, encore, le plus petit fermé S tel que A = A(S) est l'adhérence de l'ensemble des directions extrêmales pour A. En rapprochant ce résultat du Théorème 4-1, nous concluons :

COROLLAIRE 1 - Le support S_A de la mesure de surface G_A associée à un compact convexe A d'intérieur non vide est l'adhérence de l'ensemble des directions $u \in S_o$ extrêmales pour A.

Dans le même ordre d'idées, notons aussi le résultat suivant, qui nous sera utile par la suite :

COROLLAIRE 2 - Pour tout ρ tel que $0 \leq \rho < R$, on a $\varphi_A = \varphi_{A_\rho} + \rho\,\varphi_K$ sur le support S_{A_ρ} de la mesure de surface G_{A_ρ} associée à l'érosion $A_\rho = A \ominus \rho\,\check{K}$.

En effet, le dual A_ρ^* de A_ρ est l'enveloppe convexe de l'ensemble

$$D = \{\lambda u,\ u \in S_o\ ,\ \lambda(\varphi_A(u) - \rho\,\varphi_K(u)) \leq 1\}$$

soit C(D). Mais on sait que C(D) est aussi l'enveloppe convexe fermée de ses points extrêmaux, qui constituent un ensemble $\ddot{D} \subset D$. Si $y \in \ddot{D}$ est un point extrêmal de A_ρ^*, et $u = y/|y|$ sa direction, on a $u \in S_{A_\rho}$, d'après le corollaire ci-dessus, et aussi $\varphi_{A_\rho}(u) = 1/|y|$. Comme $y \in D$, on a aussi $\varphi_A(u) - \rho\,\varphi_K(u) \leq 1/|y|$, et donc l'égalité $\varphi_{A_\rho}(u) = \varphi_A(u) - \rho\,\varphi_K(u)$. Inversement, si $u \in S_{A_\rho}$, il existe une suite u_n de directions extrêmales pour A_ρ convergeant vers u, et l'égalité $\varphi_{A_\rho}(u_n) = \varphi_A(u_n) - \rho\,\varphi_K(u_n)$ passe à la limite. ∎

De ce qui précède résulte la possibilité de l'analyse et de la synthèse spectrale, au moins dans le cas de l'espace $\mathbb{R}^2$ à 2 dimensions. Si, en

effet, on désigne par S_ρ le support de la mesure de surface G_{A_ρ} associée à l'érodé $A_\rho = A \ominus \rho \check{K}$, on a $K_\rho' = K(S_\rho)$, et par suite (pour n =
2) d'après (2-2) : $K_{\rho_0} = \int_\rho^{\rho_0} K(S_\rho) \, d\rho$. Dans l'espace à 2 dimensions,
l'érodé ultime A_R, sous-dimensionnel, est au plus un segment de droite.
Dans le cas le plus usuel, $A_R = \{o\}$ est réduit à un point, et dans ce
cas on a $A = \int_0^R K(S_\rho) \, d\rho$ d'après la relation $A = A_R \oplus K_R = K_R$.
Autrement dit, dans l'espace à 2 dimensions, la connaissance des seuls
<u>supports</u> S_ρ des mesures de surface associées aux érodés A_ρ permet de
reconstituer l'ensemble initial A. Il s'agit bien là de ce que nous
avons désigné sous le nom de synthèse spectrale, c'est-à-dire de la
reconstitution de A à partir des seules propriétés d'angulosité de ses
érodés successifs A_ρ.

<u>EXEMPLE 1</u> - Dans $\mathbb{R}^2$ on a $S_\rho = S_A =$ Cste pour $0 \leq \rho < 1$ si et seulement
si A est ouvert selon K_0'.

En effet, si $A = A_{K'_0}$, on a aussi $A = A_\rho \oplus \rho K_0'$ ($0 \leq \rho < 1$) et $A_\rho =$
$A \ominus \rho \check{K}_0' = (A \ominus \check{K}_0') \oplus (1-\rho)K_0'$. Dans l'espace à 2 dimensions, cela
entraine $G_{A_\rho} = G_{A \ominus \check{K}_0'} + (1-\rho) G_{K'_0}$, et par suite $S_\rho = S_{A \ominus \check{K}_0'} \cup S_{K'_0} = S_A$.

Inversement, supposons $S_\rho = S_A$ pour $0 \leq \rho < 1$. D'après le
corollaire 2 du Théorème 4-1, on a $\varphi_A = \varphi_{A_\rho} + \rho \, \varphi_K$ sur $S_{A_\rho} = S_A$, et
aussi $\varphi_K = \varphi_{K'_\rho}$ sur S_{A_ρ} (lemme 3-1). Donc :

$$\varphi_A = \varphi_{A_\rho} + \rho \, \varphi_{K'_\rho} \quad \text{sur } S_A = S_\rho$$

Par ailleurs, $S_{K'_\rho} \subset S_{A_\rho}$ et (pyisque n = 2) $S_{A_\rho \oplus \rho K'_\rho} = S_\rho$. Le Théorème
3-1 donne donc $A = A_\rho \oplus \rho K_\rho'$. Comme $A = A_\rho \oplus K_\rho$, il en résulte $K_\rho =$
$\rho K_\rho'$, d'où l'on déduit en intégrant : $K_\rho = \rho K_0'$. Finalement, donc,
$A = A_\rho \oplus K_\rho = A_\rho \oplus \rho K_0'$ et, pour $\rho \uparrow 1$, $A = A_1 \oplus K_0'$ ∎

Dans R^2, d'après [1], A est ouvert selon K_0' si et seulement si le volume $V(A_\rho) = V(A \ominus \rho K_0')$ est un polynome de second degré en ρ sur $(0,1)$.
Comme exemple simple, on peut citer l'érosion d'un carré A par un cer-
cle $K \subset A$. Ici, K_0' est lui-même un carré contenu dans A.

EXEMPLE 2 - Dans $\mathbb{R}^2$, A est ouvert selon K si et seulement si $S_K \subset S_\rho$
pour tout $\rho < 1$. Lorsqu'il en est ainsi, on a également $S_\rho = S_A$
= Cste et $K_o' = K$.

Si A est ouvert selon K, on a $A = A_\rho \oplus \rho K$ pour $\rho < 1$, donc $S_K \subset S_A$
et $K_o' = K$. Par suite, d'après l'exemple 1, $S_K \subset S_\rho = S_A$. Inversement,
si $S_K \subset S_\rho$, on a $K_\rho' = K$ donc $K_\rho = \rho K$ et $A = A_\rho \oplus \rho K$ pour $\rho < 1$.
En faisant $\rho \uparrow 1$, il vient $A = A_1 \oplus K$, et A est ouvert selon K. ∎

5. L'EQUATION $X \ominus \check{K} = A$.

Soient A et K compacts convexes donnés dans $\mathbb{R}^n$, contenant l'origine O,
et A d'intérieur non vide. A titre d'application de ce qui précède, on
se propose de caractériser la famille des $X \in C(\mathcal{K})$ tels que $X \ominus \check{K} = A$.
Tout d'abord, il est clair que l'on doit avoir $X \supset A \oplus K$, et d'ailleurs
$A \oplus K$ lui-même est solution de l'équation proposée. Autrement, dit, $X_o =$
= $A \oplus K$ est la plus petite solution possible. Montrons, ce qui est
moins évident, qu'il existe également une plus grande solution.
D'après le corollaire 2 du Théorème 4-1, en effet, si $X \ominus \check{K} = A$ avec
$\overset{\circ}{A} \neq \emptyset$, on doit avoir $\varphi_X = \varphi_A + \varphi_K$ sur le support S_A de la mesure de
surface G_A associée a A. Or, il existe un plus grand élément $\tilde{A}$ véri-
fiant cette propriété dans $C(\mathcal{K})$, à savoir :

$$(5-1) \qquad \tilde{A} = \bigcap_{u \in S_A} E_{u, \varphi_K(u) + \varphi_A(u)}$$

Par suite, on a nécessairement $X \subset \tilde{A}$ pour toute solution X de l'équation proposée. Mais $\tilde{A}$ lui-même est une solution, et, par suite, est
la plus grande solution. En effet, par construction, on a $\varphi_{\tilde{A}} = \varphi_A +$
+ φ_K sur S_A , et donc : $\varphi_{\tilde{A} \ominus K} \leq \varphi_{\tilde{A}} - \varphi_K = \varphi_A$ sur S_A . D'après le
Théorème 3-1, cela entraine $\tilde{A} \ominus \check{K} \subset A$. Mais, par ailleurs, la relation
(5-1) qui définit $\tilde{A}$ implique manifestement $\tilde{A} \supset A \oplus K$, et donc $\widetilde{A \ominus \check{K}}$
$\supset A$. On a donc bien $\tilde{A} \ominus \check{K} = A$.

Comme $X_0 = A \oplus K$ et $\tilde{A}$ constituent la plus petite et la plus grande

solution, toute autre solution X vérifiera $X_0 \subset X \subset \tilde{A}$. Inversement,

si X est compris entre X_0 et $\tilde{A}$, X est bien une solution, puisque $A =$

$\tilde{A} \ominus \check{K} \supset X \ominus \check{K} \supset X_0 \ominus \check{K} = A$ de sorte que nous caractérisons bien ainsi

la famille de toutes les solutions de l'équation proposée.

La solution maximale vérifie $\tilde{A} \supset A \oplus K_0'$. En effet, si nous comparons

(5-1) et (4-1), il apparaît que l'on a $A = (A \oplus K)$ (S_A), soit, expli-

citement :

$$(5\text{-}2) \qquad \tilde{A} = \lim_{\varepsilon \downarrow 0} \frac{1}{\varepsilon} \left[A \ominus (\check{A} \ominus \varepsilon \, (A \oplus K) \right]$$

Mais $A \ominus \varepsilon(\check{A} \oplus \check{K}) = (1\text{-}\varepsilon) \, A \ominus \varepsilon \, \check{K}$, et $A \ominus \left[(1\text{-}\varepsilon) \, \check{A} \ominus \varepsilon \, K \right] \supset \varepsilon \, A \oplus (1\text{-}\varepsilon)$

$\left[A \frac{\varepsilon}{1-\varepsilon} \right]$. D'après (5-2), et la définition de K_0' , on en déduit $\tilde{A} \supset$

$\supset A \oplus K_0'$.

Cette inclusion est stricte, en général, pour $n \geq 3$. Mais dans l'espace

à $n = 2$ dimension, elle devient une égalité. Pour $n = 2$, en effet, on

a $G_{A \oplus K_0'} = G_A + G_{K_0'}$ et $S_{A \oplus K_0'} \subset S_A \cup S_{K_0'}$. Mais $S_{K_0'} \subset S_A$ (à cause de la

semi-continuité inférieure de l'application $G \to \mathrm{Supp}\, G$), donc en fait

$S_{A \oplus K_0'} = S_A$. L'inégalité (en fait : l'égalité) $\varphi_{\tilde{A}} \leq \varphi_A + \varphi_K \leq \varphi_A +$

$+ \varphi_{K_0'}$ sur $S_A = S_{A \oplus K_0'}$ entraine alors, d'après le Théorème 3-1, $\tilde{A} \subset$

$\subset A \oplus K_0'$, et par suite l'égalité $\tilde{A} = A \oplus K_0'$. Nous pouvons donc énon-

cer :

<u>THÉORÈME 5</u> - Soient A et K compacts convexes dans $\mathbb{R}^n$ et A d'intérieur

non vide. Pour tout $X \in C(\mathcal{K})$, on a $X \ominus \check{K} = A$ si et seulement si

$A \oplus K \subset X \subset \tilde{A}$, avec : $\tilde{A} = \bigcap_{u \in S_A} E_{u, \varphi_A(u) + \varphi_K(u)}$. Si l'on pose

$K_0' = K(S_A)$, on a $\tilde{A} \supset A \oplus K_0'$, mais non, en général, l'égalité.

Mais, dans l'espace à $n = 2$ dimensions, on a toujours $\tilde{A} = A \oplus K_0'$.

Dans cet énoncé, l'hypothèse que A est d'intérieur non vide ne peut

pas être affaiblie. Même à 2 dimensions, le théorème peut être faux

si $\overset{\circ}{A} = \emptyset$.

6. CARACTERE LOCAL DE LA CORRESPONDANCE $\varphi_A \to G_A$.

Dans le cas de l'espace à 2 dimensions, il résulte de la relation (1-4)
que la correspondance $\varphi_A \to G_A$ possède un caractère local, c'est-à-dire
que la donnée de φ_A sur un ouvert σ de la sphère unité détermine la
mesure G_A sur le même ouvert σ. Ce caractère local subsiste dans l'espace $\mathbb{R}^n$ pour $n \geq 3$, bien que la correspondance $\varphi_A \to G_A$ ne soit plus
linéaire.

Pour le voir, considérons un A compact convexe et d'intérieur non vide,
dans $\mathbb{R}^n$. Pour tout $u \in S_0$ désignons par $H_A(u)$ l'hyperplan d'appui
$H_A(u) = \{x : <u\,x> = \varphi_A(u)\}$ et par $F_A(u) = A \cap H_A(u)$ la $\underline{face}$ de A
associée à cette direction $u \in S_0$. Si σ est un $\underline{ouvert}$ de la sphère
unité S_0, on peut montrer facilement que $G_A(\sigma)$ est la mesure de la
surface de l'ensemble $F_A(\sigma) = \bigcup\limits_{u \in \sigma} F_A(u)$ telle qu'on peut la définir
à partir de considérations géométriques élémentaires. Par suite, pour
démontrer le caractère local de la correspondance $\varphi_A \to G_A$, il suffit
d'établir que $F_A(\sigma)$, pour tout σ ouvert dans S_0, ne dépend que de la
donnée de φ_A sur l'ouvert σ. Or cela résulte du lemme suivant :

$\underline{\text{LEMME DE LOCALISATION}}$ - Soit $A \in C_0(\mathcal{K})$, φ sa fonction d'appui. Pour
tout $u \in S_0$ on pose $H(u) = \{x : <u\,x> = \varphi(u)\}$ (hyperplan d'ap-
pui associé à la direction u), $F(u) = A \cap H(u)$ (face de A associée à u), $E_u = \{x : <u\,x> \leq \varphi(u)\}$, et, pour tout $\sigma \subset S_0$,
$A(\sigma) = \bigcap\limits_{u \in \sigma} E_u$. Alors, pour tout $\underline{ouvert}$ σ de la sphère unité S_0
et tout $u_0 \in \sigma$, on a $F(u_0) = A(\sigma) \cap H(u_0)$.

Comme le fermé $A(\sigma)$ ne dépend que de la donnée de φ sur l'ouvert σ ,
le lemme exprime le caractère local de la correspondance $\varphi(u_0) \to F(u_0)$,
d'où résulte aussi le caractère local de la correspondance $\varphi_A \to G_A$.
Pour démontrer le lemme, notons d'abord que l'on a nécessairement
$F(u_0) \subset A(\sigma) \cap H(u_0)$, puisque $A \subset A(\sigma)$, et il suffit donc de démontrer
l'inclusion inverse. Pour cela, nous prolongerons la fonction d'appui

φ de A, définie sur S_o, en posant $\varphi(o) = 0$ et $\varphi(y) = |y|\ \varphi(y/\,|y|)$
pour $y = 0$. On sait [3] que la fonction φ ainsi prolongée est convexe
et continue sur $\mathbb{R}^n$. On en déduit que, pour tout $x \in \mathbb{R}^n$, l'ensemble
T_x défini par : $T_x = \{y: <x,y> > \varphi(y)\}$ est un <u>cône ouvert convexe</u> ,
et que $T_x = \emptyset$ si et seulement si $x \in A$.

Si le cône T_x n'est pas vide (c'est-à-dire si $x \notin A$) sa frontière est
$\partial T_x = \{y : <x,y> = \varphi(y)\}$. Il est clair, en effet, que tout point
de la frontière de T_x est de cette forme, à cause de la continuité de
φ. Inversement, soit $y_o \in \mathbb{R}^n$ tel que $<x,\ y_o> = \varphi(y_o)$. Comme T_x n'est
pas vide, par hypothèse, il existe $y_1 \in T_x$, c'est-à-dire tel que
$<x,\ y_1> > \varphi(y_1)$ strictement. Pour $0 < \lambda < 1$, on a alors :

$$<x,\ \lambda y_o + (1-\lambda)y_1> \ >\ \lambda\varphi(y_o) + (1-\lambda)\varphi(y_1) \geq \varphi(\lambda y_o + (1-\lambda)y_1)$$

c'est-à-dire $\lambda y_o + (1-\lambda)y_1 \in T_x$. Il suffit alors de faire tendre λ
vers 1 pour obtenir $y_o \in \partial T_x$.

Soit alors $x \in A(\sigma) \cap H(u_o)$ pour un u_o appartenant à l'ouvert $\sigma \subset S_o$.
On vient de voir que, si $T_x \neq \emptyset$, c'est-à-dire si $x \notin A$, on a $x \in H(u_o)$
si et seulement si $u_o \in \partial T_x$. Donc, si x n'appartient pas à A, le voi-
sinage σ de u_o rencontre T_x. Par suite il existe un $u_1 \in \sigma$ tel que
$<x,\ u_1> > \varphi(u_1)$ strictement. Mais cela contredit $x \in A(\sigma)$. Donc
$T_x = \emptyset$, c'est-à-dire $x \in A$. Par suite, on a bien $A(\sigma) \cap H(u_o) \subset F(u_o)$ ∎.

7. INEGALITES POUR LES MESURES PERIMETRIQUES.

On a vu que les mesures de surfaces sont décroissantes pour les érosions,
soit $G_{A\ominus\check{K}} \leq G_A$. Dans l'espace à 2 dimensions, ce résultat peut être
notablement amélioré. De fait, si A et K sont suffisamment réguliers
pour que leurs mesures périmétriques admettent des densités (rayons de
courbures) continues $R_A(\theta)$ et $R_K(\theta)$, il est intuitif que le rayon de
courbure $R_{A_K}(\theta)$ de l'ouverture A_K de A selon K est majorée par le Sup

de R_A et R_K . Plus précisément, on aura $R_{A_K} = R_A$ là où le contour de A_K coïncide avec celui de A, et $R_{A_K} = R_K$ ailleurs. Celà implique, pour l'é-rodé $A_1 = A \ominus \check{K} = A_K \ominus \check{K}$, l'inégalité $R_{A_1} \leq (R_A - R_K)_+$.

Dans le cas général, les mesures périmétriques vérifient des inégali-tés analogues. Bien qu'il s'agisse d'un résultat géométriquement très intuitif, je n'ai pas trouvé de démonstration simple, dans le cas à 2 dimensions, et j'ignore si l'on a des résultats analogues à $n > 2$ dimensions. L'énoncé est le suivant :

THÉORÈME 7 - Soient A et K compacts convexes dans $\mathbb{R}^2$ et $A \ominus \check{K} \neq \emptyset$.

On pose $A_1 = A \ominus \check{K}$, $K_1 = A \ominus \check{A}_1$ et $A_K = A_1 \oplus K$. Alors :

$$(7\text{-}1) \qquad G_{A_1} \leq (G_A - G_K)_+ \; ; \quad G_{A_K} \leq G_A \vee G_K \; ; \quad G_{K_1} \geq G_A \wedge G_K$$

Compte tenu de la relation $G_{C \oplus D} = G_C + G_D$ et $A = A_1 \oplus K_1$ valables dans $\mathbb{R}^2$, ces trois inégalités sont équivalentes, et nous démontrerons la seconde, en nous appuyant sur deux lemmes (dont le premier est d'ail-leurs intéressant en lui-même).

LEMME 7-1 - Soient $K \subset A \subset C$ compacts convexes non vides dans $\mathbb{R}^2$. Si $C_A = C$, alors $C_{A_K} = C_K$. En particulier, si C est ouvert à la fois selon A et selon K, alors C est ouvert selon A_K .

En effet, si C est ouvert selon A, A est fermé selon C^c, soit $A = C \ominus (\check{C} \ominus A)$. On en déduit $A_1 = A \ominus \check{K} = C' \ominus (\check{C} \ominus A)$, avec $C' = C \ominus \check{K}$. A_1 est donc un érodé de C', et, dans $\mathbb{R}^2$, on a vu que cela implique que C' est ouvert selon A_1 , soit :

$$(7\text{-}2) \qquad\qquad C' = (C' \ominus \check{A}_1) \oplus A_1$$

Mais on a $C' \ominus \check{A}_1 = C \ominus \check{A}_K$ et $C_K = C' \oplus K$. La relation (7-2) entraine donc $C_K = (C \ominus \check{A}_K) \oplus A_1 \oplus K = C_{A_K}$. ∎

<u>LEMME 7-2</u> - Désignons par $M(a,b)$ l'ensemble des mesures $G \geq 0$ sur le cercle unité de $\mathbb{R}^2$ vérifiant :

$$\int_0^{2\pi} \cos \theta \; G(d\theta) = a \quad ; \quad \int_0^{2\pi} \sin \theta \; G(d\theta) = b$$

$(a, b$ réels). Alors : $\mathrm{Inf} \{G : G \in M(a,b)\} = 0$.

En effet, prenons $a = \rho \cos \varphi$, $b = \rho \sin \varphi$ avec $\rho \geq 0$. On peut trouver une mesure $G_1 \in M(a,b)$ concentrée sur les deux points φ, $\varphi + \pi$; et aussi une mesure $G_2 \in M(a,b)$ concentrée sur les 4 points $\varphi \pm \varepsilon$, $\varphi + \pi \pm \varepsilon$ pour ε différent de 0 et de π. Or, on a déjà $G_1 \wedge G_2 = 0$. ■

<u>DEMONSTRATION DU THEOREME 7</u>. Avec les notations du lemme 7-2, considérons l'ensemble $M(a,b)$ avec :

$$a = - \int \cos \theta \; (G_A \vee G_K) \quad ; \quad b = - \int \sin \theta \; (G_A \vee G_K)$$

Pour tout $G \in M(a,b)$, la mesure $G_C = G + G_A \vee G_K$ est positive et vérifie la condition barycentrique, donc est la mesure périmétrique d'un compact convexe C de $\mathbb{R}^2$ (Théorème de Minkowski). Comme G_C majore G_A et G_K , il en résulte (dans $\mathbb{R}^2$) que C est ouvert selon A et selon K . Donc (lemme 7-1) C est ouvert selon A_K , soit $G_{A_K} \leq G_C$. On a alors :

$G_{A_K} \leq \mathrm{Inf} \{G + G_A \vee G_K ; G \in M(a,b)\}$ et par suite $G_{A_K} \leq G_A \vee G_K$ (lemme 7-2). ■

Dans $\mathbb{R}^2$, on a $A = A_1 \oplus K_1$. En fait, $K_1 = A \ominus \check{A}_1$ <u>est le plus petit compact convexe</u> $D \supset K$ <u>tel que A soit ouvert selon D.</u>

En effet, $A_D = A$ entraine toujours $D = (D)^{A^c}$. Par suite $D \supset K$ implique $D \supset (K)^{A^c} = K_1$.

On déduit du théorème un autre résultat relatif à K_1 :

<u>COROLLAIRE</u> - <u>Soient A, K et C compacts convexes non vides dans $\mathbb{R}^2$.
Si A et K sont ouverts selon C, alors $K_1 = (K)^{A^c}$ est ouvert selon C.</u>

En effet, dans $\mathbb{R}^2$, $A_C = A$ et $K_C = K$ si et seulement si $G_C \leq G_A \wedge G_K$. D'après le Théorème, cela entraine $G_C \leq G_{K_1}$, et donc, dans $\mathbb{R}^2$, que K_1 est ouvert selon C.

BIBLIOGRAPHIE.

[1]   G. Matheron (1977) : La formule de Steiner pour les érosions
                 (à paraître dans Adv. in Appl. Prob.).

[2]   M. Minkowski (1903) : Volumen und Oberfläche. Math. Ann., Vol. 57,
                 pp. 447-495.

[3]   R.T. Rockafellar (1972) : Convex Analysis. Princeton University
                 Press.

NEEDLES AND WEDGES AS TOOLS FOR INTEGRATION
-=-

R.V. AMBARTZUMIAN

Institute of Mathematics, Erevan, Armenia
U.S.S.R.

The aim of my talk is to present some basic results from the domain of Combinatorial Integral Geometry concerning lines and planes in three dimensional Euclidean space R^3.

The subject of Combinatorial Integral Geometry presumably has direct applications in Mathematical Stereology. It is the topic of a special book <1> where among other things complete proofs of the statements we are going to formulate can be found.

The corresponding planar results were the topic of my lecture delivered at the first Buffon Bicentenary meeting which was held in Erevan (Sevan) last autumn. Results in three dimensions have since been derived starting from the planar ones. The complete account of the present state of the planar theory will soon be published <2 > in a form accessible to most listeners today. Also three dimensions is the case which is of more interest from the point of view of applications. Therefore I choose to describe here the available results for R^3, giving only minimal reference to the more extensively studied planar case.

I - BUFFONIC ROOTS

Let us reformulate the classical Buffon's problem of the needle as follows.

Instead of throwing the needle at random onto a floor ruled with a lattice of equidistant parallel lines, rather suppose that a needle v of length $|v| < 1$ is fixed (say in a disc K of unit diameter) and the lattice of parallel lines with unit separation is randomly placed on the plane. What is the probability that the line of the lattice will hit the needle ?

The position of the lattice is completely specified by that of its almost surely unique member line g_k intersecting K. For this reason to define the distribution of the position of the lattice it suffices to determine a distribution of a random line through K (that is of g_k). In these terms the "problem of the needle" is equivalent to the problem of finding the probability of the event

" g_k belongs to the set $[v]$ ".

Here $[v]$ is the subset of the space G of lines on the plane defined as

$$[v] = \{g \in G \quad : \text{the line } f \text{ intersects the needle } v\}.$$

Thus we maintain that 200 years ago Buffon was calculating the probability (or measure) of the set $[v]$. For this reason the subsets of G of the type $[v]$ have been called in < 2 > Basic Buffon.

Note that the measure on G which corresponds to the original Buffon's needle-throwing experiment is in fact proportional to the so-called invariant measure on G.

In three dimensions there are two analogues of the classical Buffon problem of the needle.

I - A lattice of equidistant parallel planes with unit separation is fixed in R^3 and a needle n of length $|n| < 1$ in placed at random in R^3. What is the probability that n will hit a plane belonging to the lattice ?

II - In R^3 a lattice of lines parallel to the Oz axis is fixed. For instance we may assume that the points of intersection of the lines from the lattice form a square lattice on the plane z = 0. A flat f (a flat is defined as a bounded convex part of a plane)with maximal diameter less then 1 is placed at random in R^3. What is the probability that f will hit a line from the lattice ?

These two problems also permit reformulations in which it is the position of the lattice which is assumed random. Thus equivalent problems I' and II' arise :

I' - Let E be the space of all planes in R^3. Define the subset $[n] \subset E$ to be the set of all planes which intersect the needle $n \subset R^3$. Find the measure (probability) of the set $[n]$.

II' - Let Γ be the space of all lines in R^3. Define the subset $[f] \subset \Gamma$ to be the set of all lines intersecting the flat f. Find the measure (probability) of the set $[f]$.

It is natural to refer to the sets of the type [n] and [f] as Basic Buffon sets in and Γ correspondingly.

II - BUFFON RINGS

We are now in a situation typically described in the first pages of many textbooks on probability and measure theory.

There is a space $X(X = G, E$ or $Γ)$ and a class B of its subsets, namely the class of X. We wish to apply the measure-theoretical operations of union $\cup$, intersection $\cap$ and complementation to the members of the class B (note that $B_1 \cup B_2$, $B_1 \cap B_2$ and B^c need not also be Basic Buffon).

In this context algebras and rings appear "almost sure".

We will consider only finite rings associated with appealing and still sufficiently general objects such as

finite sets $\{v_i\}$ of needles in R^2,
finite sets $\{n_i\}$ of needles in R^3,
finite sets $\{f_i\}$ of flats in R^3.

These rings will be called **Buffon rings** and will be denoted Br $\{[v_i]\}$, Br $\{[n_i]\}$ and Br $\{[f_i]\}$, respectively.

By definition, the ring Br $\{B_i\}$ is the ring of subsets of X generated by a collection $\{B_i\}$ of Basic Buffon subsets of X.

It is not difficult to describe the general element of Br $\{B_i\}$. The typical atom of this ring has the form

$$(\underset{i \in J}{\cap} B_i) \cap (\underset{i \in J}{\cap} B_i^c) , \qquad \begin{array}{l} \text{J is a nonempty subset of} \\ \{1,\ldots,m.\} \end{array}$$

and each member of Br $\{B_i\}$ is just a union of a number of such atoms.

Many subsets of interest in stereology associated with the structure of the sets $\{v_i\}$, $\{n_i\}$ or $\{f_i\}$ turn out to be elements of corresponding Buffon rings.

For example the set

$$\{ \gamma \in \Gamma : \gamma \quad \text{hits exactly k flats from} \{ f_i \} \ \}$$

belongs to Br $\{ [f_i] \}$ for $k > 0$. Here one may think of the f_i as the faces of a (not necessarily convex) polyhedral domain $D \subset R^3$. Further, let $\{ n_i \}$ be the set of edges of the same D. The set

$$\{ e \in E : \ e \cap D \quad \text{is a k-gon} \}$$

belongs to Br $\{ [n_i] \}$ for $k > 0$.

The introduction of Buffon rings (rather then individual subsets of X) is justified by the results stated in § 4.

III - THE COMPANION SYSTEMS

The theorems of § 4 are formulated in terms of the "companion systems".

1 - With a given set $\{ v_i \}$ of needles in R^2 associate a richer set $\{ v_i' \}$ of needles called the companion system of needles, constructed as follows.

Let $\{ P_i \}$ be the collection of endpoints of needles belonging to the original set $\{ v_i \}$. Each (unordered) pair P_i, P_j defines a needle which has P_i and P_j for its endpoints.

By definition the set $\{ v_i' \}$ of needles corresponds to all pairs P_i, P_j.

The companion systems for $\{ n_i \}$ and $\{ f_i \}$ are defined as systems (sets) of wedges, rather then needles.

A wedge is a pair (n, V), where $n \subset R^3$ is a needle and V is a domain in R^3 bounded by two planes passing through n in a vertical flat angle (fig.   ).

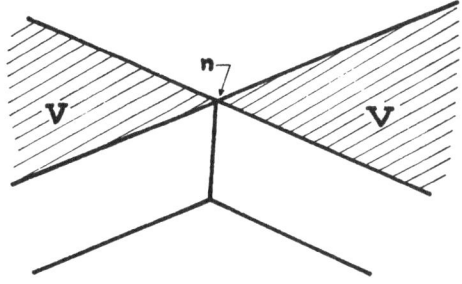

Fig. A wedge

2 - Let $\{n_i\}$ be a set of needles in R^3 and let $\{P_i\}$ be the collection of their endpoints. A wedge w, w = (n,V) ,is said to belong to the companion system iff

a) n is of the type $P_i P_j$;

b) the interior of V does not contain any points from $\{P_i\}$;

c) each plane bounding V passes throuth a point from $\{P_i\}$. The thus defined companion system obviously contains only a finite number of wedges.

3 - Now consider a set $\{f_i\}$ of flats in R^3 where, unlike the previous case, the companion system of wedges is no longer discrete.

Choose two flats, f_i and f_j. Let dl_1 and dl_2 be two linear elements on their boundaries : $dl_1 \subset \partial f_i$, $dl_2 \subset \partial f_j$. With every (i,j, dl_1, dl_2) associate two wedges, specified by an index s, s = 1,2. Both have the segment joining dl_1 and dl_2 for their common needle, the corresponding V_1 and V_2 are complementary vertical flat angles, and the two planes separating V_1 from V_2 are in fact the planes containing the mentioned needle and the elements dl_1 and dl_2.

Note, that in the case i = j (which is not excluded) one of the V_s is identically empty, while the opening of the other is π.

In this construction a wedge w is described by a discrete index I

$$I = (i,j,s)$$

and two points $l_1 \in \partial f_i$ and $l_2 \in \partial f_j$. Thus we write

$$w = W_I(l_1, l_2).$$

The collection of all such wedges composes the companion system for $\{f_i\}$.

IV - BASIC COMBINATORIAL INTEGRAL GEOMETRY

Integral Geometry has traditionnaly been interested in invariant measures in the spaces G, E and Γ. This contrasts sharply with the following theorems in which invariance assumptions are not involved.

In what follows $[P] \subset G$ denotes the bundle of lines passing through the point $P \in R^2$, $[P] \subset E$ denotes the bundle of planes through the point $P \in R^3$, μ denotes the density of a measure $m(\ )$ in Γ with respect to the invariant measure, and dg and dγ denote the elements of the invariant (normed) measure in the spaces G and Γ respectively.

THEOREM 1 - Let $m(\ )$ be a measure on G such that $M([P_2]) = 0$ for every $P \in R^2$. Let a set of needles $\{v_i\}$ be fixed in R^2. For every $A \in Br \{[v_i]\}$ the value of $m(A)$ may be calculated as a sum

$$m(A) = \sum_{v'} c_{v'}(A) \ F(v') \qquad (1)$$

where $c_{v'}(A)$ are integer coefficients which do not depend on the choice of measure $m(\ )$, and the summation is extended over the companion system of needles. The function $f(v)$ is defined on the set of needles in R^2 as

$$2 \ F_1(v) = m([v]). \qquad (2)$$

THEOREM 2 - Let $m(\ )$ be a measure on E such that $m([P]) = 0$ for every point $P \in R^3$. Let a set of needles $\{n_i\}$ be fixed in R^3. For every $A \in Br \{[n_i]\}$ the value of $m(A)$ may be calculated as a sum

$$m(A) = \pi^{-1} \sum_w c_w(A) F_2(w), \qquad (3)$$

where $c_w(A)$ are integer coefficients which do not depend on the choice of measure $m(\ )$, and the summation is extended over the companion system of wedges. The funtion $F(w)$ is defined on the set of wedges in R^3 as

$$F(w) = \int_{[n]} |V \cap e| \ m(de), \qquad (4)$$

where $|V \cap e|$ is the opening of the planar angular domain $V \cap e$.

THEOREM 3 - <u>Let</u> m( ) <u>be a measure on</u> Γ <u>possessing a density</u> μ . <u>Let a set of flats</u> $\{f_i\}$ <u>be fixed in</u> R^3. <u>For every</u> $A \in Br$ $\{[f_i]\}$ <u>the value of</u> m(A) <u>may be calculated</u> as a sum of integrals

$$m(A) = \sum_I \int_{\partial f_i} \int_{X \partial f_j} c_w(A) \sin\psi_1 \sin\psi_2 F_3(w) \, dl_1 \, dl_2, \qquad (5)$$

where $c_w(A)$ <u>are integer coefficients which do not depend on the choice of the measure</u> m( ), $w = w_I(l_1,l_2)$ <u>belongs to the companion system of wedges, and the angles</u> ψ_1 <u>and</u> ψ_2 <u>are measured between the needle</u> $n = l_1 l_2$ <u>and the elements</u> dl_1 <u>and</u> dl_2.

<u>The function</u> $F_3(w)$ <u>is defined on the set of wedges</u> $w = (n,V)$ <u>as</u>

$$F_3(w) = \frac{1}{|n|} \int_{e_\varphi \subset V} \sin\lambda_1 \sin\lambda_2 \, d\varphi \int_{[n]_\varphi} \mu(g) \, dg \qquad (6).$$

<u>Here</u> e_φ <u>is the plane through n rotated by an angle</u> φ <u>around n,</u> λ_1 <u>and</u> λ_2 <u>are the</u> <u>angles between</u> e_φ <u>and the planes bounding</u>V, $\lambda_1 + \lambda_2$ <u>= opening of V, and</u> $[n]_\varphi = \{\gamma \in [n] \; ; \gamma \subset e_\varphi$.

An algorithm for finding the integer coefficients in the equations (1), (2 and (3) is described in <1> .

We conclude with the remark that the functions F_1, F_2, F_3 completely determine the measures m( ) in the spaces G, E, Γ which generate them according to (1), (3) and (5) respectively (this is a corollary of theorems 1 - 3).

In other words, F_i-s provide an adequate language for description of measures in these spaces. The above theorems show, that, at least in the problems of calculation of the measures of the sets from Buffon rings, this language is natural.

Therefore the problem of describing the classes $\{F_1\}$, $\{F_2\}$ and $\{F_3\}$ arises naturally.

It is shown in <3> that the class $\{F_1\}$ coincides with the class of linearly - additive pseudo - metrics on the plane. However, for F_2 and F_3, the problem remains open.

In the important cases of invariant measures in G, E and the corresponding F - s are

$$F_1(v) = \text{the length of } v \text{ ;}$$

$$F_2(w) = \text{the length of n the opening of V; and}$$

$$F_3(w) = \frac{1}{2} (\sin v - v \cos v), \, v = \text{opening of V.}$$

REFERENCES

1   R.V. AMBARTZUMIAN. Combinatorial Integral Geometry (in Russian). Publishing house of the Armenian Academy of Sciences In preparation.

2   R.V. AMBARTZUMIAN. Stochastic Geometry from the standpoint of the Integral Geometry (part II). to appear in Advances in Applied Probability, 1977, December volume.

3   R.V. AMBARTZUMIAN. A note on pseudo-metrics on the plane. Z. Wahrscheinlichkeitstheire verw. Geb. 37, 145-155, 1976.

RANDOM PROCESSES OF LINEAR SEGMENTS AND GRAPHS

L. A. Santaló

SUMMARY

By a graph   G   we understand a finite set of points (vertices) together with the line segments which unites some pairs of distinct points of the set. Sets of congruent graphs are considered. The position of a graph on the plane is defined by the position of one of its vertices   P   and a rotation   φ   about   P . Assuming   P   Poisson distributed   on the plane and   φ   uniformly distributed over   0 ≤ φ < 2π, we extend to graph processes some known properties of line segment processes (Coleman [1], [2] ; Parker and Cowan [3]). We find the probability that the distance from a point chosen at random independently of the process of graphs to the nearest vertex of a graph or to the nearest graph exceeds   u .   Some of the results are also extended from   the euclidean plane to surfaces (sets of geodesic segments and sets of geodesic graphs), for instance to the sphere and to the hyperbolic plane.

1. INTRODUCTION

An oriented line segment   S   of length   s , may be defined on the plane by its origin   P(x;y)   and the angle   φ   that it makes with a fixed direction, for instance with the x-axis. If the length   s   is random variable with probability density function   f(s), so that

$$(1.1) \qquad \int_0^\infty f(s) \, ds = 1 , \qquad \int_0^\infty s \, f(s) \, ds = E(s)$$

the density for sets of uniformly distributed oriented line segments is defined by any one of the following equivalent differential forms [5]

$$(1.2) \qquad dS = f(s) \, ds \wedge dP \wedge d\phi \; = \; f(s) \, ds \wedge dG \wedge dt$$

where   dP   means the area element at   P ,   dG   is the density for oriented straight lines (corresponding to the line that contains S) and t denotes the abscissa of P on G . The densities are always considered in absolute value, so that the order of the differentials in the forms above is inmaterial. All the lengths and orientations of the segments are mutually independent.

With these assumptions, Coleman [1], [2] and Parker and Cowan [3], have considered random processes of line segments on the plane of intensity λ (mean number of points P per unit area). Though Parker and Cowan have considered more general processes, we shall assume, following Coleman, that the process of points P is a homogeneous Poisson process of intensity λ . We state some of their results:

i) The probability that the distance from a point chosen at random independently of the process of line segments to the nearest origin or end of a line segment exceeds u (0⩽ u<∞) is exp(-λH),where

$$(1.3) \quad H = 2u^2 \left(\pi - \int_0^{2u} \{arc \ cos(s/2u) - (s/2u)(1-s^2/4u^2)^{1/2}\}f(s) \ ds \right).$$

ii) The mean value of the number ν of origin or end points of the line segments that are contained in a convex set K of area F and perimeter L which is chosen at random in the plane (in Fig.1 is ν=15), is

$$(1.4) \qquad E(\nu) = 2\lambda F .$$

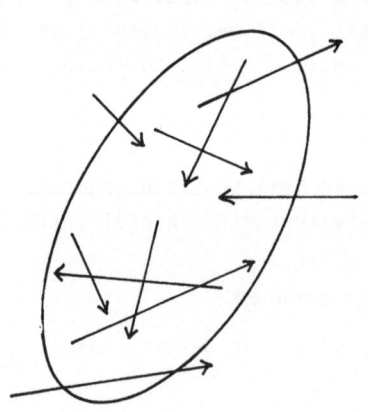

iii) The mean value of the number m of line segments which have common point with K (in Fig.1 is m=10), is

$$(1.5) \qquad E(m) = \lambda(F + \pi^{-1}E(s)L)$$

iv) The mean value of the number m* of intersection points of line segments with a rectifiable curve of length L chosen at random on the plane (uniformly distributed and independent of the process), is

$$(1.6) \qquad E(m^*) = 2\lambda\pi^{-1}E(s)L.$$

Fig.1

v) The mean value of the number N of segment-segment crossings within K (in Fig.1 is N=6), is

$$(1.7) \qquad E(N) = \pi^{-1}\lambda^2 \ (E(s))^2 F \ .$$

vi) The mean value of the total length within K of segments which intersect K , is

$$(1.8) \qquad E(\Sigma\alpha_i) = \lambda E(s)F.$$

More generally we can prove that, assuming s greater than the diameter of K ,

$$(1.9) \qquad E(\Sigma\alpha_i^n) = \pi^{-1}\lambda \left(E(s) \ I_n - \frac{n-1}{n+1} \ I_{n+1} \right)$$

where I_n are the invariants of the convex set K defined by

$$(1.10) \qquad I_n = \int \sigma^n \ dG$$

where σ is the length of the chord G $\cap$ K and the integral is extended over all the lines G of the plane.

vii) The probability that the distance from a point Q chosen at random independently of the process to the nearest line segment exceeds u $(0 \leqslant u < \infty)$ is $\exp(-\lambda H_1(u))$, where

$$(1.11) \qquad H_1(u) = \pi u^2 + 2uE(s) \ .$$

These results can be extended in three different directions: 1. Extension to random figures other than line segments; 2. Extension from the euclidean plane to other surfaces, for instance the sphere or the hyperbolic plane, 3. Extension from the euclidean plane E_2 to the euclidean space E_n .

In this paper we will be concerned with the cases 1 and 2 . The extension to E_n presents a great deal of possibilities and will be considered elsewhere.

2. FIRST EXTENSION: SETS OF RANDOM GRAPHS.

2.1 Definitions and some mean values. A graph T consists of a finite non empty set of points (vertices) together with a prescribed set of line segments (arcs) which join some pairs of vertices. Let v be the number of vertices and h the number of arcs. We consider sets of similar graphs, i.e. graphs which can be mapped one to another by a similitude, and we will denote by s the scale factor of the similitude. The lengths of the arcs are denoted by sa_1, sa_2,...,sa_h, so that the total length of the graph is $A=s(a_1+ a_2+...+a_h)$.

A graph T is defined in the plane, up to an isometry, by one of its vertices, say $P(x,y)$, the scale factor s and a rotation about P through the angle ϕ . We will consider sets of independent random graphs such that P is uniformly distributed on the plane with intensity λ (number of points P per unit area), ϕ is uniformly distributed over the range $0 \leqslant \phi < 2\pi$, and the scale factor s has a probability density function $f(s)$ which satisfies conditions (1.1). This means that the so called density for sets of graphs is the differential form (1.2), which we now write

(2.1) $$dT = f(s) \; ds \wedge dP \wedge d\phi \; .$$

The mean length of the graphs T is $E(A) = (a_1+a_2+...+a_h)E(s)$.

Applying that the expectation of the sum is the sum of expectations (provided they exist), some of the mean values of the Introduction generalize immediately to random graphs. For instance:

a) Denoting by ν the number of vertices which are contined in a convex set K of area F and perimeter L placed at random on the plane, we have

(2.2) $$E(\nu) = \lambda \nu F \; ;$$

b) The mean value of the number m^* of intersection points of a rectifiable curve of length L placed at random on the plane, with the arcs of the graphs, is

(2.3) $$E(m^*) = 2\pi^{-1} \lambda \; E(A) \; L$$

which generalizes (1.6).

c) The mean value of the total length within K of arcs that intersect with a convex set K placed at random on the plane, is

(2.4) $$E(\Sigma A_i) = \lambda \; F \; E(A) \; ,$$

which generalizes (1.8).

d) The mean value of the number N of arcs-arcs of graphs crossing within a convex set K , is

(2.5) $E(N) = \pi^{-1} \lambda^2 (E(A))^2 F$.

2.2 <u>The distribution of the distances from a given point to the nearest vertex of a random process of graphs</u>. For simplicity we shall consider now a process of isometric graphs. We will represent the lengths of the arcs by a_i , so that the total length is $A = a_1 + a_2 + \ldots + a_h$. Round each vertex of the graph T we construct

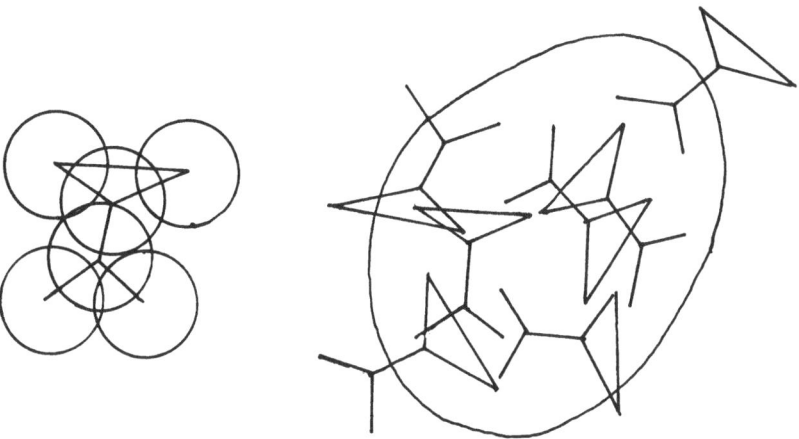

Fig.2

a disc of radius u (Fig.2) . Let D_i (i=1,2,...,v) be the set of those points which are covered exactly by i discs and let F_i denote its area. Assuming that F_1, F_2,...,F_m are $\neq 0$ and that $F_{m+1} = \ldots = F_v = 0$, we put

(2.6) $F = F_1 + F_2 + \ldots + F_m = \pi u^2 v - F_2 - 2F_3 - \ldots - (m-1)F_m$.

The functions $F_i(u)$ are characteristics of the graph. It should be interesting to know until which extent they determine T .

Consider n random graphs (with the discs included) which have the vertex P inside a large disc of radius R . The probability that a chosen point Q of the disc (sufficiently far from the boundary) be   covered exactly by r_1 sets D_1, r_2 sets D_2,..., r_m sets

D_m $(n \geqslant \Sigma r_i, \ i=1,2,\ldots,m)$ is (multinomial distribution)

$$(2.7) \quad p^{(n)}_{r_1 \ldots r_m} = \frac{n!}{r_1! \ldots r_m!(n-\Sigma r_i)!} \left(\frac{F_1}{F_0}\right)^{r_1} \ldots \left(\frac{F_m}{F_0}\right)^{r_m} \left(1 - \frac{F}{F_0}\right)^{n-\Sigma r_i}$$

where $F_0 = \pi R^2$. If n and $R \to \infty$ in such a way that $n/F_0 \to \lambda$ (number of graphs per unit area), the probability tends to the limit

$$(2.8) \quad p_{r_1 \ldots r_m} = \frac{(\lambda F_1)^{r_1}}{r_1!} \ldots \frac{(\lambda F_m)^{r_m}}{r_m!} \exp(-\lambda F)$$

which is a multiple Poisson distribution. The obtained process is called a Poisson graph process of intensity λ . Thus we have proved that

Consider a Poisson process of congruent graphs T of intensity λ . The probability that the distance from a point Q chosen at random independently of the process to i vertices of r_i graphs $(i=1,2,\ldots,m)$ does not exceed u is given by (2.8).

In particular, the probability that the distance from Q to the nearest vertex exceeds u , is

$$(2.9) \quad p_{0 \ldots 0} = \exp(-\lambda F).$$

The function $F(u)$ is in general difficult to calculate. By a direct computation, we can find:

a) If T is a <u>line</u> <u>segment</u> of length a , we have

$$(2.10) \quad F = \pi u^2 \quad \text{if} \quad a \geqslant 2u$$

$$F = 2u^2 \{\pi - \arccos(a/2u) + (a/2u)(1 - a^2/4u^2)^{1/2}\} \ \text{if} \ a \leqslant 2u$$

b) If T is a <u>rectangle</u> of sides a, b such that $b \geqslant a$, we have (Fig.3,a,b,c,d)

$$(2.11) \quad F = 4\pi u^2 \quad \text{if} \quad 2u \leqslant a ;$$

$$F = 4u^2\{\pi - \arccos(a/2u) + (a/2u)(1 - a^2/4u^2)^{1/2}\} \quad \text{if} \quad a \leqslant 2u \leqslant b ;$$

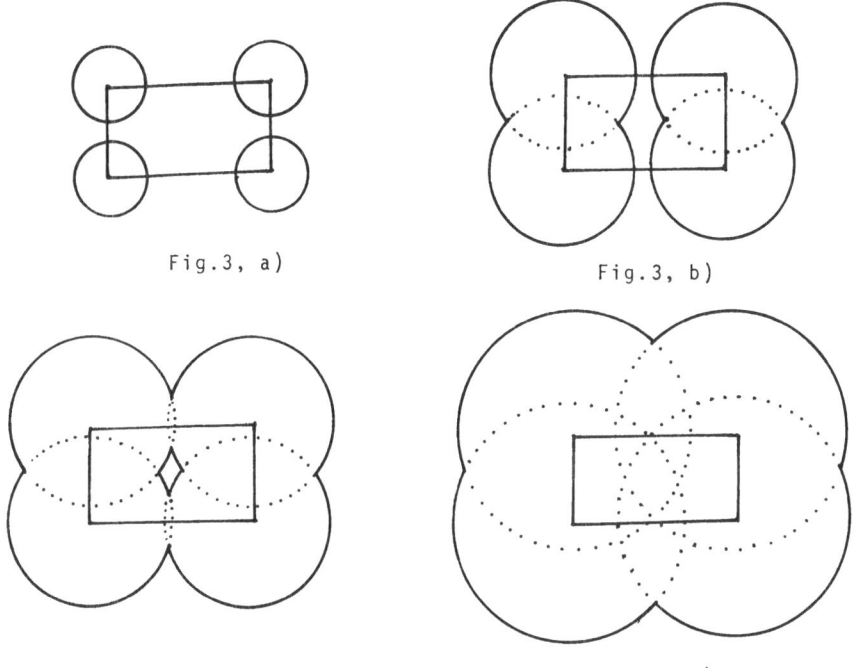

Fig.3, a)

Fig.3, b)

Fig.3, c)

Fig.3, d)

$$F = 4u^2 \{ \pi - \text{arc cos}(a/2u) - \text{arc cos}(b/2u) + (a/2u)(1 - a^2/4u^2)^{1/2}$$
$$+ (b/2u)(1 - b^2/4u^2)^{1/2} \} \quad \text{if} \quad b < 2u < (a^2 + b^2)^{1/2}$$

$$F = 2u^2 \{ (3/2)\pi - \text{arc cos}(a/2u) - \text{arc cos}(b/2u) + (a/2u)(1 - a^2/4u^2)$$
$$+ (b/2u)(1 - b^2/4u^2)^{1/2} + ab/2u^2 \} \quad \text{if} \quad (a^2 + b^2)^{1/2} < 2u .$$

c) If T is an <u>equilateral triangle</u> of side a , we have
(Fig.4, a, b, c)

$$F = 3 \pi u^2 \qquad \text{if} \qquad u \leqslant a/2$$

$$F = 6u^2 \arc \sin(a/2u) + 3au(1 - a^2/4u^2)^{1/2} \qquad \text{if } a/2 \leqslant u \leqslant a/\sqrt{3};$$

$$F = \pi u^2 + 3u^2 \arc \sin(a/2u) + (\sqrt{3}/4)a^2 + (3/2)au(1 - a^2/4u^2)^{1/2}$$
$$\text{if } a/\sqrt{3} \leqslant u .$$

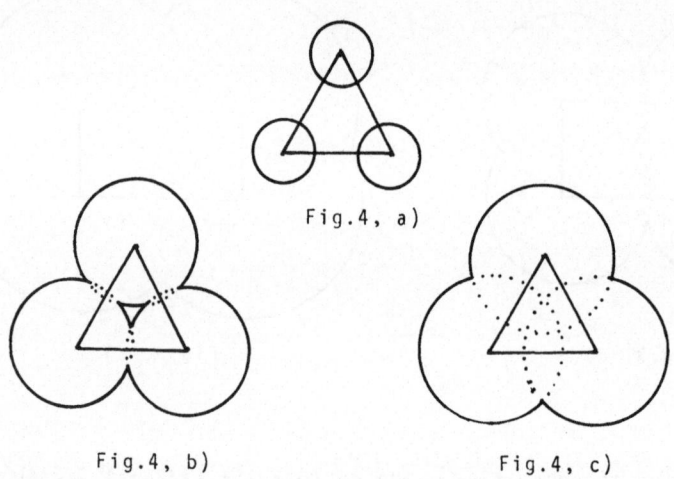

Fig.4, a)

Fig.4, b)   Fig.4, c)

2.3 <u>The distribution of the distance from a given point to the nearest graph</u>. Let $F^*(u)$ be the area of the set of points whose distance to T is $\leqslant u$. This function $F^*(u)$ is also a characteristic of T. Proceding as before we have that

The probability that the distance from a point Q chosen at random independently of the process of graphs, to the nearest graph exceeds u is $\exp(-\lambda F^*)$.

The function $F^*(u)$ must be calculated for each particular graph. For instance, by direct computation, it is easy to obtain the following results:

a) If T is a <u>line segment</u> of length a , we have

$$F*(u) = \pi u^2 + 2ua \ ;$$

b) If T is a <u>rectangle</u> of sides a, b such that b ⩾ a , we have (Fig.5, a, b):

$$F*(u) = 4(a + b)u + \pi u^2 - 4u^2 \ , \quad \text{if} \quad 2u \leqslant a \ ;$$
$$F*(u) = 2(a + b)u + \pi u^2 + ab \ , \quad \text{if} \quad a \leqslant 2u \ .$$

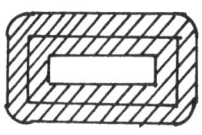

Fig.5, a)

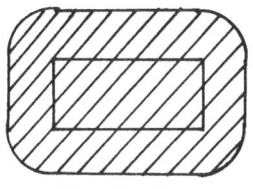

Fig.5, b)

c) If T is an <u>equilateral</u> <u>triangle</u> of side a , we have (Fig.6,a, b)

$$F*(u) = \pi u^2 - 3\sqrt{3} \ u^2 + 6au \ , \quad \text{if} \quad u \leqslant a/2\sqrt{3} \ ;$$
$$F*(u) = 3au + \pi u^2 + (\sqrt{3}/4) \ a^2 \ , \quad \text{if} \quad a/2\sqrt{3} \leqslant u \ .$$

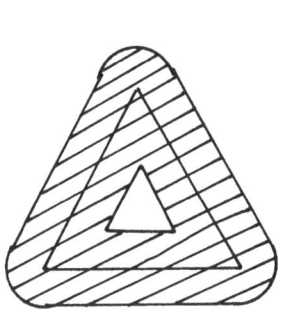

Fig.6 , a)

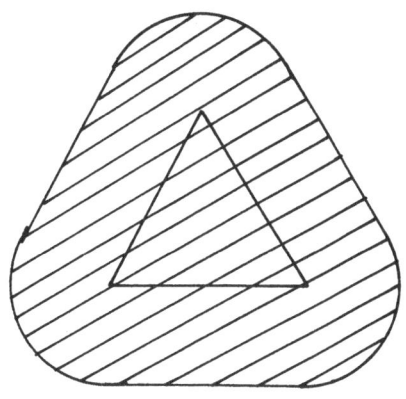

Fig.6, b)

3. SECOND EXTENSION: SETS OF GEODESIC SEGMENTS ON SURFACES.

3.1 Sets of geodesic segments which intersect a convex set.

The density for geodesic segments S on a surface Σ (riemannian space of dimension 2) is given by the same formula (1.2), where now dG stands for the density of geodesic lines of the surface (see [4]). A bounded set of points K on the surface is called convex if every geodesic has at most two points in common with the boundary ∂K, exceptimg geodesics which have an entire segment which belongs to ∂K . Let D be the diameter of K . If the surface Σ has closed geodesics and their minimal length is L_G, we will consider only segments whose maximal length s_m satisfies the inequality

(3.1) $\qquad D + s_m \leqslant L_G$.

That means that $f(s) = 0$ for $s > s_m$.

With these conditions, the measure of the set of oriented geodesic segments which intersect a fixed convex set K of area F and perimeter L , is [4]

(3.2) $\qquad \displaystyle\int_{S \cap K \neq \emptyset} dS = 2\pi F + 2E(s)L$.

and denoting by ν the number of extreme points (origin or end points) of the segments S intersecting K which are within K , we have

(3.3) $\qquad \displaystyle\int \nu\, dS = 4\pi F$.

Let S be a segment which intersect K and let α denote the length of the part of S within K . Consider the integral

(3.4) $\qquad J_1 = \displaystyle\int dG \wedge dS$

extended over all pairs (G,S) such that $G \cap S \in K$. If we first leave S fixed, we have $J_1 = 4\int \alpha\, dS$ and if we first leave G fixed and σ denotes the length of the chord $G \cap K$, we have $J_1 = 4E(s)\int \sigma\, dG = 8\pi E(s)F$ (according to well known results of Integral Geometry, see for instance [5]). Hence, we have

(3.5) $\qquad \displaystyle\int \alpha\, dS = 2\pi E(s)F$.

Consider now two geodesic segments S_1, S_2 which intersect K. Let n_{12} denote the function which is equal to 1 if $S_1 \cap S_2 \in K$ and is equal to zero otherwise. Consider the integral

$$(3.6) \qquad\qquad J_2 = \int n_{12} \, dS_1 \wedge dS_2$$

extended over all pairs such that $S_1 \cap S_2 \neq \emptyset$ $(i=1,2)$. If α_1 denotes the length of the intersection $S_1 \cap K$, integrating first with respect to S_2 we have $J_2 = 4E(s_2) \int \alpha_1 dS_1$, and according to (3.5) we have

$$(3.7) \qquad\qquad J_2 = 8\pi F \, E(s_1)E(s_2) .$$

From (3.2), (3.3), (3.5) and (3.7) we deduce:

a) The mean value $E(\nu)$ of the number of extreme points (origin or end point) within K of n geodesic segments on a surface Σ which intersect at random a convex set K of Σ, is

$$(3.8) \qquad\qquad E(\nu) = \frac{2\pi nF}{\pi F + E(s)L} .$$

b) For n geodesic segments S_i $(i=1,2,\ldots,n)$ chosen at random on the surface Σ, which intersect a convex set K, the mean value of the sum of the lengths α_i within K, is

$$(3.9) \qquad\qquad E(\Sigma \alpha_i) = \pi F \sum_{1}^{n} \frac{E(s_i)}{\pi F + E(s_i)L} .$$

If all segments have the same mean length $E(s)$, we have

$$(3.10) \qquad\qquad E(\Sigma \alpha_i) = \frac{n\pi FE(s)}{\pi F + E(s)L} .$$

c) For n geodesic segments S_i chosen independently at random on the surface Σ, which intersect a convex set K, the mean value of the number N of segment-segment crossings within K, is

$$(3.11) \qquad E(N) = 2\pi F \sum_{i<j} \frac{E(s_i)E(s_j)}{(\pi F + E(s_i)L)(\pi F + E(s_j)L)} .$$

If all segments have the same mean length $E(s)$, we have

$$(3.12) \qquad E(N) = \frac{n(n-1)\pi F(E(s))^2}{(\pi F + E(s)L)^2} .$$

3.2 Sets of segments on the unit sphere. The measure of all geodesic segments on the unit sphere has a finite value, namely

$$(3.13) \qquad \int_{Total} dS = \int f(s) \, ds \wedge dP \wedge d\phi = 8\pi^2 .$$

Hence, from the results above we can state:

a) The mean value of the number of extreme points (origin or end points) of n random segments chosen independently at random on the unit sphere, which lie within a convex set K , is

$$(3.14) \qquad E(\nu) = \frac{nF}{2\pi} .$$

b) Let K be a convex set of diameter D on the unit sphere. Consider a random segment whose maximal length satisfies the inequality $D + s_m \leqslant 2\pi$. Then, the probability that $S \cap K \neq \emptyset$, is

$$(3.15) \qquad p(S \cap K \neq \emptyset) = \frac{\pi F + E(s)L}{4\pi^2} .$$

c) The mean value of the total length within K of n random segments S_i chosen independently on the unit sphere, is

$$(3.16) \qquad E(\Sigma \alpha_i) = (4\pi)^{-1} F \sum_{1}^{n} E(s_i)$$

If all segments have the same mean length E(s) , we have

$$(3.17) \qquad E(\Sigma\alpha_i) = \frac{nE(s)F}{4\pi} .$$

d) The mean number of segment-segment crossings within K of n segments placed independently at random on the sphere is

$$(3.18) \qquad E(N) = (8\pi^3)^{-1} F \sum_{i<j} E(s_i)E(s_j) .$$

In particular, if all segments have the same mean length E(s), we have

$$(3.19) \qquad E(N) = \frac{n(n-1)(E(s))^2 F}{16\pi^3} .$$

3.3 Sets of segments on the hyperbolic plane. From (3.2), (3.5) and (3.6) we deduce:

a) If K is a convex set interior to a convex set K_0 on a given surface Σ , the probability that a random segment intersecting K_0 , also intersects K , is

$$(3.20) \qquad p = \frac{\pi F + E(s)L}{\pi F_0 + E(s)L_0} .$$

b) If we consider n random segments of the same mean length E(s) which intersect K_0 , the mean value of the total length of their intersection with K , is

$$(3.21) \qquad E(\Sigma\alpha_i) = \frac{n\pi E(s)F}{\pi F_0 + E(s)L_0} ,$$

and the mean value of the number N of segment-segment crossings within K , is

$$(3.22) \qquad E(N) = \frac{n(n-1)\pi(E(s))^2 F}{(\pi F_0 + E(s)L_0)^2} \ .$$

c) From (3.20) it follows that if there are chosen at random n segments which intersect K_0 , the probability that exactly m of them intersect K , is

$$(3.23) \qquad P_m = \binom{n}{m} \left(\frac{\pi F + E(s)L}{\pi F_0 + E(s)L_0} \right)^m \left(1 - \frac{\pi F + E(s)L}{\pi F_0 + E(s)L_0} \right)^{n-m} \ .$$

Assume that Σ is an unbounded surface of infinite area and let K_0 expand to the whole surface at the same time that $n \to \infty$ in such a way that

$$(3.24) \qquad \frac{n}{F_0} \to \lambda \quad \text{(positive constant)}.$$

Assuming moreover that under these conditions we have

$$(3.25) \qquad \frac{L_0}{F_0} \to \kappa \ ,$$

then P_m tends to the limit

$$(3.26) \qquad P_m = \frac{(\lambda H)^m}{m!} \exp(-\lambda H) \ , \qquad H = \frac{\pi F + E(s)L}{\pi + \kappa E(s)} \ .$$

It is known that $\kappa = 0$ for the euclidean plane and $\kappa = 1$ for the hyperbolic plane [6] . The obtained process is called an homogeneous Poisson segment process of intensity λ on Σ .

According to (3.26) the mean number of segments intersected by a convex set K placed at random on the surface is λH .

Using (3.21) and (3.22) we get that the mean value of the total length within K of segments that intersect with K is

$$(3.27) \qquad E(\Sigma \alpha_i) = \frac{\lambda \pi E(s)F}{\pi + \kappa E(s)}$$

and the number of segment-segment crossings within K , is

(3.28)
$$E(N) = \frac{\lambda^2 \pi (E(s))^2 F}{(\pi + \kappa E(s))^2} .$$

For the euclidean plane, $\kappa=0$, these results are the work of Parker and Cowan $\left[3\right]$. For the hyperbolic plane we must put $\kappa=1$.

d) Comsider the hyperbolic plane, $\kappa=1$. If D_r denotes the distance from a point Q chosen at random independently of the process, to the nearest r-th line segment, the probability that $D_r > u$ is equal to the probability that a disc of radius u placed at random on the plane intersect no more than r-1 line segments, that is

(3.29)
$$p(D_r>u) = \sum_{m=0}^{r-1} \frac{(\lambda H)^m}{m!} \exp(-\lambda H)$$

where H is given by (3.26) with

(3.30) $\kappa = 1$, $F = 2\pi(\cosh u - 1)$, $L = 2\pi \sinh u$.

For r=1 we have the probability that the distance from a point Q chosen at random on the hyperbolic plane to the nearest line segment is greater than u . Thus, the probability density function for the distances from Q to the nearest line segment is

(3.31)
$$2\lambda\pi(\pi + E(s))^{-1}(\pi\sinh u + E(s)\cosh u) \exp(-\lambda H)$$

with

(3.32)
$$H = \frac{2\pi}{\pi + E(s)}\left(\pi(\cosh u - 1) + E(s) \sinh u\right) .$$

REFERENCES

[1]   COLEMAN, R.   Sampling procedures for the lengths of random
                     straight lines, Biometrika, 59, 1972, 415-426.

[2]   "   The distance from a given point to the nearest
                     end of one member of a random process of linear
                     segments, Stochastic Geometry, ed.Harding and
                     Kendall, Wiley, London, 1974, 192-201.

[3]   PARKER, Ph. and COWAN, R.   Some properties of line segment
                     processes, J.Applied Probability,13,1976,96-107.

[4]   SANTALÓ, L.A. Integral Geometry on surfaces, Duke Math. Jour-
                     nal, 16, 1949, 361-375.

[5]   "   Integral Geometry and Geometric probability, En-
                     cyclopedia of Mathematics and its Applications,
                     Addison-Wesley, Reading, Mass. 1976.

[6]   SANTALÓ, L.A. and YAÑEZ, I.   Averages for polygons formed
                     by random lines in euclidean and hyperbolic
                     planes, J.Applied Probability, 9, 1972, 140-157.

Facultad de Ciencias Exactas y Naturales
Universidad de Buenos Aires
BUENOS AIRES, Argentina.

STATISTICS OF STATIONARY ORIENTED LINE
POISSON PROCESSES IN THE PLANE

-=-=-=-=-=-=-=-

A. FELLOUS[1], J. GRANARA[2], K.KRICKEBERG[1]

(1) U.E.R. de Math. PARIS V   12 Rue Cujas   75005 PARIS

(2) U.E.R. de Math. PARIS VII   2 Place Jussieu 75005 PARIS

Among the various ways in which a random mosaic of the plane
may be constructed, the Poisson lines model has been considered by many authors for
applications [see [7]] . This model presents the interesting property that the
sequence of species along any linear transect is markovian provided that these species
are independently and identically distributed in each convex cell of the mosaic
[see [8], [3a]].

Recall first that an oriented line δ in the plane (for mathematical convenience we consider oriented lines rather than unoriented lines) can be represented by two parameters, namely p, the algebr distance from δ to the origin 0
of the plane, and σ, the angle between δ and a fixed direction $\vec{O}x$.

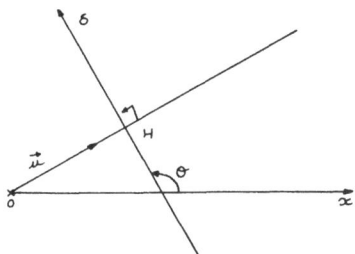

Thus, $\vec{u}$ being the unitary vector such that $(\vec{u}, \delta) = \frac{\pi}{2}$
we get $\vec{OH} = p.\vec{u}$ so that p   R, Q$\in$ [0,2 [.

We assume that the Poisson line process Z does not display any
peculiarity in the distances p of the lines, or in other words that Z is stationary
w.r.t. the group of the translations of the plane.

It is then clear that the specific features of Z are just contained in the directions Q and the orientations (which depend on both p and Q in a quite simple way) of the oriented lines.

In this case, the law of process Z is complet rage number of lines intersecting a fixed disc D of diameter 1 and whose angle lies in an interval $I \subset [0,2[$: for any interval $I \subset [0,2[$. Denoting this number by μ (I) we see that μ is a positive measure on $[0,2[$, to be called the reduced intens of the process Z, which determines the law of Z.

When $r = \mu([0,2[)$, to be called the scale factor of the process, the measure $\varpi = \frac{1}{r}. \mu$ becomes a probability measure, to be called the aniso-tropy factor, ϖ (I) being the average proportion of those lines whose angle lies in I among all the lines which intersect D.

In fact, it can be demonstrated that the average ratio ϖ (I) may actually be interpreted as a probability, as a consequence of the following general result :

Suppose that (p_1, Q_1), (p_2, Q_2),.....,(p_n, Q_n) is the (infinite) sequence of the lines of a realisation of Z, ordered in such a way that $0 \leq |p_1| \leq |p_2| \leq ... \leq |p_n| \leq ...$ Furthermore, put $\varepsilon_n = sgn (p_n)$ and $T_n = |p_n|$ (so that $p_n = {}_n. T_n$, for n = 1,2,...).

Then, the random sequence $(T_1, T_2,...,T_n,...)$ is the ordered sequence of the points of a Poisson process ζ on R_+ with intensity r.

The random sequence $(\varepsilon_1, \varepsilon_2,..., \varepsilon_n,...)$ is an infinite sample of the Bernouilli trial $(\frac{1}{2}, \frac{1}{2})$.

The random sequence $(Q_1, Q_2,..., Q_n,...)$ is an infinite sample of the law ϖ on $[0,2[$.

Moreover, these three random sequences are stochastically inde-pend nt.

This theorem gives, naturally, a very simple method for simulating such a process of lines.

On the other hand, it shows that when observing such a process, and trying to specify its law, one can work separately on the distances for estimating r, and on the angles for estimating the probabilities ϖ (I) for various I.

Let us mention briefly that one can test whether the orientations of lines are relevant or not. In fact it may happen that, although we observe oriented lines, the process does not display any peculiarity regarding the orientations - in this case it will be called disorientable. This situation arises clearly when, starting with a process of unoriented lines, one constructs a process of oriented lines by marking the orientation on each line of the previous process independently from the others and with probability $(\frac{1}{2}, \frac{1}{2})$.

It can be shown that a Poisson process of oriented lines is disorientable if and only if its anisotropy factor admits the period   .

Thus, in the situation where one can observe an a priori fixed number n of lines of the process, say the n nearest lines from the origin , the statistical problem is seen to reduce to the statistical analysis of a homogeneous Poisson process on the line (for r) [see e.g. [1]] and to sampling theory of directional data (for ϖ) [see e.g. [6]].

From a statistical theoretic point of view, the problem is quite different if one can only observe the process Z through those lines of Z which intersect an a priori fixed body K, because then the number N of lines observed is random.

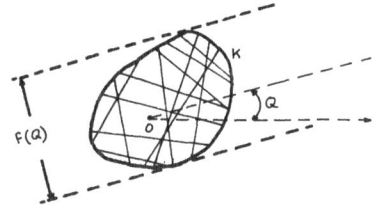

Let us suppose for convenience that K is convex and denote by f(Q) the apparent diameter of K in the direction Q.
Then, if Q_1, Q_2,...,Q_N (N random) are the random angles of the lines of Z which intersect K and if Q_1^*, Q_2^*,..., $Q_{N(I)}^*$ are those among them which lie in I(N(I) random), one can take as an estimator of $\mu(I)$ the quantity

$$\hat{\mu} (1) = \frac{1}{f(Q_1^*)} + \ldots + \frac{1}{f(Q_{N(I)}^*)} \qquad .$$

This estimator has been shown to be the minimum variance unbiased estimator of μ (I), and to have the required asymptotic properties as the size of K increases to infinity [for more details, see [2], [4]].

This is true, by a peculiar choice of I, for $\hat{r} = \hat{\mu}$ ([0,2π[) for estimation r.

Unfortunately, an unbiased estimator of ϖ (I) for I $\neq$ [0,2 π [does not exist unless the function f is constant as in the case of a disc or the following classical example :

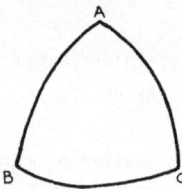

From each edge of an equilateral triangle ARC, draw the portion of the circle joining the two other edges, with radius the common distance c between these three points.

Now, whatever be K, suppose that f $\equiv$ c, a known constant. Then $\hat{\mu}$ (I) = $\frac{N(I)}{c}$ and $\hat{r} = \frac{N}{c}$. Put $\hat{\varpi}$ (I) = $\frac{\hat{\mu}(I)}{\hat{r}} = \frac{N(I)}{N}$. This random quantity - which is not defined for N = 0 - is clearly the "naïve" estimator of the average ratio ϖ (I) and has been shown to satisfy the same requirements as $\hat{\mu}(I)$.

Finally, let us mention that this study has been performed in the more general frame of stationary cox processes of marked k-flats in d-dimensional Euclidean space (1< k < d)   [see [5] , [3b]].

Corresponding results hold for more general parameters, this increase in generality going together with that of the mathematical complexity of the various expressions which appear.

Bibliography

[1] COX D.R. and LEWIS P.A.W. : The Statistical Analysis of series of events. METHUEN 1966.

[2] FELLOUS A. : Estimation dans les modèles de processus de Cox isotropes de droites et d'axes du plan.
Thesis (3e cycle) PARIS V June 1976.

[3a]   FELLOUS A. et GRANARA J. : "Une caracterisation de processus de Poisson".
         Zeitschrift für Wahrscheinlichkeitstheorie - To appear·

[3b]   FELLOUS A. et GRANARA J. : "Statistique des processus de Cox stationnaires
         de $k_{-p}a_{ns}$ marqués sur un espace Euclidien de dimension d". To be published.

[4]   GRANARA J. : "Compléments à la théorie des processus ponctuels et statistique
         des processus de Cox stationnaires d'hyperplans orientés". Thesis (3° cycle)
         PARIS V Déc. 76.

[5a]   KRICKEBERG K. : "Moments of point processes". Stochastic Geometry, Ed. E.F. Har-
         ding et D.G. Kendall, Wiley 1974, p.89

[5b]   KRICKEBERG K. : "The Cox process", Sympos. Math. Calcola Proba. Teor.
         Turbolenza (1972) pp. 151-167.

[5c]   KRICKEBERG K. : "Fundamentos del Analysis Estadistica del processes puntuales",
         CIENES, Santiago de Chile, 1973

[5d]   KRICKEBERG K. : "Statistical Analysis of Hyperplane processes", International
         conference on Probability Theory and Mathematical Statistics,Vilnius 1973,
         pp. 335-338.

[6]   MARDIA K.V. : Statistics of directional data. Academic Press, 1972.

[7]   PIELOU E.C. : An Introduction to Mathematical Ecology.Wiley Intersciece, 1969.

[8]   SWITZER P. : "A random set process in the plane with a Markovian property"
         Ann. Math. Statist. 36 : 1859-1863.

AUTOMATED CYTOLOGY

F. MEYER

General Introduction

Under this title we group two different papers dealing with two successive steps of a more general research project sponsored by the Ministry for Scientific Research of Western Germany. The objective is to construct a system for an automatic screening of cervical smears in order to permit an early detection of uterine cancer.

Not all cells of a slide are interesting for the diagnosis; Thus you first have to rapidly identify the interesting cells without being misled by the presence of many types of artefacts such as cell-overlappings or dust particles. This is the subject of the first paper, entitled "Automatic cell finding".

The first step increases the probability of any of the retained cells of being cancerous. However, for the final diagnosis of the slide, a further and finer analysis has to be done on each of the detected cells. The second paper entitled "Contrast extraction and Chromatin analysis" describes how to characterize the cancer alteration of the chromatin of the nuclei.

Automatic Cell Finding

Abstract : This program performs a prescreening of cervical smears, in order to detect cancer. The problem consists of extracting the key cells from a background full of uninteresting cells, artefacts, overlappings, edge perturbations, etc. Specific iterations of image transformations lead to a decision tree which filters the important cells.

Material and Cytological background

Dr. PLOEM et AL invented a new fluorescent double staining technique for cervical smears and implemented it on a staining machine (ref. 4). It is sufficient to change a filter in the microscope to obtain successive and highly contrasted masks of the cytoplasm and of the nuclei. With epi-illumination, the Feulgen staining of the nucleus gives an acceptable estimation of the DNA in the cells; in transmitted light, the chromatin structures of the cell appear nicely

and the measurement of the absorption of light gives a very accurate
estimation of the DNA.

It is thus possible, by using several openings at two or
three grey levels of the fluorescence image, to detect, in real time,
the cells with an increased value of DNA. Dr PLOEM proved that such
cells are present in all suspect slides and that it is therefore sufficient to look for these cells. Unfortunately in this first step one
detects not only single cells but also overlapping cells and other
artefacts, and it is precisely one of the tasks of the C.M.M. to discriminate between cells and artefacts.

Method

In classical pattern recognition, one first builds a large
training set grouping all types of cells which occur in a slide,
then one measures a set of parameters on each cell ; a multivariate
analysis on the results gives the best discriminators between the
different classes. On the real slide one first has to find the particle on which to make the measurements. This is mostly done by a rou-
tine which finds the first connex particle of the field. Then one makes the measurements on this cell : the discriminators determined in
this learning phase permit one to classify the cell or the artefact.
One proceeds in the same manner on cell after cell. To summarize, one
first filters the field in order to get a single particle on which
parameters will be measured.

In contrast, we propose a different method. We want to
avoid working on a single particle and measuring parameters as long
as possible. To achieve this, we have to improve the first step of
filtering the image. We construct several non-linear filters, each of
them filtering a different artefact. In this way we have no single
measurement to make in the first step.

We construct, for each type of artefacts, a special image-
dependent non-linear spatial filter composed of iterative image transformations. Let us take a pedagogical example and assume that each
cell is convex and that a non-convex particle is therefore an artefact. Now it is well known that a closing of size R closes all concavities of a radius smaller than R.

Fig. 1

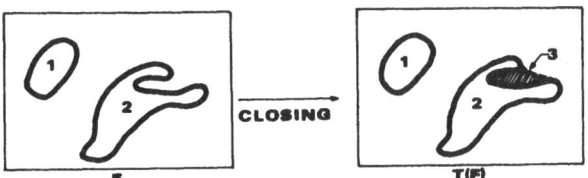

If we take the difference between the sets T(F) and F, particle 1 vanishes, but particle 2 is marked by the concavity 3. It is then sufficient to reconstruct all marked particles and to eliminate them, in order to retain only the convex particles. The iteration "closing - set difference - reconstruction of the marked particles - elimination of these "constitutes a filtering of the initial image F. We remark that this time we do not have to work particle by particle, but that the whole field is treated simultaneously.

<u>Flowsheet</u>

In the same manner, after the storage of the primary information (we store three binary images) :

a) the mask of the cytoplams
b) the mask of the nuclei
c) the mask of the bright chromatin particles (See Fig. 2)

We do the following successive operations :

1   Detection of high DNA-valued particles by working on images b and c

2   Elimination of too long or too large particles

3   Elimination of the degenerate particles (whithout any cytoplasm)

4   Elimination of the particles cutting the edges of the mask

5   Edge analysis of the particles. Elimination of the particles with strong local concavities. This step detects many clusters and all kinds of dust artefacts.

6   Partition of the cells in two classes, according
     their nucleo cytoplasmic ratio

7   Only during the last step does an individual analysis
     of the nuclei take place. Here, the single cells are
     separated from the double overlapping cells. These
     individual cells have no strong concavities, and thus
     have not been detected in an earlier step.

Results

        The best way to judge the results is to see on the TV
screen how the program works on each new field. As this is not possible here, we will show you a few pictures taken of the TV screen.

Fig. 2.1 - Shows the image as it appears in transmitted light. The
nuclei appear dark with a visible chromatin pattern.

Fig. 2.2 - The background appears in black, the cytoplasm in dark
grey, the nuclei in light grey and the chromatin in white. These are
the three binary images we get from the fluorescence images of the
cytoplasms and of the nuclei.

Fig. 2.3 - Shows the end of the procedure. The nuclei which have
been chosen for a further analysis appear in white and are superimposed on the original picture 2.1

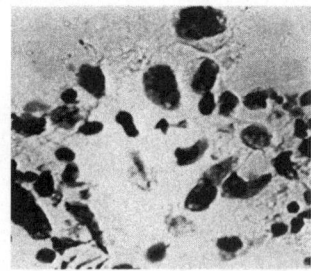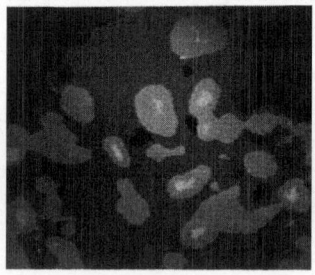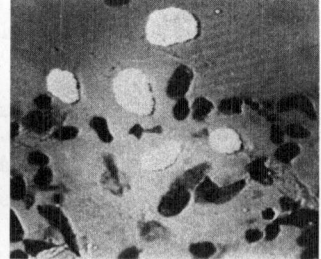

CONTRAST EXTRACTION AND CHROMATIN ANALYSIS
-=-

Abstract

Chromatin measurements : One wants to characterize the hetero-chromatin in the nucleus. The method has to combine morphometry and densitometry. Masks of the heterochromatin are automatically generated independently of the grey levels. The classical densitometric histograms are then computed inside the masks.

Introduction

This paper has two targets. The first is to present a new contrast algorithm for the Fontainebleau prototype of Texture Analyser. The second is to show, within the vast range of applications in Biology, how well such an algorithm is suited for the difficult problem of chromatin analysis of the nuclei in cervical smears, in order to give a fine diagnosis of malignancy (See. ref. 1,2 for the importance of the cancer associated chromatin alterations).

I - METHOD

A/ Preliminary definitions

All image analysis classically deals with two kinds of pictures, the grey tone images and the binary images, and continuously bridges the gap between the first and the second by an image transformation called thresholding For each grey tone image, we can define a grey tone function $g(x,y)$, representing the grey value of the point (x,y). Such a function takes the value 0 for the white points, 1 for the black points, and all values between 0 and 1 according the darkness of the point. To threshold an image at level i means to extract all points with a grey value higher than i.

Fig. 3

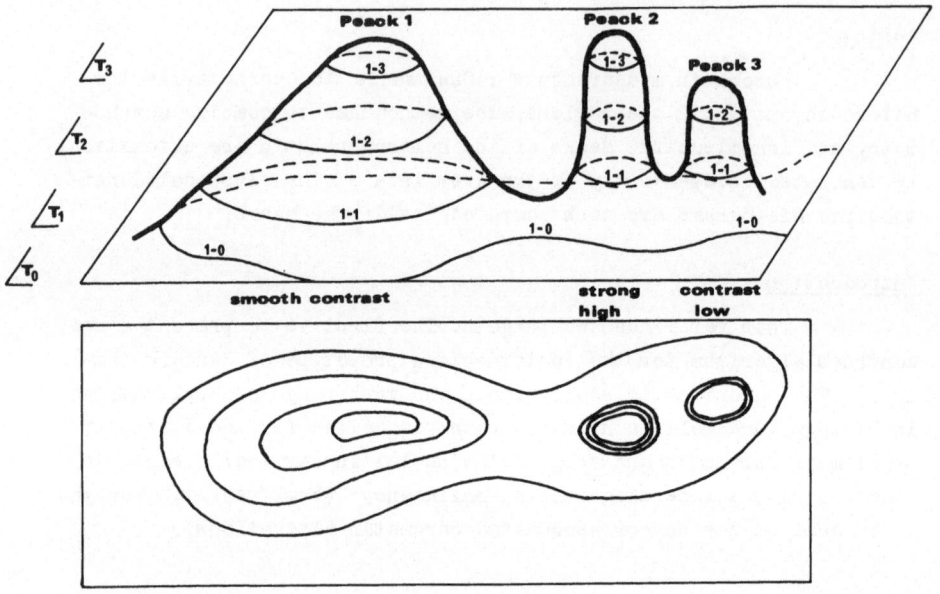

Let's suppose we wish to automatically detect the strong peaks of contrasts 2 and 3 without detecting the smooth peak 1. For that we compare the images given by the thresholds T_1 and T_2. For each pak, we will retain the image given by the threshold T_2 only if the image given by the threshold T_1 is not too large. So we will retain the images 2-2 and 3-2 for peaks 2 and 3 but not the image 1-2 because 1-1 was much larger than 2-1 and 3-1. Practically we detect the size of a given pattern in a image by the morphological transformation "opening" (see ref. 3 by J. SERRA).

It is easy to see that

1/ If the size of the opening increases, the detected contrasts become smoother.

2/ If the difference of grey levels between the 2 thresholds increases, the detected contrasts become higher : For example, if in fig. 3, we had used the grey levels 1 and 3 the low contrast peak 3 would have been eliminated.

II - APPLICATION TO CHROMATIN ANALYSIS

A/ **Material**   See § Material in the paper "Automated Cytology"

B/ Construction of a binary mask for the chromatin

We note that on picture **4** there exist both clear and dark chromatin particles. Therefore we perform the contrast extraction at several grey levels and take the union of the binary images of contrast we obtained at each level. Thus, according to the strength of the contrast extraction we obtain the binary mask of chromatin (in white) superimposed upon the original image (fig. 4A, 4B, 4C)

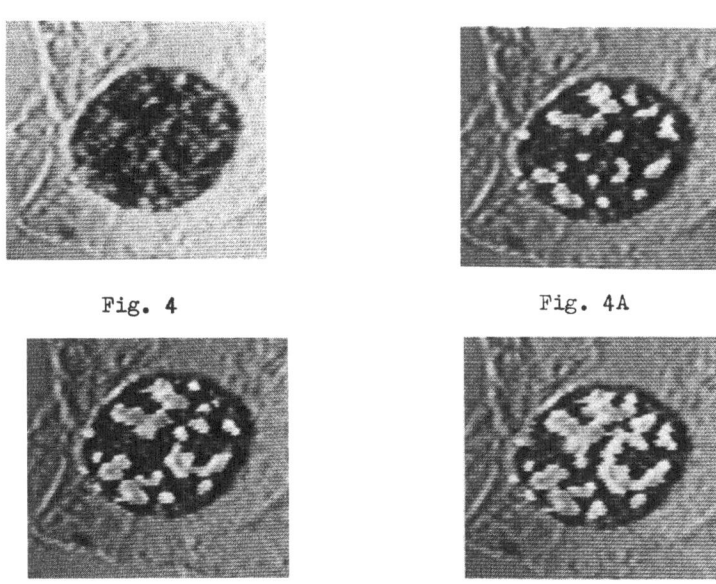

Fig. 4   Fig. 4A

Fig. 4B   Fig. 4C

C/ Measurements

The Feulgen staining has the particularity that the integral of the light within the mask of the nucleus gives a relative estimation of the DNA of the cell. This value, when divided by the value measured on a standard cell of 2n-DNA, produces an absolute estimation of DNA.

If we now measure the integral of light inside the above-constructed chromatin mask, we have an estimation of the amount of chromatin represented by the heterochromatin (HCH). We can also measure the light inside the chromatin situated in the border of the cell (Bord).

Fig. 5

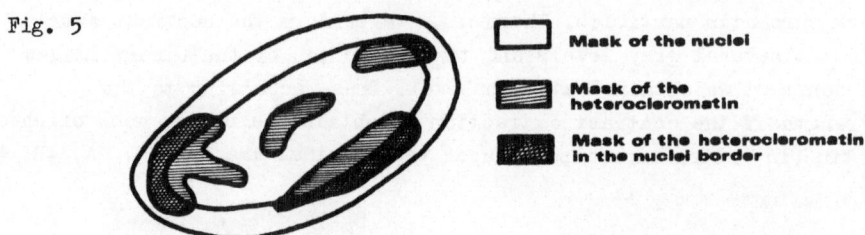

Mask of the nuclei

Mask of the heterocleromatin

Mask of the heterocleromatin in the nuclei border

We can then compute, such parameters as

$$\frac{HCH}{ADN} , \quad \frac{HCH \times HCH}{ADN} , \quad \frac{Bord}{HCH} , \quad \frac{Bord}{ADN} ,$$

which are good descriptors of the chromatin in the cell.

D/ Results

The first results concern about 100 cells from 5 different slides. These cells were first diagnosed by a cytologist, then machine-scanned. On all severe dysplasia and carcinoma in situ cells, we made not one false negative diagnosis and on all negative and inflammatory cells, we made not one false positive diagnosis. Of course much more cells must be measured, but the results here and now look very promising.

III - CONCLUSION

The contrast-extraction algorithm seems to be well adapted to biological problems. Often the staining does not allow one to precisely visualise the structures which the biologist wish to analyse. No satisfactory threshold can be found. Nevertheless, the eye sees contrasted structures in the image. It is now possible to extract these contrasted features and to visualize them in a binary image to which all the classical tools of Mathematical Morphology can then be applied.

Références

1 - E. SPRENGER, W. SANDRITTER : DNA Content and Chromatin Pattern
    Analysis on Cervical Carcinoma in Situ in Acta Cytologica,
    Vol. 17 : 27-31, 1973.

2 - KIEFER G. , KIEFER R. and SANDRITTER W. Die Dichtverteilung des
    Chromatins innerhalb des Zellkerns. Verk. Deutsch. Ges. Path.
    55 : 621, 1971

3 - SERRA J. : One, two, three... Infinity
    Proceedings of the Buffon Symposium.

4 - TANKE H.I., PILOUW-MULDER M.J. and PLOEM J.S.
    An automated System for a Feulgen - Protein Staining Procedure.
    To be published in Stain Technology.

RANDOM PROCESSES SIMULATIONS
ON THE TEXTURE ANALYSER

-=-=-=-=-=-=-=-

S. BEUCHER

CENTRE DE MORPHOLOGIE MATHEMATIQUE, 35 Rue St-Honoré 77305 FONTAINEBLEAU
France

The prototype Texture Analyser at Fontainebleau permits very swift simulations of many random processes, be they point processes or random set processes. We shall illustrate, with some examples of general processes, the possibilities of this apparatus. Three kinds of processes are presented here :

- The Poisson point process
- Cluster Processes ("shooting" scheme)
- Hard-Core processes (repulsion models)

With various combinations of these three fundamental processes, one may induce more complex realizations of random sets containing two or three phases.

Poisson point process

A Poisson point process, with intensity θ, is very easy to simulate on a bounded set A. The simulation is divided in two stages :

One chooses n, the number of points of the set A from a Poisson distribution with mean $\theta \cdot x \ |A\ |$. ($|A\ |$, Lebesgue measure of A).

One distributes the n points at random over the set A, which in the present case, is the rectangular field of the texture analyser.

It is possible, starting from a Poisson realization, to simulate random sets processes by centering in every point of the Poisson process one realization from a non-stationary random set, called the primary grain X'. Two particular realizations of such processes, the Boolean Schemes, are shown in figure 1.

Cluster Processes

When the primary grain of a Boolean Scheme is a cluster of N points distributed independantly around each of the Poisson process points according to a distribution law F, one is simulating a cluster process, also called a "schooting scheme", ("Schema du Tireur" - Matheron G.) or a Neymann-Scoot process (fig. 2).

Once such a point process is achieved, the subsequent simulation of random sets is a simple task (fig. 3).

Hard - Core Models

These models were defined by Matern B. (1960). One samples a Poisson process of intensity θ , and deletes any point which is within R of any other point. Figure 4 illustrates this kind of process.

Mixtures of models

These models are obtained by mixing simpler processes as shown in Figure 5.

Analyser program and applications

The afore-mentioned simulations do not consume large amounts of computer time (two or three minutes for the longest ones). Biases introduced by edge effects are eliminated by suppressing a strip around the sampling field, whose width depends on the parameters of the processes. Balls are approximated by hexagons (one could use dodecagons to increase the accuracy) ; Hence, only three needle directions can be defined on the hexagonal grid of the analyser, o, $\pi/3$, 2 $\pi/3$. Thus the orientation of needles is chosen at random among these three values.

Because of the rapidity of the simulations, the production of a large number of runs for these models for various parametric values is easy. In figure 6, an example of a covariance function calculated from the simulation of a hard-core model is schown.

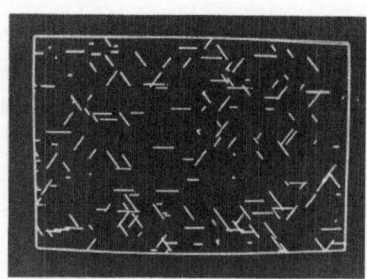

Figure 1 : Boolean Scheme

The primary grain is a needle, whose length is a random variable chosen from a uniform probability distribution, and oriented at random on the plane

$$\theta \times |A| = 250$$

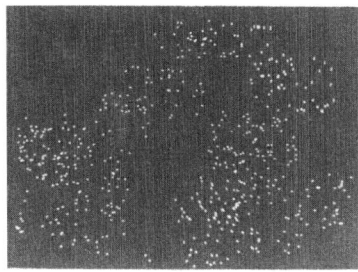

Figure 2 : Cluster Process (Schēma du tireur)

This realization is obtained by sampling in a circle of radius R centered on each point of the parent Poisson process of intensity θ , a daughter Poisson process of intensity θ . Then, the number N of points in the primary grain is a random variable following a Poisson law of mean $\theta' \pi R^2$, and the distribution function of these points is uniform over the R-Cercle.

R = 30

Mean of the Parent process : $\theta |A| = 20$

Average number of points in a cluster : 50

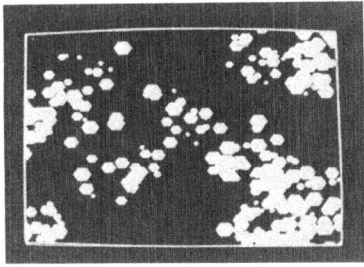

Figure 3 : Cluster Random Sets
R = 30 ; $\theta |A| = 15$; Ave(N) = 30

314

Figure 4 - Hard-Core Model

This sample corresponds to the first model defined by Matern.
One takes into account every point of the parent Poisson process, whether or not
it has been already deleted.

Average number of Parent process : θ |A| = 200

Repulsion distance   R = 5.

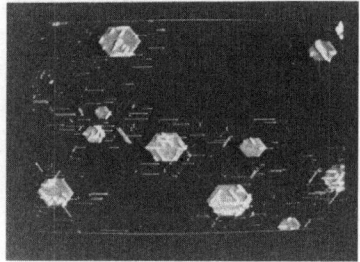

Figure 5 : Three-phase boolean scheme

Boolean scheme (the primary grain is a ball) and cluster process
(the needles are centered on the points of this second process)

θ . |A| = 20

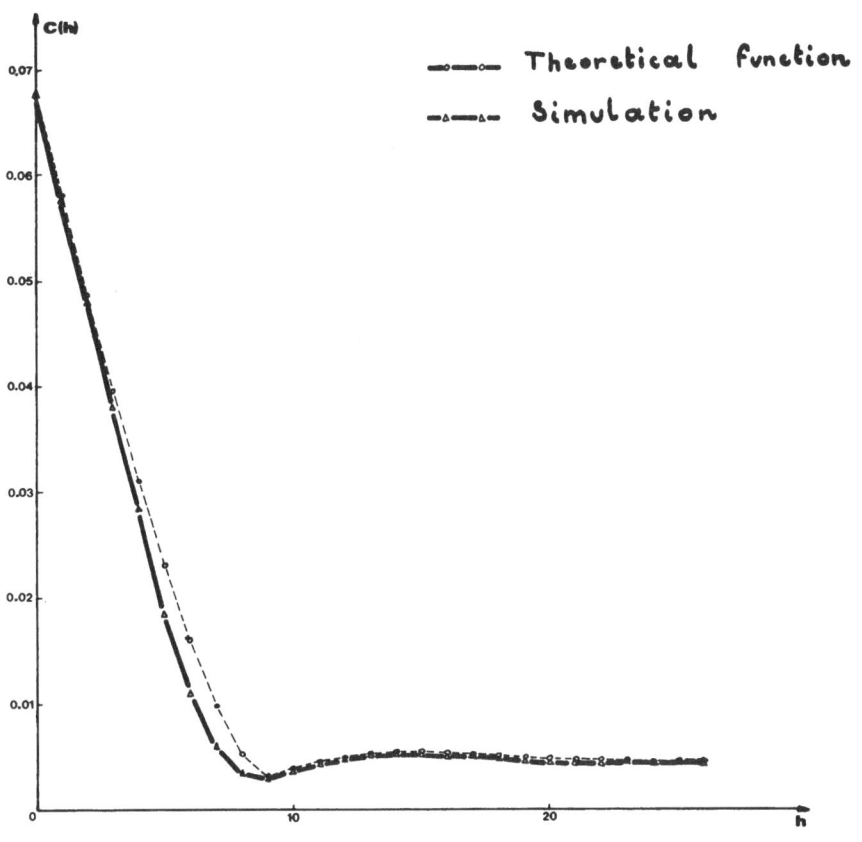

<u>Figure 6</u> : <u>Covariance function for the Hard-Core model</u>

On each point of the Hard-core model of intensity λ and Repulsion distance R, we center a ball of fixed radius r with $2r < R$ (non-overlapping balls). The theoretical formula is given by :

$$C(h) = \alpha\lambda \; K_r(h) + \lambda^2 \int_{R^n} K_r(z-h)k(|z|)dz$$

with $\quad \alpha = \exp[-\lambda C_n R^n]$, where C_n is the volume of the unit ball in R^n

$$K(z) = \begin{cases} 0, \text{ if } |z| < R \\ \exp\left[-\lambda(2\,K_R(0) - K_R(z))\right] & \text{if } |z| \geq R \end{cases}$$

K_r and K_R, are the geometric covariograms of balls with radii r and R (resp.) in R^n.

BIBLIOGRAPHY

1   Ch. LANTUEJOUL : Les modèles probabilistiques en morphologie
                mathématique. C.M.M. Fontainebleau, Avril 1975

2   B. MATERN : Spatial Variation
                Meddelanden frän Statens Skogeforskningsinstitut (1960)

3   G. MATHERON : Schémas aléatoires de germes. C.M.M. Fontainebleau

4   B.D. RIPLEY : Modelling spatial Patterns. Times Maths. 1977.

CONDITIONAL PROBABILITIES SIMULATION
-=-=-=-=-=-=-=-=-=-=-=-=-=-=-=-=-=-=-

Problem introduction

Let us consider the picture on figure 1.This image presents a great variety of grey tones. One digitalizes this image by partitioning the grey scale into n classes. Here four grey levels were chosen, so every picture point belongs to one class level, denoted (1), (2), (3) and (4). One wants to estimate the conditional probability that a picture point belongs to level (i), when its neighbowring points belong to given levels. In this case, only the six vertex points of an hexagon centered on the point will be considered. These points are at distance 1 from the central point (fig. 2).

Probleme solution

Let us take a numerical example. Suppose one wants to know the probability for a picture point to be at level (3), given that the six vertices of the size 5 hexagon belong, respectively, to levels (2), (3), (4), (4), (2) and (1). We estimate this probability by the formula :

$$p^* = \frac{n}{m}$$ n, number of configurations
   in the image

m, number of configurations

(central point not taken into account).

This estimator p^* will be meaningful only when the number m of configurations is great enough. This would be expected, even in a first approach, since there are at least 45.000 patterns in the image.

Using the texture analyser, one can easily look for such configurations as (2) - (3) - (4) - (4) - (2) - (1). The results are shown in figures 3 et 4.

As can be judjed from the figure, the value of m is very low. Hence, the calculated estimator p^* will not be meaningful.

The result obtained from other configurations are the same. For example, with :

$$2 \hexagon 4 \quad , \text{ we obtain } \quad m = 10.$$

In the fact, with six conditioning points and four grey levels, there are 4^6 = 4096 different possible configurations, that is, an average value of ten possible events for a given configuration. Actually, one notices that, in many cases, the actual number of configurations is lower than this average value.

When the number of conditioning factors is greater than a very low value (here six points, four grey classes), the number of events for a given configuration is too low to give a meaningful statistical inference for the conditional law.

BIBLIOGRAPHY

MATHERON G. : Le choix et l'usage des modèles Topo-probabilistes.
              C.M.M. Fontainebleau - 1976.

Figure 1 : Scanning electron microscope image
of a clay sample (x 3000)

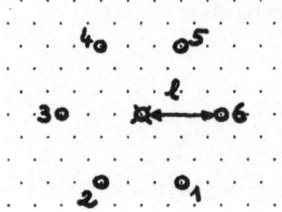

Figure 2 : Sampling grid of the image points

Figure 3 : Picture points surrounded by the
outline (2) - (3) - (4) - (4) - (2) - (1)

at distance 5

four points with proper configurations

Figure 4 : intermediate result

The six vertices of the hexagon are tested successively.

Image is shown after the three first vertices have been processed.

COMPUTATION OF THE HISTOGRAMS OF THE NUMBER OF EDGES AND NEIGHBOURS OF CELLS IN A TESSELLATION

C. LANTUEJOUL

ABSTRACT

The spatial random structure studied is a population of curvilinear polygons partitioning the space. The object of the paper is to provide an automatic method to compute the histograms of the number of edges and neighbours per polygon. The method consists of three steps : first, polygon boundaries are skeletonized ; then, a procedure to compute the number of edges and polygon neighbours is outlined. Finally, we develop a statistical method given by MILES to compensate edge effects in planar sampling. Practical applications are made on cells of a sample of sintered alumina.

I - INTRODUCTION AND NOTATIONS.

Throughout this paper, we are concerned with a structure common to histology, metallography and many other fields. It is an aggregate of non-overlapping cells which, along with their boundaries, cover the whole space. Each cell of the aggregate is a curvilinear polygon. Edges are common to two grains, vertices to three. An aggregate of cells verifying such properties is called a <u>tessellation</u> (cf. Figure 1).

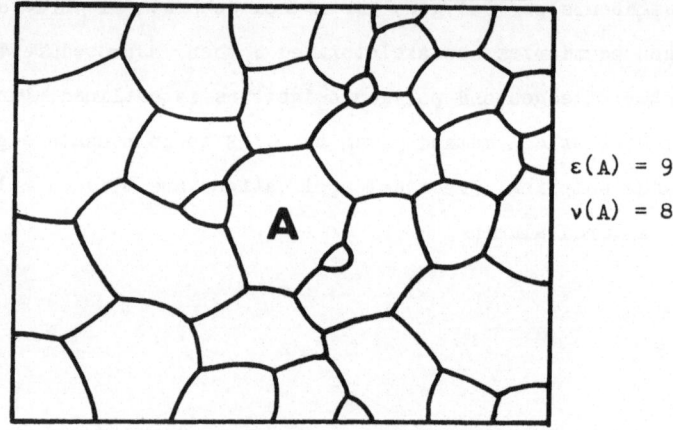

$$\varepsilon(A) = 9$$
$$\nu(A) = 8$$

<u>Figure 1</u> : Example of a tessellation

Let A be an arbitrary cell of the tessellation. The number of edges and neighbours of A are respectively denoted by $\varepsilon(A)$ and $\nu(A)$. Note that these two numbers are not necessarily equal, for two adjacent grains may have more than one common edge. The objective of this paper is to provide a method for computing the histograms of the number of edges and neighbours per cell. The paper is divided into three parts. At first, the concept of skeletonization is introduced. Then, as a consequence, the procedure to compute the number of edges and neighbours of a cell is given. And lastly, we use MILES' formulae for compensating for edge effects to obtain the histograms.

2 - SKELETONIZATION.

When working on any automatic scanning instrument, cells must be widely separated, in order not to be connected on the thresholded picture. However, our measurement procedure can yield unbiased results if, and only if, cell boundaries are as narrow as possible. So, just before making measurements, we skeletonize cell boundaries. The skeletonization consists in dilating cells to a maximal size, or in reducing cell boundaries to a minimal width, without connecting adjacent cells (cf. Figure 2).

More precisely, let us denote by A_1, ...,A_n,... the initial cells. The skeletonization transforms each cell A_n into a cell B_n, which is the set of the points nearer to A_n than to any other cell A_p.

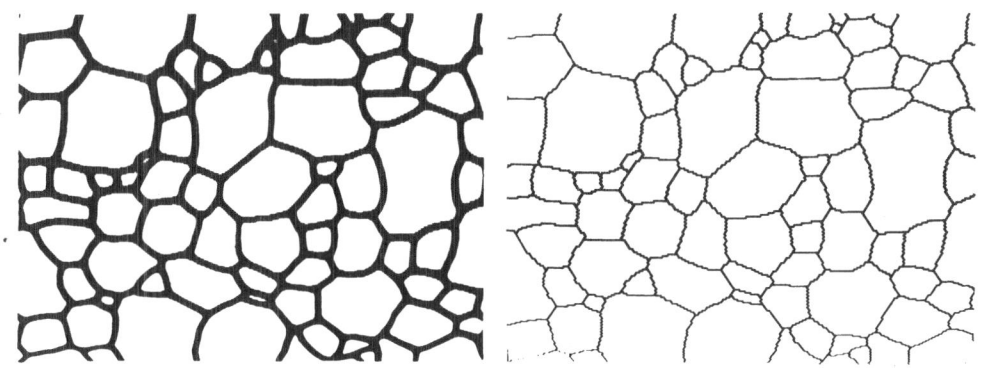

Figure 2 : Cells before and after skeletonization

The skeletonization can be automatically performed with the texture analyzer devised and built in the Mathematical Morphology Center in 1975 (1, 2). Various skeletonization algorithms have been conceived by many authors (3, 4, 5, 6). All of these are iterative.

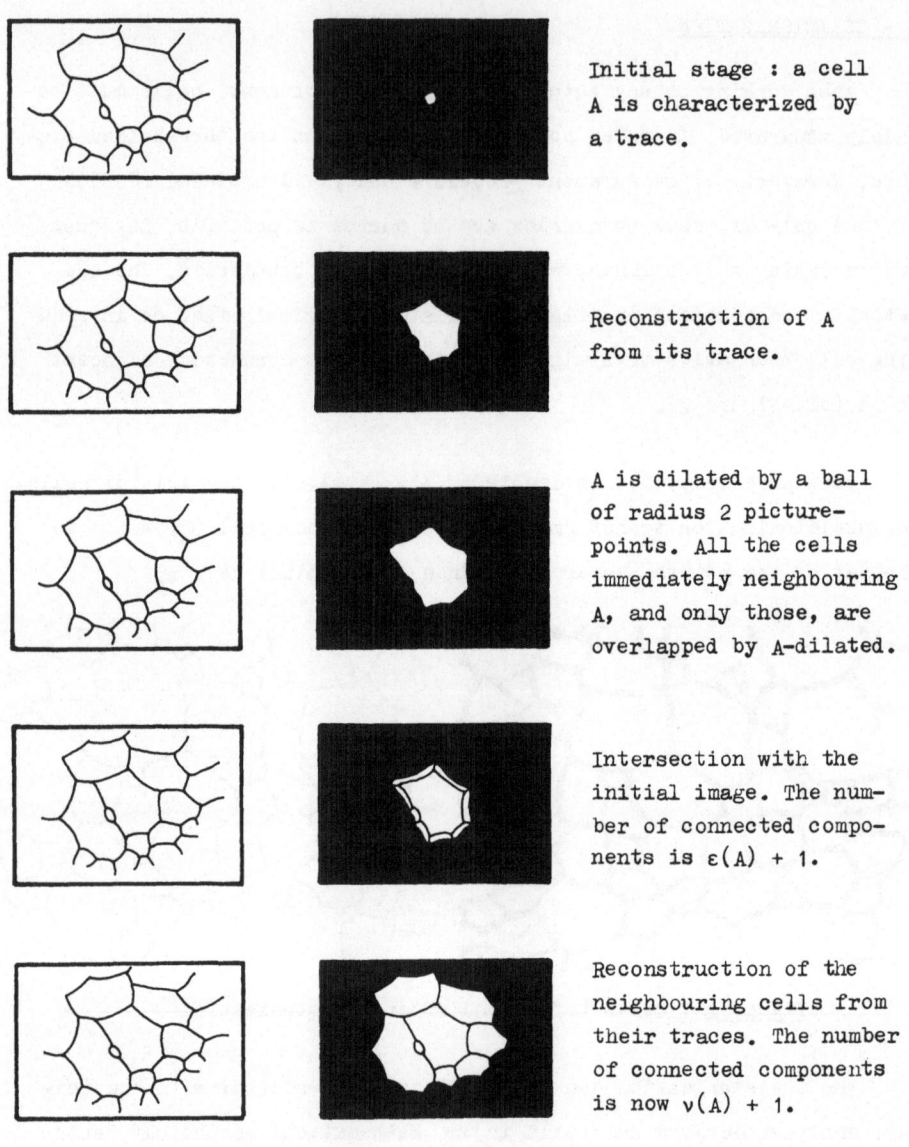

Initial stage : a cell A is characterized by a trace.

Reconstruction of A from its trace.

A is dilated by a ball of radius 2 picture-points. All the cells immediately neighbouring A, and only those, are overlapped by A-dilated.

Intersection with the initial image. The number of connected components is $\varepsilon(A)$ + 1.

Reconstruction of the neighbouring cells from their traces. The number of connected components is now $\nu(A)$ + 1.

<u>Figure 3</u> : Computation of the number of edges and neighbours of a given cell.

3 - COMPUTATION OF THE NUMBER OF EDGES AND NEIGHBOURS OF A CELL.

A procedure is outlined for computing the number of edges and neighbours of a given cell. It rests on the fact that cell boundaries are only one picture point wide because of the skeletonization. For such measurements, the basic transformations used are dilation, reconstruction of a cell from a trace, and intersection. For more clarity and brevity, the procedure is described in Figure 3 with pictures and explanations.

4 - COMPENSATION FOR EDGE EFFECTS.

Measurements are made on a sample of sintered alumina. A little more than 1000 cells are taken into account.

Let f_ε be the experimental histogram of the number of edges per cell (cf. Figure 4). Note that there are almost no two-edged cells. As a consequence, the numbers of neighbours and edges of a given cell are almost surely equal, and the corresponding histograms are the same.

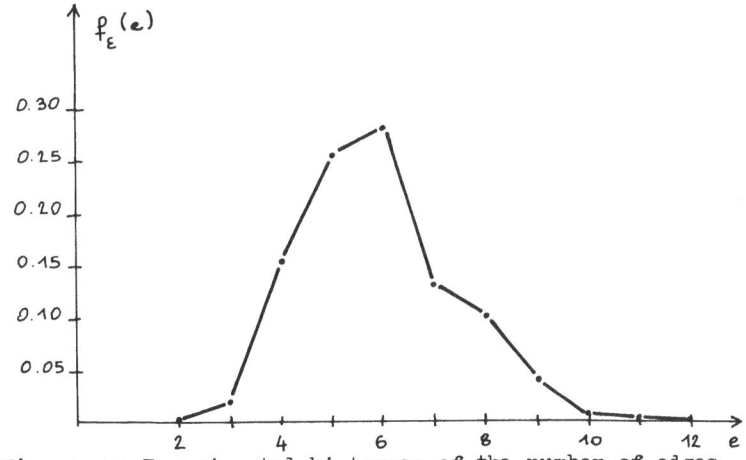

Figure 4 : Experimental histogram of the number of edges per cell.

The mean value is $E(\varepsilon) = 5.895$ whereas the ergodic mean value is 6.

Measurements are made on the cells falling within the frame. The cells cutting the edges are not taken into account. But the largest cells have the best chance of cutting the edges of the frame. In other words, the natural histogram is biased.

MILES proposed a procedure to compensate for the edge effect bias in planar sampling (7). In order to present it, let us assume that measurements are made in various frames which are identical, up to a translation, to a set D. Let us denote by $A_1,...,A_N$ all the cells enclosed within the measurement frames. For a cell A, let $M(A)$ be the surface area of the set composed of all points x where it is possible to set a figure D containing A. If measurements are made on frames far enough from each other for cells located within different frames to be independent, MILES showed that the quantity

$$\frac{\sum_{\{i = 1,...,N \,:\, \varepsilon(A_i) = e\}} \frac{1}{M(A_i)}}{\sum_{i=1,N} \frac{1}{M(A_i)}} \qquad (1)$$

is an asymptotically unbiased estimator of the conditional probability $P\{\varepsilon(A) = e \mid M(A) \neq 0\}$. If the cells A are rather small with respect to the frame, $P\{M(A) \neq 0\} = 1$ and the condition "$M(A) \neq 0$" disappears.

Practically speaking, the frame D is a rectangle. Then it is easy to see that $M(A) = M(A*)$, where A* is the smallest rectangle enclosing A, with edges parallel to those of D. Let us denote by (L,H) and (ℓ,h) the respective sizes of the rectangles D and A*. Then, we have $M(A*) = (L-\ell)(H-h)$.

On Figure 5, one can compare the experimental histogram f_ε (solid line) with the histogram f_ε^* corrected by MILES's procedure.

329

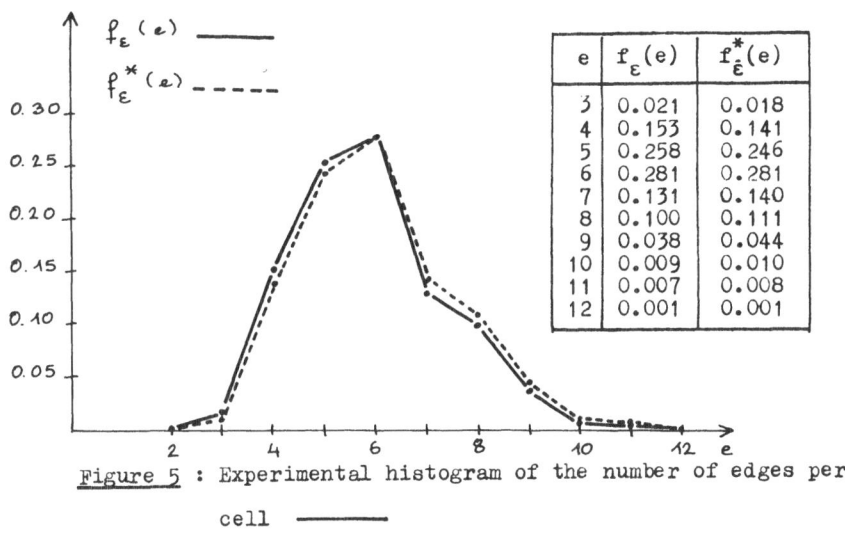

e	$f_\varepsilon(e)$	$f_{\hat\varepsilon}^*(e)$
3	0.021	0.018
4	0.153	0.141
5	0.258	0.246
6	0.281	0.281
7	0.131	0.140
8	0.100	0.111
9	0.038	0.044
10	0.009	0.010
11	0.007	0.008
12	0.001	0.001

<u>Figure 5</u> : Experimental histogram of the number of edges per

cell ———

Histogram corrected by MILES' procedure. — — — —

Clearly, large values are underestimated in f_ε. The new mean value is now $E^*(\varepsilon) = 5.998$. It appears that few measurements are sufficient to obtain a practically unbiased result, even though the estimator is only asymptotically unbiased.

BIBLIOGRAPHY

1. J.C. KLEIN, Engineer Doctorate Thesis, Nancy (1975).
2. H. DIGABEL, Internal Report, CMM C-65, Fontainebleau (1975).
3. K. PRESTON Jr., I.E.E.E. Trans on Comp. C-20, 9 (1971) 1007-
4. C. LANTUEJOUL, Engineer Doctorate Thesis (to be published)
5. H. DIGABEL, Internal Report, CMM, Fontainebleau (to be published)
6. R.E. MILES, Stochastic Geometry (Ed. D.G. Kendall and E.F. Harding) Wiley and Sons, London (1974) 228-

AN INDIRECT METHOD FOR COUNTING PARTICLES

IN 3-D AND 2-D SPACES

H. DIGABEL

In histological pictures, direct counts of structures (cells of a given type...) are often impossible to perform on automatic devices. This stems from both theoretical and practical reasons : the thickness of slides allows different cells to overlap in 2-d projected images, even though all cells are separate in 3-d space, and the heterogeneity of staining makes few projected cells appear as connex and holeless objects after thresholding operations. It is therefore necessary to use indirect methods of counting.

1 - DEFINITION OF THE PROBLEM.

Let us have a brief look at the practical problem. We assume that we have 3-d cells of known geometrical description (e.g. known diameters distribution in the case of spheres). These cells, embedded in a matrix (intersticial tissue), are sliced and projected (usually by transmission microscopy) into a 2-d image. Let X be this planar set.

Figure 1 : Cross section of the formation of X, cells being spheres of diameter D : H = h + D

The number of sets in X per unit test area, N_A, is the number of cells located in a slide of thickness H, usually larger than h, for cells can be centered outside the slide and still be seen. We assume that H is known, and that we can compute, or measure - by selecting non-overlapping projections - the mean area of the projection of the intersection of a cell and the slide. Let $\overline{A}$ be that area.

To estimate the density of cells in 3-d space, N_V, we have to measure N_A, for :

$$N_V = \frac{N_A}{H}$$

Hence, to solve our problem -the determination of N_V - it is sufficient to estimate N_A, and we shall give methods to achieve that, using only $\overline{A}$. It must be pointed out that the computation of N_A is a 2-d problem.

2 - THEORETICAL BACKGROUND.

The relationship between N_A and $\overline{A}$ can be easily given in two extreme cases (A_A standing for the area of set X per unit test area).

2.1 - If there are no overlapping in X, it is obvious that

(F1) $\qquad A_A = N_A \cdot \overline{A}$

2.2 - If the different parts of X are convex, and randomly and independently located, it can be proved that [1] :

(F2) $\qquad 1 - A_A = \exp(- N_A \cdot \overline{A})$

2.3 - If we are in the case of 2.2, and if there are few particles (N_A small) or if the cells are small ($\overline{A}$ small), there is a very low probability of overlapping occurring in X. Therefore we might be in a case equivalent to 2.1. This is fortunately true, for, using the expansion

$$\exp(- N_A \cdot \overline{A}) = 1 - N_A \overline{A} + \frac{(N_A \overline{A})^2}{2!} + \ldots$$

and (F2), we have :

$$N_A \cdot \overline{A} \simeq A_A \qquad \text{which is (F1).}$$

2.4 – We have assumed that $\overline{A}$ is known. In both cases, we can estimate N_A using only area measurements, which are very stable parameters and not sensitive to noise (although the direct determination of the number of cells is a very unstable measurement). This is a very important factor in practice.

3 – SIMULATIONS.

Actual histological pictures are not random pictures. Nevertheless, overlappings often occur in projected slides : in this case, the use of (F1) is biased, for N_A is underestimated (to have a biased estimation, the overlappings area might be beasured twice for determining A_A). To test if (F2) would give better results, we have used simulations.

3.1 – Principle.

We build a synthetic picture on the texture analyzer. The total number N of cells settled in the measuring mask is measured, and their mean area $\overline{A}$ is computed. Then, we measure the percentage area of the cells, and compute N_A using (F1) and (F2). Multiplying these numbers by the area of the mask, we deduce two estimations of the number of cells, N_1 and N_2, and compare them with N.

For simplicity's sake, all the cells are hexagonal. The randomness lies in the location of their centers and in the distribution of their sizes.

3.2 – Centers location process.

This process should not be a Poisson process, otherwise (F2) is true and the simulation is useless. The choice of our simulation technique is a function of several parameters :

. We build a Poisson point process P, the mean number of points in the mask being n_1.

. An intersection between P and another set S selects only some points of P. The remaining points P' are not a realization of a Poisson point process. S can be either a regular set (a union of rectangles denoted by R) or an irregular set (obtained by the dilation of a Poisson process, denoted I) chosen independently of P.

. All the points of P' having a neighbour at a distance less than d are then destroyed (which gives a Matern's 1st hard-core model [2]).

This process of center locations is denoted by $[n_1,R,d]$ or $[n_1,I,D]$.

3.3 - Size distribution of cells.

Four laws can be chosen for the size of cells : constant (C), Poisson distribution (P), exponential distribution (E), or a more complex law, denoted by S, simulating the distribution of intersections of spheres and planes, conditionnally to intersections larger than a given circle. The first three laws depending on one parameter - the mean value m of the radii - are denoted by $[C,m]$, $[P,m]$ or $[E,m]$. The distribution S, depending on two parameters is denoted $[S,R,h]$, where R is the radius of spheres, and $\sqrt{R^2-h^2}$ is the radius of the smallest intersection allowed. $\overline{A}$ is computed from the size distribution of cells.

3.4 - Definition of the cell process.

The size of the mask in which the process is simulated is 200 × 200 points. All the sizes are expressed in mesh units. The simulated process is defined as soon as the center location process and size distribution are given. An example of such a process could be :

$$([100,I,4] / [E,3])$$

where [100, I,4] is the center location process, and [E,3] the size distribution of cells.

Some realizations of simulations can be seen on Figure 2.

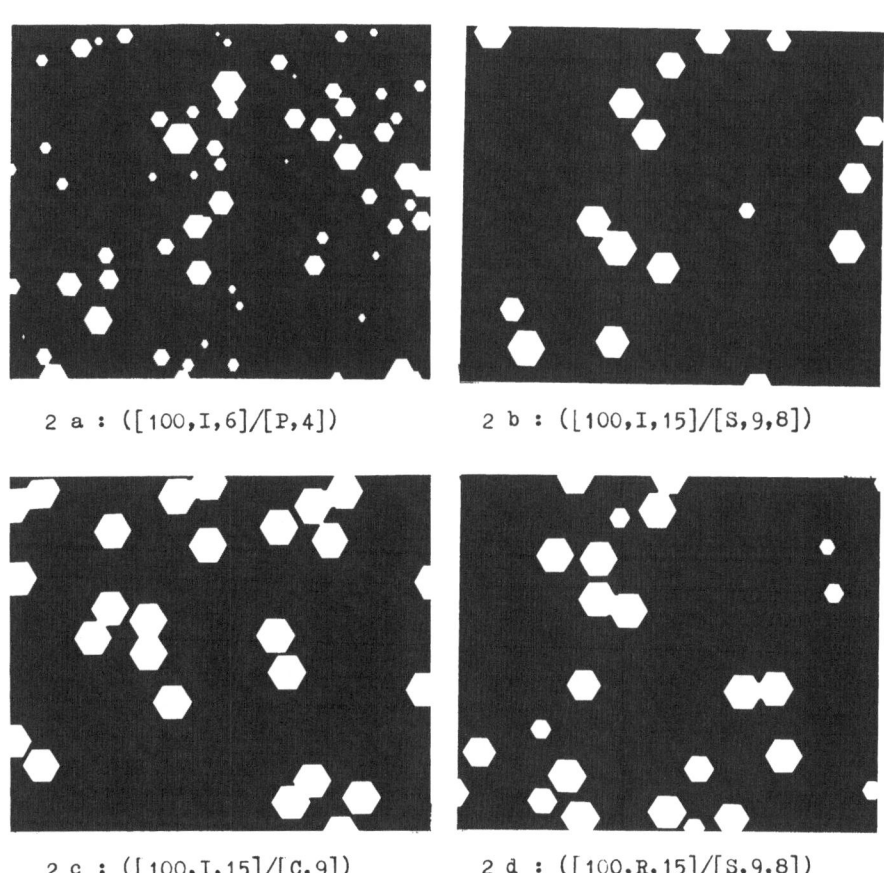

2 a : ([100,I,6]/[P,4])   2 b : ([100,I,15]/[S,9,8])

2 c : ([100,I,15]/[C,9])   2 d : ([100,R,15]/[S,9,8])

4 - RESULTS.

4.1 - Experimental fluctuations.

We have chosen two sets S, one R and the other I (§ 3.2).

For each of these sets, ten realizations of the same process gave the
following results, where E1 is the percentage of relative error when
using (F1) and E2 when using (F2).

CELL PROCESS	N	N1	N2	E1	E2
([100,1,15]/[C, 9])	22	22	23	0.0	-4.5
([100,1,15]/[C, 9])	8	9	9	-12.5	-12.5
([100,1,15]/[C, 9])	17	16	17	5.8	0.0
([100,1,15]/[C, 9])	13	11	11	15.3	15.3
([100,1,15]/[C, 9])	20	20	22	0.0	-10.0
([100,1,15]/[C, 9])	25	22	24	12.0	4.0
([100,1,15]/[C, 9])	12	11	11	8.3	8.3
([100,1,15]/[C, 9])	16	16	17	0.0	-6.2
([100,1,15]/[C, 9])	20	17	18	15.0	10.0
([100,1,15]/[C, 9])	26	22	24	15.3	7.6
([100,R,15]/[C, 9])	12	10	10	16.6	16.6
([100,R,15]/[C, 9])	12	10	11	16.6	8.3
([100,R,15]/[C, 9])	15	15	15	0.0	0.0
([100,R,15]/[C, 9])	19	18	19	5.2	0.0
([100,R,15]/[C, 9])	23	24	26	-4.3	-13.0
([100,R,15]/[C, 9])	18	19	20	-5.5	-11.1
([100,R,15]/[C, 9])	19	20	21	-5.2	-10.5
([100,R,15]/[C, 9])	15	11	11	26.6	26.6
([100,R,15]/[C, 9])	14	12	12	14.2	14.2
([100,R,15]/[C, 9])	9	9	10	0.0	-11.1

The fluctuations of estimations and errors are very important.
some values of E1 can even be negative : this is due to non ergodicity,
although A_A was measured in an eroded frame to avoid edge effects. So
we decided to compute the average, or equivalently the sum, of estimations over ten realizations of each simulation.

4.2 - <u>More significant results</u>.

The following results have been obtained.

Number	Cell Process	N	N1	N2	E1	E2
1	([100,I,8]/[C,10])	365	318	370	6,9	-1,3
2	([100,I,15]/[C,9])	179	166	176	7,3	0,2
3	([100,R,15]/[C,9])	156	148	155	5,1	0,6
4	([100,I,6]/[P,4])	738	738	793	0	-7
5	([50,R,6]/[E,4])	386	362	386	6,2	0
6	([100,I,15]/[S,9,8])	186	177	186	4,8	0
7	([100,R,15]/[S,9,8])	191	173	182	9,4	4,7

4.3 - Interpretation of experimental data.

The results of simulations 4 and 5 are not significant, for we would have to average on a very large number of simulations to obtain good results. We have not used this averaging techniques for those cases which show a large dispersion in the cell sizes : such dispersions are the cause of the fluctuations in the estimations, and create unrealistic histological images (Fig. 2 a).

The first simulation gives good results, but there are too many overlappings, due to the relative size of hard-core parameters (8) and of cells (10).

The remaining simulations have two common points : the size distribution of cells is plausible, and the choice of parameters leads to realistic pictures, with not too many overlappings. There are even many realizations in which no overlappings occur. In these four simulations, the use of (F2) is better then the use of (F1) : the estimation given by (F2) seems to be robust.

4.4 - Why ?

We have already stated that (F1) leads to biased estimations. If there are many overlappings, (F2) is good, for the picture is closed to the union of randomly located cells. If there are few overlappings, $N_A \bar{A}$ is usually small : the first order approximation of the exponential is good and (F2) gives the same estimation as (F1). In other words, the estimator given by (F2) is never too bad, and the estimating errors seem to cancel each other out when computed over several fields.

5 - CONCLUSION.

To measure the number of overlapping particles in the plane, the use of (F2) (§ 2.2) seems to give reliable results on simulations. It requires only area measurements, which are stable, and the knowledge of the mean area of a cell. If a short stereological study is performed, it is possible to deduce a number of particles in 3-d from this 2-d estimation.

It would be very interesting to try these manipulations on various histological samples to verify the robustness of this method.

BIBLIOGRAPHY.

[1] G. MATHERON - Eléments pour une Théorie des Milieux Poreux,
                    Masson, Paris, 1967.

[2] B. MATERN - Spatial Variation, Almaenna Foerlaget - Stockholm,
                    1960.

[3] H. DIGABEL, J.C. KLEIN - Manuel d'Utilisation de l'Analyseur de
                    Textures - Internal Report, C.M.M.

Bio-mathematics

Managing Editors: K. Krickeberg, S. A. Levin

Editorial Board: H. J. Bremermann, J. Cowan, W. M. Hirsch, S. Karlin, J. Keller, R. C. Lewontin, R. M. May, J. Neyman, S. I. Rubinow, M. Schreiber, L. A. Segel

Volume 1:
Mathematical Topics in Population Genetics
Edited by K. Kojima
1970. 55 figures. IX, 400 pages
ISBN 3-540-05054-X

"...It is far and away the most solid product I have ever seen labelled biomathematics."
American Scientist

Volume 2: E. Batschelet
Introduction to Mathematics for Life Scientists
2nd edition. 1975. 227 figures. XV, 643 pages
ISBN 3-540-07293-4

"A sincere attempt to relate basic mathematics to the needs of the student of life sciences."
Mathematics Teacher

M. Iosifescu, P. Tăutu
Stochastic Processes and Applications in Biology and Medicine

Volume 3
Part 1: **Theory**
1973. 331 pages.
ISBN 3-540-06270-X

Volume 4
Part 2: **Models**
1973. 337 pages
ISBN 3-540-06271-8

Distributions Rights for the Socialist Countries: Romlibri, Bucharest

"... the two-volume set, with its very extensive bibliography, is a survey of recent work as well as a textbook. It is highly recommended by the reviewer."
American Scientist

Volume 5: A. Jacquard
The Genetic Structure of Populations
Translated by B. Charlesworth, D. Charlesworth
1974. 92 figures. XVIII, 569 pages
ISBN 3-540-06329-3

"...should take its place as a major reference work.."
Science

Volume 6: D. Smith, N. Keyfitz
Mathematical Demography
Selected Papers
1977. 31 figures. XI, 515 pages
ISBN 3-540-07899-1

This collection of readings brings together the major historical contributions that form the base of current population mathematics tracing the development of the field from the early explorations of Graunt and Halley in the seventeenth century to Lotka and his successors in the twentieth. The volume includes 55 articles and excerpts with introductory histories and mathematical notes by the editors.

Volume 7: E. R. Lewis
Network Models in Population Biology
1977. 187 figures. XII, 402 pages
ISBN 3-540-08214-X

Directed toward biologists who are looking for an introduction to biologically motivated systems theory, this book provides a simple, heuristic approach to quantitative and theoretical population biology.

Springer-Verlag
Berlin
Heidelberg
New York

A
Springer
Journal

Journal of

Mathematical Biology

Ecology and Population Biology
Epidemiology
Immunology
Neurobiology
Physiology
Artificial Intelligence
Developmental Biology
Chemical Kinetics

Edited by H.J. Bremermann, Berkeley, CA; F.A. Dodge, Yorktown Heights, NY; K.P. Hadeler, Tübingen; S.A. Levin, Ithaca, NY; D. Varjú, Tübingen.

Advisory Board: M.A. Arbib, Amherst, MA; E. Batschelet, Zürich; W. Bühler, Mainz; B.D. Coleman, Pittsburgh, PA; K. Dietz, Tübingen; W. Fleming, Providence, RI; D. Glaser, Berkeley, CA; N.S. Goel, Binghamton, NY; J.N.R. Grainger, Dublin; F. Heinmets, Natick, MA; H. Holzer, Freiburg i. Br.; W. Jäger, Heidelberg; K. Jänich, Regensburg; S. Karlin, Rehovot/Stanford CA; S. Kauffman, Philadelphia, PA; D.G. Kendall, Cambridge; N. Keyfitz, Cambridge, MA; B. Khodorov, Moscow; E.R. Lewis, Berkeley, CA; D. Ludwig, Vancouver; H. Mel, Berkeley, CA; H. Mohr, Freiburg i. Br.; E.W. Montroll, Rochester, NY; A. Oaten, Santa Barbara, CA; G.M. Odell, Troy, NY; G. Oster, Berkeley, CA; A.S. Perelson, Los Alamos, NM; T. Poggio, Tübingen; K.H. Pribram, Stanford, CA; S.I. Rubinow, New York, NY; W.v. Seelen, Mainz; L.A. Segel, Rehovot; W. Seyffert, Tübingen; H. Spekreijse, Amsterdam; R.B. Stein, Edmonton; R. Thom, Bures-sur-Yvette; Jun-ichi Toyoda, Tokyo; J.J. Tyson, Blacksbough, VA; J. Vandermeer, Ann Arbor, MI.

Springer-Verlag
Berlin
Heidelberg
New York

Journal of Mathematical Biology publishes papers in which mathematics leads to a better understanding of biological phenomena, mathematical papers inspired by biological research and papers which yield new experimental data bearing on mathematical models. The scope is broad, both mathematically and biologically and extends to relevant interfaces with medicine, chemistry, physics and sociology. The editors aim to reach an audience of both mathematicians and biologists.